Control for Fractional-Order Systems and Fractional-Order Control

This book provides systematic guidance and theoretical basis for control theories and applications of fractional-order control systems. *Control for Fractional-Order Systems and Fractional-Order Control: Theories and Applications* describes the background, mathematical preliminaries, and recent progress with related illustrated examples, and proposes suggested future trends.

The authors describe control schemes based on the adaptive backstepping technique and sampled data for nonlinear uncertain fractional-order systems and linear fractional-order systems or multi-agent systems. Secure control for linear fractional-order systems subject to cyber attacks is also studied. In addition, the fractional-order adaptive backstepping control problem for high-order uncertain nonlinear integer-order systems or multi-agent systems is also investigated as employing fractional-order controllers in integer-order systems can help increase additional control design degrees of freedom and enable superior control performance. This book contains both control for fractional-order systems in linear or nonlinear forms even with uncertainties and cyber attacks, and fractional-order control schemes. Rigorous analyses, mathematical rules, and practical examples are presented throughout, providing insights for future applications of fractional-order control systems. The resulting improved control is vital for defending against denial-of-service attacks and developing fractional-order control systems.

This book is an invaluable reference for advanced undergraduate and graduate students and researchers in the fields of control science and engineering, fractional-order control schemes, control for fractional-order systems, and related topics. To understand this book, readers should have a knowledge of basic linear and nonlinear control theory.

Xinyao Li received her PhD from Nanyang Technological University, Singapore, in 2023. She is an associate professor at Guangdong Polytechnic Normal University, China. Her research interests include adaptive control, fractional-order systems control theory and applications, fractional-order control, distributed systems, and cyber-physical systems.

Xiaolei Li is an associate professor at Yanshan University, China. His research interests include secure localization and formation control of underwater robotic systems and cyber security of cyber-physical systems. He has published more than 80 peer-reviewed papers, including some highly cited ESI papers.

Changyun Wen is currently a full professor at Nanyang Technological University, Singapore. His research interests include control systems and applications, cyber-physical systems, smart grids, and complex systems and networks. He is an IEEE fellow and a fellow of the Academy of Engineering, Singapore.

Control for Fractional-Order Systems and Fractional-Order Control

Theories and Applications

Xinyao Li
Xiaolei Li
Changyun Wen

CRC Press
Taylor & Francis Group
Boca Raton London New York

CRC Press is an imprint of the
Taylor & Francis Group, an informa business

Designed cover image: Getty

First edition published 2026
by CRC Press
2385 NW Executive Center Drive, Suite 320, Boca Raton FL 33431

and by CRC Press
4 Park Square, Milton Park, Abingdon, Oxon, OX14 4RN

CRC Press is an imprint of Taylor & Francis Group, LLC

ISBN: 978-1-041-02639-6 (hbk)
ISBN: 978-1-041-02654-9 (pbk)
ISBN: 978-1-003-62037-2 (ebk)

DOI: 10.1201/9781003620372

Typeset in CMR10 font
by KnowledgeWorks Global Ltd.

Publisher's note: This book has been prepared from camera-ready copy provided by the authors.

Contents

Preface

As a generalization of integer-order calculus, fractional-order one is a branch of the mathematical field with over three centuries of history. Owing to its infinite memory and hereditary properties, fractional-order calculus possesses significant advantages in accurately modeling and characterizing dynamic properties of many real-world phenomena or systems arising from various fields such as physics, biology, mechanics, chemistry, finance, and engineering. In recent years, fractional-order calculus has attracted increasing attention in different areas, particularly in control engineering, yet compared with the widely studied traditional integer-order case, the research on fractional-order control systems is still at its beginning stages.

In this book, control schemes based on the adaptive backstepping technique and sampled-data are developed respectively for nonlinear and linear fractional-order systems. Secure control for linear fractional-order systems subject to cyber attacks will also be studied. In addition, the fractional-order adaptive backstepping control problem for high-order nonlinear integer-order systems or multi-agent systems (MASs) will also be investigated as employing fractional-order controllers in integer-order systems can help increase additional control design degrees of freedom and enable superior control performance.

First of all, the adaptive backstepping smooth control problem for a class of commensurate fractional-order nonlinear systems in strict feedback form with system uncertainties and unknown time-varying external disturbance is studied. As one of the most prevalent techniques applied in controlling nonlinear systems, backstepping control demonstrates a recursive systematic design procedure. For fractional-order systems, the time derivative of composite functions, which is required for the backstepping controller design process, can not be easily obtained by the chain rule because of the specific mathematical characteristics of fractional-order calculus. To overcome this obstacle, a novel approach for approximating the fractional-order time derivatives of virtual control signals, which are composite functions about system states, is proposed, and adaptive laws are designed to estimate the bounds of approximation errors. Besides, instead of using a noncontinuous control signal, an auxiliary function is employed to obtain a smooth control input that is still able to achieve asymptotic tracking in the presence of arbitrarily bounded external disturbances. It is proved that the obtained closed-loop system is globally asymptotically stable under the assumption that the approximation errors between the actual values of fractional derivatives of virtual control signals and their approximated values remain bounded. Subsequently, in order to relax such an assumption, we propose an alternative adaptive backstepping smooth control strategy, under which the resulting closed-loop fractional-order system is guaranteed to be stable in the sense that all the closed-loop signals are uniformly ultimately bounded and the output tracking error can be driven toward a small region around the origin. Moreover, a quantizer backstepping-based control scheme is proposed on the basis of the aforementioned results, enriching the control for fractional-order nonlinear systems with uncertainties and external disturbances.

Furthermore, it is also of great importance yet challenging to investigate the incommensurate fractional-order system, which is a universal form and could be found in various practical scenarios such as fractional-order phase-locked loop (PLL) and chaotic systems. Nevertheless, few results on the backstepping control of such systems have been reported.

Consequently, we construct a smooth adaptive backstepping controller for a class of incommensurate fractional-order nonlinear systems with uncertainties and time-varying disturbance to achieve that the system output tracks a predefined signal asymptotically while ensuring the global boundedness of all closed-loop signals.

In practice, control schemes are commonly implemented digitally so that digital control signals are generated, resulting in sampled-data control. Therefore, the consensus control problem for linear fractional-order MASs in single-integrator form over directed topology is investigated using sampled-data control. Unlike the classical integer-order derivatives or integrals, fractional-order ones are non-local operators and own infinite memory and hereditary properties. As a result, a novel distributed control scheme with the consideration of these unique properties of fractional-order calculus is proposed to achieve asymptotic consensus and ensure global stability in the sense that all the signals in the closed-loop systems are globally bounded.

Hereafter, for linear single-integrator fractional-order systems, a finite-dimensional sampled-data-based control scheme, which is computationally tractable and more applicable in practical usage, is developed to guarantee the global stability of the corresponding closed-loop system in the sense that the closed-loop states can converge to an ultimate bound around the origin with control signals remaining bounded. A guideline on selecting the control signal storage space for the finite-dimensional controller to satisfy preset control accuracy is also provided. Then a new sampled-data controller is designed for a more general class of linear fractional-order systems whose solution contains the Mittag-Leffler function, one of the fundamental functions in fractional-order calculus which does not obey the basic exponentiation identity. It is ensured that the resulted closed-loop system is asymptotically stable with the hereditary and infinite memory characteristics of fractional-order calculus being taken into account. In addition, the security control for fractional-order systems subject to denial-of-service (DoS) attacks will be studied, which provides a novel method for defending against attacks while simultaneously ensuring stable control for systems in fractional order.

Thereafter, the fractional-order adaptive backstepping control problem for a class of high-order nonlinear integer-order systems with uncertainties and unknown external disturbance is addressed. Different from the existing works that are only for second-order integer systems, the considered class of uncertain nonlinear integer-order systems can be of arbitrary order. Based on the Lyapunov stability theorem, we provide theoretical proof to assure the global stability of the closed-loop system in the sense that all the closed-loop signals are uniformly ultimately bounded and the system output error converges to an arbitrarily small tunable bound by adjusting certain designs parameters. Then, a prescribed performance-based (PPB) fractional-order adaptive backstepping control scheme is further proposed in this book for nonlinear high-order integer-order systems, which guarantees the output can stay within predefined bound constantly and track target trajectory asymptotically, showing improvement from the previous results that could only maintain the maximum value of the transient system output error within a tunable limit. Furthermore, this book studies the distributed quantized tracking control problem with prescribed output performance requirements for uncertain nonlinear integer high-order MASs with both DoS attacks and external unknown disturbances under data-rate-constrained communication networks.

In all the problems we considered above, rigorous theoretical analysis and demonstrative examples are provided to verify and illustrate the efficiency of the established results. The rich content of this book provides systematic guidance and a theoretical basis for the theories and applications of control for fractional-order systems and fractional-order control. This book can provide a reference for researchers in the fields of control science and engineering, fractional-order control schemes, control for fractional-order systems, etc. Understanding this book requires knowledge of basic linear and nonlinear control theory.

This book was supported by the National Natural Science Foundation of China under Grant 62403147. We express our deep gratitude to our beloved families who have made it possible to write this book. Xinyao Li is very grateful to her parents, Juewei Li and Weifeng Gao, and her husband, Jie Fang, for their care, understanding, and constant encouragement. Xiaolei Li is greatly indebted to his parents, Jingyun Li and Fengxiang Zhao, his wife, Jiange Wang, and his child Ziyi Li, for their constant support throughout these years. Changyun Wen is greatly indebted to his wife, Xiu Zhou, and his children Wen Wen, Wendy Wen, Qingyun Wen, and Qinghao Wen for their constant invaluable support and assistance throughout these years.

Xinyao Li
Guangdong Polytechnic Normal University, China

Xiaolei Li
Yanshan University, China

Changyun Wen
Nanyang Technological University, Singapore

Authors

Xinyao Li earned a BEng in automation from the University of Electronic Science and Technology of China in 2017, an MS in electrical engineering from the National University of Singapore, Singapore, in 2018, and a PhD in control science and engineering from Nanyang Technological University, Singapore, in 2023. She is currently an Associate Professor at the School of Automation, Guangdong Polytechnic Normal University, Guangzhou, China. Her research interests include adaptive control, fractional-order systems control theory and applications, fractional-order control, distributed systems, and cyber-physical systems. She received the Best Paper Award of the 16th Conference on Industrial Electronics and Applications ICIEA 2021 and the 2022 National Award for Outstanding Self-financed Chinese Students Abroad in 2023.

Xiaolei Li earned a BEng and PhD in control engineering from Yanshan University, China, in 2012 and 2018, respectively. From 2018 – 2022, he was a Research Fellow at the School of Electrical and Electronic Engineering, Nanyang Technological University, Singapore. He joined the Yanshan University in 2022, where he is currently an Associate Professor. He received the Best Paper Award of the 16th Conference on Industrial Electronics and Applications ICIEA 2021. His current research interests include secure localization and formation control of underwater robotic systems and cyber security of cyber-physical systems.

Changyun Wen earned a BEng in automatic control from Xi'an Jiaotong University, China, in 1983 and PhD in control science and engineering from the University of Newcastle, Australia, in 1990. From 1989 – 1991, he was a Postdoctoral Fellow at the University of Adelaide, Australia. Since 1991, he has been with Nanyang Technological University, Singapore, where he is currently a Full Professor. His main research activities are in the areas of control systems and applications, cyber-physical systems, smart grids, complex systems and networks.

He is a Fellow of the IEEE and a Fellow of the Academy of Engineering, Singapore, was a member of the IEEE Fellow Committee from 2011–2013, and a Distinguished Lecturer of the IEEE Control Systems Society from 2010–2013. He is currently the co-Editor-in-Chief of *IEEE Transactions on Industrial Electronics*, an Associate Editor of *Automatica*, and an Executive Editor-in-Chief of *Journal of Control and Decision*. He also served as an Associate Editor of *IEEE Transactions on Automatic Control* from 2000–2002, *IEEE Transactions on Industrial Electronics* from 2013–2020, and *IEEE Control Systems Magazine* from 2009–2019. He has been actively involved in organizing international conferences as a general chair (including general chair of IECON 2020 and IECON 2023), TPC chair (e.g., Chinese Control and Decision Conference since 2008), etc. He was the recipient of a number of awards, including the Prestigious Engineering Achievement Award from the Institution of Engineers, Singapore in 2005, and the Best Paper Award of IEEE Transactions on Industrial Electronics in 2017.

Acronyms and Symbols

Acronyms

CRONE	Commande Robuste d'Ordre Non Entier (Robust Non-Integer-Order Control in French)
DoS	Denial-of-Service
FO	Fractional-Order
GL	Grünwald-Letnikov
IO	Integer-Order
ISS	Input-to-State Stable
MAS	Multi-Agent System
MIMO	Multi-Input-Multi-Output
PID	Proportional Integral Derivative
PLL	Phase-Locked Loop
PPB	Prescribed Performance-Based
RL	Riemann-Liouville
SISO	Single-Input-Single-Output

Symbols

$\mathbb{R}$	set of reals
$\mathbb{R}_0$	set of non-negative reals
$\mathbb{R}^+$	set of positive reals
$\mathbb{R}_{>p}$	set of positive reals greater than a number p
$\mathbb{R}^n$	real-valued vectors of dimension n
$\mathbb{R}^{m\times n}$	real-valued $m \times n$ matrices
$\mathbb{N}$	set of positive integers
$\mathbb{N}_0$	set of non-negative integers
$\mathbb{N}_{>p}$	set of positive integers greater than a number p
$\mathbb{Z}$	set of integers
$\mathbb{C}$	set of complex numbers
$Re(x)$	real part of the number x
$[x]$	rounding of the number x
$\arg(x)$	argument of the number x
$\|x\|_\infty$	maximum value of $\|x\|$ over time interval $t \in [0, \infty)$ $(x \in \mathbb{R})$
$f : S_1 \to S_2$	function f mapping a space S_1 into a space S_2
$L^1[t_0, t_1]$	$\{f(t) \| \int_{t_0}^{t_1} \|f(s)\| \mathrm{d}s < +\infty\}$
$C([t_0, t_1], \mathbb{R})$	$\{f(t) \| f$ is continuous, $f : [t_0, t_1] \to \mathbb{R}\}$
$C^k([t_0, t_1], \mathbb{R})$	$\{f(t) \| f$ with continuous derivatives up to $k \in \mathbb{N}$ orders, $f : [t_0, t_1] \to \mathbb{R}\}$
$\dot{g}(t)$	first-order derivative of $g(t)$ with respect to time
$g^{(p)}(t)$	pth-order derivative of $g(t)$ $(p \in \mathbb{R})$ with respect to time
$g^{(0)}(t)$	function $g(t)$ itself
$f'_u(u)$	partial derivative of a function f with respect to u
I	identity matrix with an appropriate dimension
I_n	identity matrix with dimension $n \times n$
$x^T(A^T)$	transpose of a vector x or a matrix A
A^{-1}	inverse of a matrix A
$\\|\cdot\\|$	Euclidean norm of a vector or matrix
$\lambda_{\max}(A)$	maximum eigenvalue of a matrix A
max	maximum
min	minimum
$\to$	tends to
$\mathrm{o}(x)$	infinitesimal of x
$q!$	factorial of a number q
$rank(A)$	rank of a matrix A
sup	supremum

1

Introduction

This chapter summarizes the research background of the control for fractional-order systems and the fractional-order control, as well as gives the basic mathematical preliminaries for fractional-order calculus and literature review related to the content of this book. Finally, the major contributions and the organization of the book are presented.

1.1 Background

As a branch of mathematics with a history of more than 300 years, fractional-order calculus can be regarded as the extension of the traditional integer-order calculus that contains fractional derivatives and integrals.

Over the past three centuries, without the support of disciplines such as physics and mechanics, fractional-order calculus has been studied by mathematicians as a pure theory in the field of mathematics with well-established mathematical theory [5, 6, 7, 8]. In [6], a good review of fractional-order calculus is given from a purely mathematical perspective, and the historical development of such discipline is summarized year-by-year up to 1975 in [9]. The research boom in fractional-order calculus began in the 1970s, and the main reason was that researchers found out that fractal geometry, power-law phenomena, memory processes, and other related phenomena or processes can be well described and characterized by fractional-order calculus. In 1960, Professor Manabe from Japan first extended the theoretical work of fractional-order calculus into the field of systems control [10], which marked the beginning of the research of fractional-order calculus in science and engineering fields. Based on fractional-order derivatives, a dissipation model is proposed in [11]. In [12], the structure of the fractional-order proportional integral derivative (PID) controller and its corresponding applications are given. Professor Oustaloup developed the CRONE control, which is one of the most widely applied fractional-order control methods currently, and successfully applied this theory to the suspension control systems in automobiles [13]. In [14], the nonlinear motion control systems with fractional-order control applications are discussed. A tuning scheme for the fractional-order PID controller of the permanent magnet synchronous motor speed servo is shown in [15]. For integer nonlinear systems, fractional sliding mode control methods and fractional backstepping control schemes are respectively investigated in [16, 17, 18, 19] and [20, 21, 22, 23, 24]. Since about 20 years ago, there have been some monographs devoted to fractional-order calculus and its applications in various professional fields. Among these works, there are many influential ones, including [8] that focuses on fractional-order differential equations and control, [25] about bioengineering and [26] from the field of physics. Some books have also been published in the field of fractional-order control in recent years, such as [27], [7], [14], [28], [29], and [30] issued in 2010, 2011, 2012, 2014, 2017, and 2019, respectively.

DOI: 10.1201/9781003620372-1

Recently, extensive studies have been conducted on fractional-order systems due to the fact that with the non-local property of fractional-order derivative, fractional-order calculus is more appropriate to describe phenomena or processes with hereditary and infinite memory characteristics compared to integer-order one [31]. The voltage-current relation of a semi-infinite lossy transmission line [32, 33] and the diffusion of heat through a semi-infinite solid [8, 34] are typical examples of fractional-order systems. Applications of fractional-order systems can be found in many other fields. For instance, using fractional-order dynamic models to manifest the electrical characteristics of fuel cells can provide higher accuracy compared with applying integer-order ones [35]. In [36] and [37], fractional-order techniques are adopted to model lithium-ion batteries. A fractional-order full-cell model that owns clear physical meaning is constructed in [38]. A fractional-order capacitor with order within (1, 2) can be found in [39]. Furthermore, fractional-order difference equations are employed to depict the behaviors of biological systems; see for examples [40, 41, 42]. The drugs release and adsorption, and the cardiac tissue-electrode interface are described with fractional-order derivatives respectively in [43] and [44]. A fractional-order model for the premotor neurons in the vestibulo-ocular reflex is given in [45]. In [46], the relaxation and retardation functions of the Maxwell model are formulated with fractional-order derivatives. For the viscoelasticity of polymer materials, it is suggested in [26] and [47] that the rheological constitutive equation can be more precisely described by fractional-order differential equations. Moreover, the chaotic nonlinear systems with fractional-order dynamics have lately attracted increasing attention in research community, such as fractional-order Lorenz system [48, 49], fractional-order Chua's system [50, 51], fractional-order Liu system [52, 53], and fractional-order Chen system [54, 55]. Also, phase-locked loop (PLL) developed in fractional-order can be found in [56, 3, 57]. A fractional-order dynamic model for crowds of pedestrians that considers the temporal and spatial influences and long-range interactions is investigated at three different scales in [58]. Some other science and engineering applications of fractional-order calculus can be found summarized in [59].

Apart from fractional-order single systems, the control problems of the fractional-order multi-agent systems (MASs) have also been broadly investigated. The consensus control problem of fractional-order MASs is first studied in [60], in which the sufficient conditions on the network structure and the fractional-order of MASs are provided for achieving consensus. The relationship between the number of agents and the fractional-order to ensure consensus is also presented. After doing comparisons between MASs of fractional-order and integer-order, this work comes up with a conclusion that there exists a time instant $t = \delta$ such that, over the time interval $(0, \delta)$, the consensus convergence rate of MASs with fractional-order within (0, 1) is faster than the one with integer-order. Based on [60], a consensus protocol with switching order is proposed in [61] to effectively increase the convergence rate of fractional-order MASs over an undirected graph. Fractional-order MASs consensus control subject to communication delays is investigated in [62]. Observer-based leader-following consensus problem is addressed in [63], in which the observers of the second-order leader designed for the followers are fractional-order systems with order within (0, 2). In [64], the consensus control problem for fractional-order MASs involving input saturation and external disturbances is discussed. For incommensurate fractional-order MASs, whose orders of fractional derivation, in contrast to the commensurate case, are not integer multiples of a base order [27], the consensus problem is settled in [65] under system nonlinearities, uncertainties, and external disturbances. By carefully selecting a piecewise Lyapunov function, the convergence of the consensus tracking error of incommensurate fractional-order nonlinear MASs under directed switching topologies to an adjustable residual set within finite time can be ensured in [66]. In consideration of the control for fractional-order systems and the corresponding stability analysis, the stability theorem for linear time-invariant fractional-order systems is available in [67], and some lemmas, which are important for constructing Lyapunov functions to analyze the stability of fractional-order nonlinear systems, are proposed in [68, 69, 70].

With regard to the fractional-order control [71, 72, 73, 74], fractional calculus possesses the benefit of extending the adjustment range of design parameters, leading to improved control performance and enabling the superiority of the control method over integer ones [75]. Several recent studies have focused on the application of fractional controllers to integer systems in order to improve control performance, for instance, fractional-order PID control [12, 76, 77, 78] and fractional sliding mode control [79, 80, 19].

The present book is devoted to studying the control for fractional-order systems and fractional-order control with theories and applications. In particular, in this book, control schemes based on the adaptive backstepping technique and sampled-data are developed respectively for nonlinear and linear fractional-order systems. Secure control for linear fractional-order systems subject to cyber attacks will also be studied. In addition, the fractional-order adaptive backstepping control problem for high-order nonlinear integer-order systems or MASs will also be investigated, as employing fractional-order controllers in integer-order systems can help increase additional control design degrees of freedom and enable superior control performance. Rigorous stability analysis for the developed control strategies will be presented with demonstrative examples provided to verify the theoretical results.

1.2 Mathematical Preliminaries

1.2.1 Special Functions in Fractional-Order Calculus

1.2.1.1 Gamma Function

Definition 1.1 ([8, 29]) *The Gamma function is defined by the integral*

$$\Gamma(z) = \int_0^\infty e^{-t} t^{z-1}\,\mathrm{d}t, \tag{1.1}$$

which will converge in the right-half of the complex plane $Re(z) > 0$.

Euler's Gamma function is one of the most important basic functions of the fractional-order calculus. One of its significant properties is that it satisfies the following equation:

$$\Gamma(z+1) = \int_0^\infty e^{-t} t^{z}\,\mathrm{d}t = (-e^{-t}t^z)|_{t=0}^{t=\infty} + z\int_0^\infty e^{-t} t^{z-1}\,\mathrm{d}t = z\Gamma(z). \tag{1.2}$$

Apparently from (1.2), it can be deduced that $\Gamma(1) = 1$, $\Gamma(n+1) = n\Gamma(n) = n(n-1)\Gamma(n-1) = \cdots = n!$ and $\Gamma(-n) = \infty$ for $n \in \mathbb{N}_0$. It can also be noticed that the Gamma function summarizes the computation of $n!$ where n can be a non-integer and even a complex number.

Definition 1.2 ([8, 29]) *The Gamma function can also be represented by the limit*

$$\Gamma(z) = \lim_{n\to\infty} \frac{n^z n!}{z(z+1)\cdots(z+n)},\ Re(z) > 0, \tag{1.3}$$

from which it can be seen that the Gamma function has poles at $z = -n$, $n \in \mathbb{N}_0$.

1.2.1.2 Mittag-Leffler Function

Just as the exponential function e^z plays an essential role in integer-order calculus, the Mittag-Leffler function, especially the Mittag-Leffler with two parameters, plays a very important role in the theory of fractional-order calculus.

Definition 1.3 ([8, 29]) *The Mittag-Leffler function with one parameter, which is first introduced by Swedish mathematician Mittag-Leffler in 1903, is denoted as*

$$\mathrm{E}_{\alpha}(z)=\sum_{k=0}^{\infty}\frac{z^k}{\Gamma(\alpha k+1)}. \tag{1.4}$$

The convergent condition for this infinite series is Re(α) >0 for all $z\in\mathbb{C}$.

It can be easily checked that the exponential function is a special case of the one-parameter Mittag-Leffler function since when $\alpha=1$ in (1.4), it gives

$$\mathrm{E}_1(z)=\sum_{k=0}^{\infty}\frac{z^k}{\Gamma(k+1)}=\sum_{k=0}^{\infty}\frac{z^k}{k!}=e^z. \tag{1.5}$$

Definition 1.4 ([8, 29]) *The Mittag-Leffler function with two parameters is defined by*

$$\mathrm{E}_{\alpha,\,\beta}(z)=\sum_{k=0}^{\infty}\frac{z^k}{\Gamma(\alpha k+\beta)},\ \alpha>0,\ \beta>0,\ z\in\mathbb{C}. \tag{1.6}$$

If $\beta=1$ in (1.6), then the two-parameter Mittag-Leffler function returns back into the one-parameter one, i.e.:

$$\mathrm{E}_{\alpha,\,1}(z)=\mathrm{E}_{\alpha}(z). \tag{1.7}$$

Definition 1.5 ([8, 29]) *The integer-order derivatives of the Mittag-Leffler function can be computed as*

$$\left(\frac{\mathrm{d}}{\mathrm{d}t}\right)^m\left(t^{\beta-1}\mathrm{E}_{\alpha,\,\beta}(t^\alpha)\right)=t^{\beta-m-1}\mathrm{E}_{\alpha,\,\beta-m}(t^\alpha),\ m\in\mathbb{N}. \tag{1.8}$$

Definition 1.6 ([8]) *The integration of the Mittag-Leffler function is given by*

$$\int_0^z\mathrm{E}_{\alpha,\,\beta}(\lambda t^\alpha)t^{\beta-1}\,\mathrm{d}t=z^\beta\mathrm{E}_{\alpha,\,\beta+1}(\lambda z^\alpha),\ \beta>0. \tag{1.9}$$

Practically, a more general form of (1.9) is shown as follows [8]:

$$\frac{1}{\Gamma(\nu)}\int_0^z(z-t)^{\nu-1}\mathrm{E}_{\alpha,\,\beta}(\lambda t^\alpha)t^{\beta-1}\,\mathrm{d}t=z^{\beta+\nu-1}\mathrm{E}_{\alpha,\,\beta+\nu}(\lambda z^\alpha),\ \beta,\,\nu>0. \tag{1.10}$$

Moreover, a special case of (1.10) that will be used in the proof of stability analysis in Section 2.3 is displayed here:

$$\frac{1}{\Gamma(\alpha)}\int_0^z(z-t)^{\alpha-1}e^{\lambda t}\,\mathrm{d}t=z^\alpha\mathrm{E}_{1,\,\alpha+1}(\lambda z),\ \alpha>0. \tag{1.11}$$

Lemma 1.1 ([8]) *If $0<\alpha<2$, β represents an arbitrary complex number and real number μ satisfies that $\frac{\pi\alpha}{2}<\mu<\min\{\pi,\pi\alpha\}$, then for any integer $n\geq1$,*

$$\mathrm{E}_{\alpha,\,\beta}(z)=-\sum_{j=1}^{n}\frac{z^{-j}}{\Gamma(\beta-\alpha j)}+\mathrm{o}(|z|^{-1-n}), \tag{1.12}$$

where $\mu\leq|\arg(z)|\leq\pi$ and $|z|\to\infty$.

1.2.2 Fractional-Order Derivatives and Integrals

In this subsection, three commonly used types of fractional-order derivatives and integrals in the areas of mathematics and engineering are introduced.

1.2.2.1 Grünwald-Letnikov (GL) Fractional-Order Derivative and Integral

The GL derivative is obtained based on the integer-order backward difference generalization, and it is usually used for numerical computation.

Definition 1.7 ([8, 29]) *The αth-order GL derivative of a function with initial time t_0 is given by*

$$\begin{aligned} {}_{t_0}^{GL}\mathcal{D}_t^{\alpha} f(t) &= \lim_{h\to 0}\frac{1}{h^{\alpha}}\sum_{j=0}^{[(t-t_0)/h]}(-1)^j\binom{\alpha}{j}f(t-jh) \\ &= \frac{1}{\Gamma(-\alpha)}\int_{t_0}^{t}(t-\tau)^{-\alpha-1}f(\tau)\,\mathrm{d}\tau,\ \alpha\in\mathbb{R}^+, \end{aligned} \tag{1.13}$$

where

$$\binom{\alpha}{j} = \frac{\alpha(\alpha-1)(\alpha-2)\cdots(\alpha-j+1)}{j!} = \frac{\Gamma(\alpha+1)}{\Gamma(j+1)\Gamma(\alpha-j+1)}. \tag{1.14}$$

Definition 1.8 ([8]) *The αth-order GL integral of function $f(t)$ is defined as*

$$\begin{aligned} {}_{t_0}^{GL}\mathcal{D}_t^{-\alpha} f(t) &= \lim_{h\to 0}\frac{1}{h^{-\alpha}}\sum_{j=0}^{[(t-t_0)/h]}(-1)^j\binom{-\alpha}{j}f(t-jh) \\ &= \frac{1}{\Gamma(\alpha)}\int_{t_0}^{t}(t-\tau)^{\alpha-1}f(\tau)\,\mathrm{d}\tau,\ \alpha\in\mathbb{R}^+. \end{aligned} \tag{1.15}$$

1.2.2.2 Riemann-Liouville (RL) Fractional-Order Derivative and Integral

The nth-order integral of function $f(t)$ is given as

$${}_{t_0}I_t^n f(t) = \int_{t_0}^{t}\frac{f(\tau)}{(t-\tau)^{1-n}(n-1)!}\,\mathrm{d}\tau,\ n\in\mathbb{N}. \tag{1.16}$$

By replacing the integer n with a real number $\alpha > 0$, we then can obtain the definition of a fractional-order integral.

Definition 1.9 ([8, 29]) *The formula of the RL fractional integral of a function with $\alpha \in \mathbb{R}^+$ is*

$${}_{t_0}^{RL}\mathcal{D}_t^{-\alpha} f(t) = \frac{1}{\Gamma(\alpha)}\int_{t_0}^{t}\frac{f(\tau)}{(t-\tau)^{1-\alpha}}\,\mathrm{d}\tau \tag{1.17}$$

where the initial time equals t_0 and $f(t)$ is an integrable function.

Conducting enough times of fractional integrations firstly, then taking a sufficient number of integer differentiations will give the RL fractional derivative.

Definition 1.10 ([8, 29]) *The RL fractional derivative of a function is given by*

$${}_{t_0}^{RL}\mathcal{D}_t^{\alpha} f(t) = \frac{\mathrm{d}^m}{\mathrm{d}t^m}\,{}_{t_0}I_t^{m-\alpha} f(t) = \frac{1}{\Gamma(m-\alpha)}\frac{\mathrm{d}^m}{\mathrm{d}t^m}\int_{t_0}^{t}\frac{f(\tau)}{(t-\tau)^{\alpha-m+1}}\,\mathrm{d}\tau \tag{1.18}$$

where ${}_{t_0}I_t^{\alpha}$ means the fractional integral of order α and $m-1<\alpha<m$, $m\in\mathbb{N}$.

If $f(t)$ is $(p+1)$-times continuously differentiable in the interval $[t_0, t]$ with an integer number p satisfying $p > \alpha - 1$, then after doing integration by parts repeatedly, equation (1.13) can be expressed as [8]:

$$\begin{aligned} {}_{t_0}^{GL}\mathcal{D}_t^{\alpha} f(t) =& \sum_{k=0}^{p} \frac{f^{(k)}(t_0)(t-t_0)^{-\alpha+k}}{\Gamma(-\alpha+k+1)} \\ &+ \frac{1}{\Gamma(-\alpha+p+1)} \int_{t_0}^{t} (t-\tau)^{p-\alpha} f^{(p+1)}(\tau)\,\mathrm{d}\tau. \end{aligned} \tag{1.19}$$

Besides, under the same assumption, (1.18) can be calculated as follows [8]:

$$\begin{aligned} {}_{t_0}^{RL}\mathcal{D}_t^{\alpha} f(t) =& \frac{\mathrm{d}^m}{\mathrm{d}t^m} \frac{1}{\Gamma(m-\alpha)} \int_{t_0}^{t} \frac{f(\tau)}{(t-\tau)^{\alpha-m+1}}\,\mathrm{d}\tau \\ =& \frac{\mathrm{d}^m}{\mathrm{d}t^m} \left[\sum_{k=0}^{p} \frac{f^{(k)}(t_0)(t-t_0)^{m-\alpha+k}}{\Gamma(m-\alpha+k+1)} \right. \\ &\left. + \frac{1}{\Gamma(m-\alpha+p+1)} \int_{t_0}^{t} (t-\tau)^{m-\alpha+p} f^{(p+1)}(\tau)\,\mathrm{d}\tau \right] \\ =& \sum_{k=0}^{p} \frac{f^{(k)}(t_0)(t-t_0)^{-\alpha+k}}{\Gamma(-\alpha+k+1)} + \frac{1}{\Gamma(-\alpha+p+1)} \int_{t_0}^{t} (t-\tau)^{p-\alpha} f^{(p+1)}(\tau)\,\mathrm{d}\tau \\ =& {}_{t_0}^{GL}\mathcal{D}_t^{\alpha} f(t). \end{aligned} \tag{1.20}$$

Remark 1.1 ([8]) *For a function $f(t)$ that has $p+1$ continuous derivatives over time interval $[t_0, t]$, its GL fractional derivative coincides with the RL fractional derivative.*

1.2.2.3 Caputo Fractional-Order Derivative and Integral

Among some practical problems, such as the viscoelasticity theory in which fractional-order models can better describe the properties of the materials, initial conditions are necessary for solving fractional-order differential equations. Although the RL type fractional derivative and integral play an important role in the development and application of fractional calculus in mathematics, to get the initial conditions for RL fractional differential equations, the limit values of RL fractional derivative at $t \to t_0$ should be calculated as follows:

$$\lim_{t\to t_0} {}_{t_0}^{RL}\mathcal{D}_t^{\alpha-1} x(t) = c_1,\ \lim_{t\to t_0} {}_{t_0}^{RL}\mathcal{D}_t^{\alpha-2} x(t) = c_2,\ \cdots,\ \lim_{t\to t_0} {}_{t_0}^{RL}\mathcal{D}_t^{\alpha-m} x(t) = c_m, \tag{1.21}$$

where $m-1 < \alpha < m$, $m \in \mathbb{N}$ and $c_1, c_2, \cdots, c_m$ are preset constants [8].

Even though mathematically the initial value problem for the RL derivative is solvable, it lacks a practical explanation since the physical meanings of these initial conditions are unknown yet. On the contrary, fractional differential equations with Caputo derivatives have initial values that are in-line with integer-order differential equations, i.e., $x(t_0), \dot{x}(t_0), \ldots, x^{(m-1)}(t_0)$, which contain specific physical interpretations [8]. Therefore, Caputo-type fractional systems are frequently employed in practical analysis.

The Caputo fractional derivative is obtained by doing enough times of integer-order differentiations first and then a sufficient number of fractional integrations.

Definition 1.11 ([8, 29]) *The Caputo fractional derivative of a function is defined as*

$${}_{t_0}^{C}\mathcal{D}_t^{\alpha} f(t) = {}_{t_0}I_t^{m-\alpha} \frac{\mathrm{d}^m}{\mathrm{d}t^m} f(t) = \frac{1}{\Gamma(m-\alpha)} \int_{t_0}^{t} \frac{f^{(m)}(\tau)}{(t-\tau)^{\alpha-m+1}}\,\mathrm{d}\tau \tag{1.22}$$

where $m-1 < \alpha < m$, $m \in \mathbb{N}$.

From the above equation, we can observe that the Caputo derivative of a constant is 0.

Remark 1.2 *Since the Caputo derivative empowers the utilization of initial values of traditional integer-order derivatives with known physical meaning, as mentioned earlier, Caputo fractional-order systems with order within* $(0, 1]$ *are studied in this book. Additionally, all the fractional-order systems under consideration in this book stay in steady-state until the control signal is imposed, i.e., for time* t *before the initial time instant* t_0*, the value of system state* $x(t)$ *remains constant which is denoted as* $x(t_0)$*, or in other words, the class of fractional-order systems under initial condition* $x(t_0)$ *is studied. Consequently, based on the fact that the derivative of constant* $x(t_0)$ *under the Caputo definition is zero, the initial condition will not cause any effect related to the initialization issue [81] of fractional-order systems described in state space form, as shown below, that for any* $t_i \leq t_0$,

$$\begin{aligned}
{}_{t_i}^{C}\mathcal{D}_t^{\alpha} x(t) =& \frac{1}{\Gamma(1-\alpha)} \int_{t_i}^{t} \frac{x^{(1)}(\tau)}{(t-\tau)^{\alpha}} \mathrm{d}\tau \\
=& \frac{1}{\Gamma(1-\alpha)} \int_{t_i}^{t_0} \frac{x^{(1)}(t_0)}{(t-\tau)^{\alpha}} \mathrm{d}\tau + \frac{1}{\Gamma(1-\alpha)} \int_{t_0}^{t} \frac{x^{(1)}(\tau)}{(t-\tau)^{\alpha}} \mathrm{d}\tau \\
=& {}_{t_0}^{C}\mathcal{D}_t^{\alpha} x(t).
\end{aligned} \tag{1.23}$$

Theorem 1.1 ([8, 29]) *The Laplace transform formula of the Caputo fractional derivative is*

$$\int_0^{\infty} e^{-st} \{{}_0^{C}\mathcal{D}_t^{\alpha} f(t)\} \mathrm{d}t = s^{\alpha} F(s) - \sum_{k=0}^{m-1} s^{\alpha-k-1} f^{(k)}(0), \tag{1.24}$$

where $m - 1 < \alpha < m$*,* $m \in \mathbb{N}$*, and* $F(s)$ *stands for the Laplace transform of* $f(t)$*.*

Property 1.1 ([8]) *It follows from the definition of the Caputo fractional derivative that, for any* p*,* $q \in (0, 1)$*, the semigroup property holds*

$$ {}_{t_0}^{C}\mathcal{D}_t^{p} \left[{}_{t_0}^{C}\mathcal{D}_t^{q} f(t) \right] = {}_{t_0}^{C}\mathcal{D}_t^{q} \left[{}_{t_0}^{C}\mathcal{D}_t^{p} f(t) \right] = {}_{t_0}^{C}\mathcal{D}_t^{p+q} f(t). \tag{1.25}$$

Property 1.2 ([8]) *Fractional differentiation is a linear operation, i.e., for functions* $f(t)$ *and* $g(t)$,

$$ {}_{t_0}^{C}\mathcal{D}_t^{\alpha} [af(t) + bg(t)] = a {}_{t_0}^{C}\mathcal{D}_t^{\alpha} f(t) + b {}_{t_0}^{C}\mathcal{D}_t^{\alpha} g(t) \tag{1.26}$$

holds for any constants a *and* b*.*

Definition 1.12 ([29]) *The Caputo fractional integral of function* $f(t)$ *is given by*

$$ {}_{t_0}^{C}\mathcal{D}_t^{-\alpha} f(t) = \frac{1}{\Gamma(\alpha)} \int_{t_0}^{t} \frac{f(\tau)}{(t-\tau)^{1-\alpha}} \mathrm{d}\tau. \tag{1.27}$$

By comparing (1.20) and (1.22), the following remark can be made.

Remark 1.3 ([8, 29]) *The relationship between the definitions of Caputo and RL fractional derivatives can be depicted as*

$$ {}_{t_0}^{C}\mathcal{D}_t^{\alpha} f(t) = {}_{t_0}^{RL}\mathcal{D}_t^{\alpha} f(t) - \sum_{k=0}^{p} \frac{f^{(k)}(t_0)(t-t_0)^{k-\alpha}}{\Gamma(-\alpha+k+1)}, \ p+1 = m \in \mathbb{N}. \tag{1.28}$$

From (1.15), (1.17), and (1.27) we can observe that the definitions of these three types of fractional integrals are unified:

Definition 1.13 ([6]) *Let function $f(t)$ be piecewise continuous on $J' = (0, \infty)$ and integrable on any finite subinterval of $J = [0, \infty)$. Then for $t > t_0$ and $\alpha \in \mathbb{R}^+$, the fractional integral of $f(t)$ of order α is defined as follows*

$$ {}_{t_0}I_t^\alpha f(t) = \frac{1}{\Gamma(\alpha)} \int_{t_0}^t \frac{f(\tau)}{(t-\tau)^{1-\alpha}} \mathrm{d}\tau. \tag{1.29} $$

Lemma 1.2 ([70]) *Assume $x(t) \in \mathbb{R}^n$ is a smooth function, $n \in \mathbb{N}$, then for $\forall t \geq t_0$,*

$$ \frac{1}{2} {}_{t_0}^{C}\mathcal{D}_t^\alpha [x^T(t)Px(t)] \leq x^T(t)P {}_{t_0}^{C}\mathcal{D}_t^\alpha x(t), \quad 0 < \alpha < 1 \tag{1.30} $$

where $P \in \mathbb{R}^{n\times n}$ is a positive definite constant matrix.

Lemma 1.3 ([70, 82]) *If the Caputo fractional derivative ${}_{t_0}^{C}\mathcal{D}_t^\alpha f(t)$ is integrable, then*

$$ {}_{t_0}I_t^\alpha {}_{t_0}^{C}\mathcal{D}_t^\alpha f(t) = f(t) - \sum_{k=0}^{m-1} \frac{f^{(k)}(t_0)}{k!}(t-t_0)^k \tag{1.31} $$

where $m-1 < \alpha < m$, $m \in \mathbb{N}$. Particularly, for $0 < \alpha \leq 1$, it follows from (1.31) that

$$ {}_{t_0}I_t^\alpha {}_{t_0}^{C}\mathcal{D}_t^\alpha f(t) = f(t) - f(t_0). \tag{1.32} $$

Lemma 1.4 ([83]) *If $x(t) \in \mathbb{R}^n$ satisfies*

$$ {}_{t_0}^{C}\mathcal{D}_t^\alpha x(t) = f(x,t), \; x(t_0) = x_0, \tag{1.33} $$

where $0 < \alpha < 1$ and $f(x,t) \in L^1[t_0, T]$, then it also satisfies the Volterra fractional integral

$$ x(t) = x_0 + \frac{1}{\Gamma(\alpha)} \int_{t_0}^t (t-\tau)^{\alpha-1} f(x(\tau), \tau) \mathrm{d}\tau, \tag{1.34} $$

with $t_0 \leq t \leq T$ and vice versa.

Lemma 1.5 ([83]) *For the Caputo-type fractional-order linear differential equation*

$$ {}_{t_0}^{C}\mathcal{D}_t^\alpha x(t) = \mu x(t) + g(t), \; x(t_0) = x_0, \tag{1.35} $$

where $\mu \in \mathbb{R}$, $\alpha \in (0,1)$, and $g(t) \in C([t_0, t_1], \mathbb{R})$, its solution is expressed as

$$ x(t) = x_0 \mathrm{E}_\alpha\big(\mu(t-t_0)^\alpha\big) + \int_{t_0}^t (t-\tau)^{\alpha-1} \mathrm{E}_{\alpha,\alpha}\big(\mu(t-\tau)^\alpha\big) g(\tau) \mathrm{d}\tau, \;\; t_0 \leq t \leq t_1. \tag{1.36} $$

Property 1.3 ([8]) *The fractional-order derivative of a composite function $f(h(t))$ is*

$$ \begin{aligned} {}_{t_0}^{C}\mathcal{D}_t^\beta f(h(t)) = & \sum_{q=1}^{\infty} \binom{\beta}{q} \frac{(t-t_0)^{q-\beta}}{\Gamma(q-\beta+1)} \sum_{i=1}^{q} f^{(i)}(h(t)) \sum \\ & \frac{q!}{a_1! \cdots a_{q-i+1}!} \left(\frac{h^{(1)}(t)}{1!}\right)^{a_1} \cdots \left(\frac{h^{(q-i+1)}(t)}{(q-i+1)!}\right)^{a_{q-i+1}} \\ & + \frac{(t-t_0)^{-\beta}}{\Gamma(1-\beta)} f(h(t)), \end{aligned} \tag{1.37} $$

where non-negative integers a_r $(r = 1, \cdots, q-i+1)$ satisfy

$$ \sum_{r=1}^{q-i+1} a_r = i \;\; and \;\; \sum_{r=1}^{q-i+1} r a_r = q. \tag{1.38} $$

Lemma 1.6 ([84]) *Assume the function $f(t)$ on the real half axis $\mathbb{R}^+ \to \mathbb{R}^+$ is uniformly continuous and* $\lim_{t\to\infty} {}_{t_0}I_t^{\alpha} f(t) = 0$ *with* $0 < \alpha < 1$ *and* $\forall t > t_0 > 0$, *then* $\lim_{t\to\infty} f(t) = 0$.

Lemma 1.7 ([85]) *If $x(t) : [t_0, \infty) \times \mathbb{R}^n \to \mathbb{R}^n$ is uniformly continuous,* ${}_{t_0}I_t^{\alpha} |x(t)|^p \leq K$ *for* $\forall t > t_0 > 0$ *and* $0 < \alpha < 1$ *with constants* p, $K > 0$, *then* $\lim_{t\to\infty} x(t) = 0$.

Lemma 1.8 ([68]) *Let $z = 0$ be an equilibrium point for a fractional nonautonomous system ${}_{t_0}^{C}\mathcal{D}_t^{\alpha} z(t) = g(z,t)$ with initial condition $z(t_0)$ and $\alpha \in (0,1)$. Assume that there exists a Lyapunov function $W(t, z(t))$ and $\mathcal{K}$-functions ξ_j with $j = 1, 2, 3$ satisfying $\xi_1(\|z\|) \leq W(t, z(t)) \leq \xi_2(\|z\|)$ and ${}_{t_0}^{C}\mathcal{D}_t^{\beta} W(t, z(t)) \leq -\xi_3(\|z\|)$, where $0 < \beta < 1$, then the equilibrium point of the fractional system is asymptotically stable.*

As mentioned in Remark 1.2, the Caputo fractional-order derivative is considered in this book. Thus in the rest of this book, the Caputo-type fractional derivative of a function $g(t)$, i.e., ${}_0^{C}\mathcal{D}_t^{\alpha} g(t)$, will be expressed as ${}_0\mathcal{D}_t^{\alpha} g(t)$ for simplification.

1.3 Literature Review

1.3.1 Adaptive Backstepping Control for Nonlinear Fractional-Order Systems

Backstepping control is one of the most popular techniques employed in controlling nonlinear systems that applies a systematic design procedure by recursively building Lyapunov functions and virtual control signals step by step [86]. It is an important control approach for integer-order nonlinear systems with a triangular structure. Numerous results have been established to design controllers for uncertain nonlinear integer-order systems based on a backstepping scheme, see for examples [86, 87, 88, 89, 90, 91] and references therein. Recently, the backstepping control strategy has been extended to the control of nonlinear fractional-order systems. As is well known, the time derivatives of virtual control signals are needed when designing controllers for nonlinear high-order systems using the backstepping approach. With the help of the chain rule, such derivatives can be easily calculated in integer-order systems. Nevertheless, the fractional derivative of a composite function like virtual control signal is the sum of an infinite number of terms [9], which is hard, if not impossible, to be expressed in a concise closed-form compared to integer-order case. As a result, this is a challenge for fractional-order backstepping control problem.

Backstepping control strategy is applied to the fractional-order nonlinear system in [92], in which the nonlinear function only appears in the last equation of the considered system. In this case, the design of virtual control signals is straightforward without computing their fractional derivatives using the chain rule. For a more general class of uncertain nonlinear fractional-order systems, because of the lack of mathematical tools to compute fractional derivatives of composite functions succinctly, it remains a difficult task to apply the backstepping scheme to such systems. An approach to transform the backstepping control problem for fractional-order systems to integer-order is proposed in [93, 94, 95] based on the frequency distributed model [96], and under such transformation, the stability analysis for the fractional-order closed-loop systems is carried out with the integer-order Lyapunov method. Backstepping-based adaptive control scheme is developed in [97, 98, 99] for the stabilization problem of fractional-order chaotic systems without providing an analysis of the boundedness or smoothness of the designed controller. In [100, 101, 102], the fractional derivatives of virtual control signals, which as mentioned before are composite functions that contain system states and estimated parameters, are directly counteracted in the subsequent design steps during the backstepping-based controller design procedure, where the boundedness of closed-loop signals is validated by simulation studies and the corresponding theoretical proof still needs to be further explored. Bounded functions are constructed in [103] to approximate the virtual control signals inside a sufficiently large open set to avoid computing their fractional derivatives. By assuming that the fractional derivatives of virtual controllers remain bounded in compact sets, adaptive backstepping dynamic surface control schemes are proposed for uncertain fractional-order nonlinear systems in [104, 105, 106], where the fractional derivatives of virtual controllers are handled by estimating their upper bounds based on neural network technique. In [107, 108, 109, 1], instead of calculating the fractional derivatives of virtual control signals, fuzzy logic systems are employed to approximate such fractional derivatives together with systems uncertainties. Even though neural network techniques or fuzzy logic systems can be applied to handle the complicated fractional derivatives of virtual control signals when designing adaptive backstepping control law, global stability of the resulting closed-loop system is hard to be achieved since such two techniques are applicable within certain compact sets and the selection and tuning of membership functions depend heavily on experience and knowledge.

For dealing with disturbances in fractional-order nonlinear systems, nonlinear disturbances observers are proposed in [110] and [111] to counteract the effects caused by unknown disturbances. Nevertheless, strong assumptions that the disturbances have to be constants or must have bounded derivatives are needed to achieve satisfactory control performances. Adaptive controllers with sign functions are introduced in [1, 112, 113, 114] to deal with disturbances while ensuring the stabilization or tracking error converges to zero. Notably, the designed control signals in these papers are noncontinuous because of the non-continuity of the sign function.

Furthermore, the limited bandwidth available for data transmission in digital control systems makes the development of quantized control strategies essential from both a theoretical and practical standpoint. Over the past few years, the issue of stabilizing uncertain integer-order linear or nonlinear systems with adaptive quantized control inputs has gained considerable attention, see for examples [115, 116, 117, 118] and references therein. While there has been extensive research on adaptive quantized control for integer-order nonlinear systems, research on fractional nonlinear systems remains sparse. In [119] and [120], adaptive quantized control schemes are designed for fractional nonlinear systems, with stability analysis being conducted using the integer-order Lyapunov method on the basis of the frequency distributed model [96]. In [121], an issue focused on quantized control input is conducted to analyze the synchronization problem of fractional-order neural networks with delays, wherein the nonlinearities of the systems should be globally Lipschitz continuous and bounded.

The above studies merely aim at controlling the commensurate nonlinear fractional-order systems. For incommensurate nonlinear fractional-order systems, few results about the backstepping control of such a class of systems have been made. An approach that transforms the incommensurate fractional-order systems to integer-order ones based on frequency distributed model is presented in [93, 94, 122, 123, 124, 2] to implement the backstepping control scheme. However, only the stability analysis based on the transformed systems rather than the original fractional-order systems is provided in these works. An adaptive controller is proposed in [125] to achieve stabilization of a three-dimensional incommensurate fractional-order chaotic system with an assumption that its first subsystem is asymptotically stable without any control effort. In [126], the stabilization of a class of incommensurate fractional-order chaotic systems is realized via an impulsive control approach. Nevertheless, neither system uncertainties nor external disturbances are considered in both mentioned works, while the asymptotic stability of incommensurate fractional-order chaotic systems in the presence of uncertainty and disturbance is guaranteed by utilizing a discontinuous adaptive sliding mode controller, as presented in [127].

1.3.2 Sampled-Data Control for Linear Fractional-Order MASs and Systems

In consideration of constructing control signals in digital controllers, which have been widely used in practice, sampled-data control shows its potential in reducing costs and being more practical in implementation, compared with continuous control that requires control signals to be updated continuously. On the basis of H_∞ optimal theory, a sampled-data control scheme is developed for continuous-time linear integer-order systems in [128]. For nonlinear integer systems, the sampled-data control problem is addressed in [129]. In [130], a sampled-data-based controller is proposed for a class of integer-order networked control systems where the network-induced time delays are considered. [131] concerns the stability of linear time-varying integer systems based on sampled-data control strategy, and the consensus problem for second-order integer MASs is studied in [132] under a sampled-data control approach.

Different from integer-order systems that have been intensively studied for decades, the sampled-data control for fractional-order systems has only received attention in recent years. The consensus problem of the fractional-order MASs with directed graph based on sampled-data control is investigated in [4, 133, 134, 135, 136]. In [4], the consensus with or without a leader of the fractional-order MASs with single-integrator form dynamics is studied and the necessary and sufficient conditions for the systems to achieve consensus are given while the system dynamics in double-integrator form are considered among [133, 134, 135, 136]. The necessary and sufficient conditions for achieving consensus with the consideration of network structure, sampling period, and the fractional order are presented in [133]. Leaderless consensus problem of fractional-order MASs with intermittent sampled-data control scheme is addressed in [135] to reduce the updating rate and working time of controllers. Then based on [135], a consensus method for systems under input time delay is proposed in [136]. In [137], the consensus problem of linear fractional-order MASs is studied over a communication topology whose coupling structure is not necessary to be Laplacian. Event-triggering sampled-data control of fractional-order MASs is proposed in [138], where the networked topology among the agents is assumed to be undirected and the consensus of nonlinear fractional-order MASs via event-triggered control is investigated in [139]. For the containment control of linear fractional-order double-integrator MASs, control schemes based on sampled data are presented in [73, 140, 141].

From the definitions of fractional-order integrals and derivatives reviewed in Subsection 1.2.2, it can be observed that they can be treated as weighted integrals. Therefore compared with traditional integer-order ones, fractional-order calculus has hereditary and infinite memory properties, which makes the study of fractional-order control systems meaningful. Due to such properties, different from integer-order, the initial values and also the "history" of the variables over the entire time interval of integration are essential and cannot be ignored when solving fractional-order differential equations. However, after dividing the whole time interval into sampling intervals, instead of considering these properties when designing sampled-data control methods for fractional-order MASs, the aforementioned existing related literature treats the solutions of fractional-order equations at the beginning of each sampling interval as the new initial values for this interval. In this way, the closed-loop fractional-order systems are transformed into discrete-time integer-order systems in which the evolution of the one-step-ahead states merely relies on the values of current states and current control input.

By taking the hereditary and infinite memory properties into account, all the historical values of inputs should be involved for solving fractional differential equations at the point where the derivative is calculated. This implies that the present control input depends on all its previous control values. Nevertheless, it will be a computationally increasingly demanding task to implement such a control law in practice since as time goes infinite, the number of stored input data will increase dramatically. Hence utilizing only a finite number of previous inputs to compute the present control signal could be more applicable in practice and meaningful for fractional-order systems. For discrete-time fractional-order systems, a finite-dimensional approximation of the fractional-order systems by truncating the original infinite-dimensional fractional-order operators is introduced in [142, 143, 144, 145] to carry out finite-dimensional control for fractional-order systems. However, only the practical stability of the truncated approximated finite-dimensional systems rather than the actual systems is analyzed in these studies. In [146, 147, 148], the finite-dimensional approximation approach is improved by adopting an additive uncertainty term to address the approximation error. In such a way, the practical asymptotic stability of the original systems can be derived. Yet for continuous-time fractional-order systems, to the best of our knowledge, no work has addressed finite-dimensional sampled-data control for such systems. Additionally, as far as we know for a more general class of linear fractional-order systems rather than those in single-integrator form, no work about sampled-data control that takes into account the unique properties of fractional-order calculus has currently been reported.

With regard to the security control for systems subject to cyber attacks, some controllers have nowadays been created to mitigate the impact of denial-of-service (DoS) attacks [149, 150, 151] on integer-order systems [149, 152, 153, 154, 155]. In [154, 155], event-triggered resilient cooperative control schemes are proposed for continuous-time MASs under DoS attacks, such that the controlled MASs can achieve secure consensus exponentially. As for discrete-time MASs in the presence of aperiodic sampling and random DoS attacks, a distributed output-feedback control scheme is developed to reach output consensus by assuming that the sampling process is nonuniform and the consecutive attack duration is upper-bounded [153]. Despite this, there are a limited number of published security control studies for fractional-order systems in the literature. [156] and [157] analyze the control problem for continuous-time fractional-order MASs and complex networks that are vulnerable to DoS attacks, respectively. By truncating the fractional operator, which is infinite-dimensional, [142, 143, 144, 145] propose the finite-dimensional control approaches for fractional-order systems in the discrete-time domain based on the truncated approximated finite-dimensional systems. Further improvement has been made in [146, 147, 148], where the finite-dimensional approximation errors were considered in designing a controller by treating them as the additive uncertainty terms, thus ensuring practical asymptotic stability of the actual fractional-order systems. [158] studies the control of discrete-time fractional-order MASs under DoS attacks, disregarding the nonlocal characteristics of fractional-order calculus when designing the control scheme. To date, there has been limited in-depth exploration of DoS attacks and their effects on discrete-time fractional-order networked systems.

1.3.3 Adaptive Fractional-Order Backstepping Control for Nonlinear Integer-Order Systems and MASs

In comparison with the widely-studied integer-order controllers, adopting fractional-order controllers to integer-order systems can increase additional control design degrees of freedom and enable superior control performance [75]. Nowadays, numerous studies such as fractional-order PID control [12, 77, 159] and fractional-order sliding mode control [16, 17, 18, 19] have been investigated with the aim of further improving control performance by incorporating fractional-order controllers into integer-order systems.

As already mentioned in Subsection 1.3.1 that the backstepping technique, which demonstrates a step-by-step design procedure by constructing the Lyapunov function and virtual control signal at each step, is widely applied in controlling integer-order nonlinear systems in strict feedback form, see [86, 87, 88] for examples. Recently, fractional-order calculus is studied by combining it with the backstepping controller design procedure for nonlinear integer-order uncertain systems. In this regard, existing works on adaptive backstepping fractional-order control for nonlinear integer-order systems can be found in [20, 21, 22, 23, 24]. However, the results are only applicable to second-order integer-order systems. Besides, since derivatives of variables are contained in the designed control inputs in [20, 21, 22], the implementation of such controllers might cause some practical issues, as the differential operation is non-causal and will amplify measurement noises in the system. In addition, in [21] and [23], the absolute value of the error of the system output to a given reference input is required to be continuously greater than a positive limit from the initial time up till a specific non-zero time instant to ensure the stability of the closed-loop system.

The majority of existing control techniques applied to uncertain nonlinear integer-order systems with adaptive backstepping tracking control schemes focus primarily on system stability and the steady-state performance of output tracking errors. For instance, [160] and [161] proposed adaptive backstepping-based control schemes for uncertain nonlinear systems with predefined output tracking time requirements, and [162] investigates an adaptive fuzzy backstepping control of nonlinear systems with pre-given tracking time and steady-state accuracy. Nevertheless, for systems to operate safely and stably in practice, the design of the control method should also take the transient performance of the overall closed-loop system into account. With regard to this issue, certain outcomes [163, 164, 165, 166, 167, 168] concerning fractional control for integer-order systems have been reported in the study of ensuring system predefined performance. The paper [163] explores the study of adaptive fractional control for discrete-time integer-order nonlinear systems with predetermined performance requirements. For systems in the continuous-time domain, [164] and [165] investigate fractional prescribed performance sliding mode control for nonlinear integer-order systems, where system output in [164] is ensured to remain ultimately uniformly bounded with its transient path staying in specified limits. Further, the authors in [165] propose a new method by directly replacing the integer derivative of output tracking error in the dynamic model with its fractional derivative without theoretical explanation to achieve such that not only the system transient output trajectory can be restricted within predefined bounds but also the system output converges to zero. An adaptive fractional backstepping-based prescribed performance controller is developed in [166] to assure the output error of integer-order nonlinear systems goes towards zero asymptotically apart from ensuring its transient performance, wherein the employment of sign function in the design of controller leading to the occurrence of chattering in the control signals. [167] and [168] put forth prescribed performance fractional sliding mode control strategies for multiple integer-order nonlinear systems. However, the inclusion of integer derivatives or fractional derivatives of variables in the controllers designed in [164] and [166, 167, 168] can make the implementation of such controllers impractical, as the non-causal differential operation could amplify system measurement noises or disturbances. Furthermore, the aforementioned studies [163, 164, 165, 166, 167, 168] merely focus on nonlinear systems of integer degree no higher than two.

In the area of distributed control for MASs, the effectiveness of tracking in MASs heavily hinges on the ability of agents to exchange information efficiently and reliably. The transmission capability of a practical digital network is limited in accordance with the fundamental principle of communication, resulting in a restricted manageable amount of data that can be transmitted between agents during each time interval. Therefore, the investigation of the distributed tracking problem for MASs within the constraints of limited data rates imposed by communication becomes of significant importance from both a theoretical and practical standpoint. To model data-rate constraints for MASs, a quantization-encoding-decoding procedure with a finite number of quantization levels can be introduced, as described in [169], where the leaderless consensus problem for first-order MASs is studied. Further research on data-rate-constrained MASs with higher-order dynamics using quantized encryption-decryption schemes is conducted in recent works such as [170] and [171], where the leaderless fault-tolerant consensus problem is studied in [170], and [171] investigates the leaderless differentially private consensus issue. Besides, the open nature of the digital communication network also makes it susceptible to malicious cyber attacks, for instance, DoS attacks [149, 150, 151], which can disrupt information availability, causing deterioration of tracking performance and even potentially destabilize MASs. Despite this, there have been few comprehensive studies exploring the control schemes that address both data-rate restrictions and DoS attacks in either single systems [172] or MASs [173, 174]. Moreover, the control methods investigated in such aforementioned literature merely focus on linear systems that do not possess uncertainties. Nevertheless, it is noted that nonlinearity and uncertainty typically exist inherently within the majority of physical systems. The consensus control problem under data-rate constraints and DoS attacks is addressed in [175] for uncertain nonlinear MASs, where the nonlinearities are required to be locally Lipchitz. For the more general nonlinear systems with non-Lipchitz property, for example, nonholonomic mobile robots [176] and hypersonic vehicles [177], the backstepping technique [86] provides a well-structured approach for handling the tracking control issue in such nonlinear systems. In recent times, a notable development of distributed tracking control approaches utilizing adaptive backstepping techniques [178, 179] has emerged for accomplishing output tracking for high-order uncertain nonlinear MASs. However, the above distributed adaptive backstepping approach-related outcomes are established assuming an ideal network communication scenario in MASs without limitations on data rate and influence of DoS attacks. Note that the signals received by the agents in the nonlinear MASs subjected to DoS attacks and data-rate constraints formulated as a quantized-encoding-decoding process will be discontinuous, yet the high-order derivative of signals should be necessary for applying the backstepping-based control scheme for high-order nonlinear MASs, leading to the non-applicability of such control method to solve the distributed tracking problem for uncertain nonlinear MASs under the influence of DoS attacks and data-rate constraints. On the other hand, the existing works mentioned above are merely concerned with the steady-state performance of the controlled systems, while in practical circumstances, in order to operate and track the reference effectively and reliably for MASs, the development of the control method should consider not only the agent output steady-state performance but also its transient performance. [180] studies the tracking control problem for MASs to drive the transient and steady-state tracking error to the prescribed bound in the presence of DoS attacks, where the traditional integer control method is applied without considering the communication constraints among agents. Fractional calculus offers a distinct advantage by widening the scope of design control parameters [80] that can be adjusted, producing enhanced control performance and establishing superiority over integer-order approaches. Recently, a groundbreaking approach has been developed by merging fractional-order calculus with a backstepping adaptive controller for the control of nonlinear integer uncertain systems with external disturbances [21] that outperforms the conventional integer ones, offering improved control performance in both transient and steady states. In terms of assuring the prescribed performance of the system, specific results have been made in [165, 167, 168] regarding fractional control on second-order integer MASs. Nonetheless, the direct incorporation of differential operations into the controllers in [167] and [168] can pose practical implementation challenges as system disturbances or measurement noises could be magnified, resulting from the non-causal characteristics of differential operations. Moreover, the impact of restricted data transmission capacity in contemporary digital control systems underscores the significance of devising quantized control methods from both a theoretical and practical standpoint. In recent times, attention has been paid to the control problem for uncertain nonlinear systems with adaptive quantized control signals, see for instances [118] and [181], where traditional integer quantized control strategy is developed in [118] while taking into account the influence of DoS attacks, and in [181], a prescribed-performance based quantized control scheme is designed for nonlinear systems with delays. Nevertheless, to the authors' knowledge, no research has been conducted on quantized adaptive fractional backstepping control for uncertain nonlinear MASs, much less on applying this control to such systems with data-rate-constrained communication and DoS attacks, all while maintaining the prescribed output performance.

1.4 Major Contributions of the book

In recent years, fractional-order calculus has attracted increasing attention in different areas, particularly in control engineering, yet compared with the widely-studied traditional integer-order case, the research on fractional-order control systems is still at its beginning stage. Fractional-order control systems contain control for fractional-order systems and fractional-order control for systems which can be both integer and fractional order. Nowadays, to the best of the authors' knowledge, a book containing both the control for fractional-order systems in linear or nonlinear forms even with uncertainties and cyber attacks, and fractional-order control schemes is yet to be available. Besides, some researchers new to fractional-order control systems might confuse fractional-order control with the control for fractional-order systems, and some mathematical rules established in integer-order calculus needed for control laws development cannot be directly extended to fractional-order cases, which will be explained in our book. Thus, this book would be beneficial for the development of fractional-order control systems.

The major results of this book are summarized as follows.

1. The adaptive backstepping control problem for a class of uncertain single-input-single-output (SISO) commensurate fractional-order nonlinear systems in the strict feedback form with external disturbance is studied in Chapter 2. Due to the lack of proper mathematical tools, for fractional-order systems, the time derivative of composite functions, which is required for the backstepping controller design procedure, can not be easily obtained by the chain rule. To overcome this difficulty, a novel approach for approximating the fractional-order time derivatives of virtual control signals is proposed, and adaptive laws are designed to estimate the bounds of approximation errors. Moreover, an auxiliary function is employed to obtain a smooth control input for achieving asymptotic output tracking in the presence of arbitrary bounded disturbances. It is shown that the resulting closed-loop system is globally asymptotically stable under the assumption that the approximation errors between the actual values of fractional derivatives of virtual control signals and their approximated values are globally bounded. Afterward, to relax such an assumption, an alternative adaptive backstepping smooth control law is proposed, under which the obtained closed-loop fractional-order system is ensured to be stable in the sense that all the closed-loop signals are uniformly ultimately bounded and the output tracking error can be driven to a small region around the origin.
2. The development of backstepping-based adaptive control for a class of nonlinear fractional-order systems that encounter systematic uncertainties, time-varying

external disturbances, and quantized control input is explored in Chapter 3. A novel analog chain rule computation method is proposed to overcome the difficulty in handling the fractional derivatives of composite functions (i.e. virtual control signals) encountered in intermediate recursive steps of the backstepping procedure with rigorous stability analysis being provided. A quantified adaptive controller is strictly crafted to guarantee the stability of the system, with all closed-loop signals staying within bounds and the system output tracking error converging to an adjustable residual set asymptotically, leveraging the direct fractional Lyapunov method and such an approach. The simulated studies confirm the efficacy of the investigated control strategy.

3. Investigating the incommensurate fractional-order system is also of great significance yet challenge. A smooth adaptive backstepping controller for incommensurate fractional-order nonlinear systems with uncertainties and time-varying disturbance to achieve that the system output asymptotically tracks a predefined signal, while ensuring global stability of the resulted closed-loop system in the sense that all the closed-loop signals are bounded is to be developed in Chapter 4. Besides, by virtue of a smooth controller, continuous control effort can be well implemented on the actual actuators, thereby avoiding chattering and also potentially extending the lifespan of devices.
4. The consensus control of linear fractional-order MASs in single-integrator form is first studied in Chapter 5 via sampled-data control, which is an approach that possesses various benefits compared with continuous control such as low cost and being more practical in implementation. A novel scheme to design a distributed controllers with rigorous analysis is presented by utilizing the unique hereditary and infinite memory properties of fractional-order calculus. Different from existing research works where the evolution of one-step-forward states in the closed-loop systems only depends on the current control input and current states, which further can be analyzed using methods for traditional discrete-time integer-order systems, our resulting fractional-order system is rigorously analyzed by considering time evolution of the system states which relies on their initial values and also all the previous control signals. In this chapter, we also overcome a new challenge, namely the establishment of the global boundedness of the proposed distributed controller containing the weighted sum of all the historical control inputs. It is shown that with our proposed control scheme, the closed-loop system is globally stable, and asymptotic consensus for fractional-order MASs is realized. Subsequently, a finite-dimensional sampled-data-based control scheme for linear single-integrator form fractional-order systems is developed to guarantee the global stability of the corresponding closed-loop system in the sense that under any initial condition, the closed-loop states converge to an ultimate bound around the origin with control signals remain bounded. A guide for the finite-dimensional controller on selecting the control signal storage space to fulfill preset control accuracy is also provided.
5. Based on the sampled-data control method, the stabilization problem for a general class of linear fractional-order systems whose solution contains the Mittag-Leffler function that does not obey the basic exponentiation identity is addressed in Chapter 6. Note that since the existence of the Mittag-Leffler function in the solution expression of the fractional differential equation, it is impracticable to directly extend the control design procedure in Chapter 5 to this case. By considering the infinite memory and hereditary characteristics of fractional-order calculus, a novel sampled-data controllers that guarantees the corresponding closed-loop system to be asymptotically stable is investigated.

6. In Chapter 7, the security control problem of discrete-time fractional-order networked systems under DoS attacks is considered. A practically applicable finite-dimensional control strategy will be developed for fractional-order systems that possess nonlocal characteristics. By employing the Lyapunov method, it is theoretically proved that under the proposed controller, the obtained closed-loop fractional system is globally input-to-state stable (ISS), even in the presence of DoS attacks. Finally, the effectiveness of the designed control method is demonstrated by the numerical example.
7. The fractional-order adaptive backstepping control problem for a class of high-order nonlinear integer-order systems with uncertainties and unknown external disturbance is considered in Chapter 8. Fractional-order calculus is integrated within the conventional backstepping controller design procedure to obtain enhanced control performance. In contrast to the existing works that are only applicable to second-order integer systems, the considered class of uncertain nonlinear integer-order systems is allowed to be arbitrary order. The theoretical proof is provided based on the Lyapunov stability theorem to ensure the global stability of the closed-loop system in the sense that all the closed-loop signals are uniformly ultimately bounded and the output tracking error to a reference signal converges to an adjustable arbitrarily small range by tuning certain design parameters.
8. In Chapter 9, an adaptive fractional backstepping controller is designed for integer-order nonlinear uncertain systems of arbitrary order with prescribed performance subject to unknown time-varying disturbances. The integration of fractional adaptive backstepping control procedure and prescribed performance bound technique, which describes the convergence rate and largest overshoot of the output error, into a unitary framework can result in an enhanced control scheme with high precision. Based on Lyapunov analysis, it is theoretically demonstrated that under the proposed control scheme, all the closed-loop signals remain bounded and the output tracking error with respect to the reference signal, which is strictly confined within the specified performance bound for all time, will asymptotically converge to zero. Simulation studies are presented to show the effectiveness of the proposed fractional adaptive prescribed performance-based (PPB) control strategy. Additionally, this chapter presents a scheme that assures perfect output tracking and provides a transient output error trajectory that can be confined to a range with predefined convergence rate and maximum overshoot, which shows a significant improvement from Chapter 8 where fractional adaptive backstepping control is developed for high-order nonlinear integer-order uncertain systems, yet could only maintain the maximum value of the transient system output error within a tunable limit.
9. Chapter 10 studies the distributed quantized tracking control problem with prescribed output performance requirements for uncertain nonlinear integer high-order MASs with both DoS attacks and external unknown disturbances under data-rate-constrained communication networks. A hierarchical control framework including a quantized resilient distributed observer layer and a decentralized quantized adaptive fractional-order controller layer is proposed for solving the considered control problem. Specifically, quantized resilient distributed observers are designed at the observer layer for each agent to guarantee the availability of observing the reference system under the influence of data-rate-constrained communication and DoS attacks. Based on the designed observers, smooth reference trajectories with the existence of high-order derivatives, which ensure the applicability of the backstepping technique, are then constructed at the decentralized

controller layer, where a novel adaptive fractional-order PPB backstepping quantized control strategy is proposed for high-order integer uncertain systems, resulting in an enhanced control scheme with the integration of fractional calculus. It is theoretically proven that under data-rate constraints and DoS attacks, the output tracking error of each agent in relation to the reference signal will constantly strictly stay within the predefined performance limit under the proposed control scheme with all closed-loop signals remaining bounded. Simulation studies are provided as evidence to establish the efficacy of the proposed distributed quantized fractional adaptive PPB control strategy.

1.5 Organization of the Book

This book is organized as follows.

Chapter 1 first introduces the background and some basic mathematical preliminaries for fractional-order calculus, which would be useful for the analysis required in subsequent chapters. Literature review of recent related works, including adaptive backstepping control for nonlinear fractional-order systems, sampled-data control for linear fractional-order MASs and systems, and adaptive fractional-order backstepping control for nonlinear integer-order systems and MASs, is also conducted. In Chapter 2, novel adaptive backstepping smooth control methods for commensurate nonlinear uncertain fractional-order systems will be proposed. A quantizer backstepping-based control scheme will be proposed in Chapter 3, enriching the control for fractional-order nonlinear systems with uncertainties and external disturbances. Chapter 4 will demonstrate an adaptive backstepping control scheme with a smooth control signal for incommensurate uncertain nonlinear fractional-order systems. Sampled-data control for linear fractional-order MASs in the form of single integrator and finite-dimensional sampled-data control for the linear single-integrator form fractional-order systems will be investigated in Chapter 5 with considering the unique hereditary and infinite memory properties of fractional-order calculus. Chapter 6 will introduce a sampled-data control scheme developed for a more general class of linear fractional-order systems than those in the single-integrator form. The security control for fractional-order systems subject to DoS attacks will be studied in Chapter 7, which provides a novel method for defending against attacks while simultaneously ensuring stable control for systems in fractional order. In Chapter 8, an adaptive fractional-order backstepping-based controller for high-order uncertain nonlinear integer-order systems will be presented. Then, a PPB fractional-order adaptive backstepping control scheme will be further proposed in Chapter 9 for nonlinear high-order integer-order systems, showing improvement from the previous results in Chapter 8 that could only maintain the maximum value of the transient system output error within a tunable limit. The distributed quantized tracking control problem with prescribed output performance requirements for uncertain nonlinear integer high-order MASs with both DoS attacks and external unknown disturbances under data-rate-constrained communication networks will be investigated in Chapter 10. Finally, the book will be summarized in Chapter 11 with some interesting future topics provided.

2

Adaptive Backstepping Control for Commensurate Uncertain Nonlinear Fractional-Order Systems with External Disturbance

In this chapter, the adaptive backstepping-based smooth control schemes for commensurate fractional-order nonlinear systems involving both parametric uncertainties and external unknown time-varying disturbances are proposed. As is well-known, when designing controllers for high-order nonlinear systems using backstepping technology, the time derivatives of virtual control signals are required and for integer-order systems, such derivatives can be easily obtained through the chain rule. Nevertheless, this is not the case for fractional-order systems. A novel approach to approximate the fractional-order time derivatives of virtual control signals is proposed in this chapter to handle such difficulty encountered in applying backstepping control scheme to fractional-order systems and corresponding adaptive laws are designed to estimate the bounds of approximation errors. Furthermore, an auxiliary function instead of the sign function is used in the designed controller to compensate for the effect induced by the unknown external disturbance in order to get rid of the chattering phenomenon that appears in the control signal. The fractional directed Lyapunov method is used for the stability analysis of the corresponding closed-loop system.

2.1 Introduction

The adaptive backstepping control problem for a class of uncertain SISO commensurate fractional-order nonlinear systems in the strict feedback form with external disturbance is studied in this chapter. As mentioned in Subsection 1.3.1 that due to the lack of proper mathematical tools, for fractional-order systems, the time derivative of composite functions, which is required for the backstepping controller design procedure, can not be easily obtained by the chain rule. To overcome this difficulty, a novel approach for approximating the fractional-order time derivatives of virtual control signals is proposed in Section 2.3 and adaptive laws are designed to estimate the bounds of approximation errors. Moreover, an auxiliary function is employed to obtain a smooth control input for achieving asymptotic output tracking in the presence of arbitrary bounded disturbances. It is shown that the resulting closed-loop system is globally asymptotically stable under the assumption that the approximation errors between the actual values of fractional derivatives of virtual control signals and their approximated values are globally bounded. Afterward, to relax such an assumption, an alternative adaptive backstepping smooth control law is proposed in Section 2.4, under which the obtained closed-loop fractional-order system is ensured to be

DOI: 10.1201/9781003620372-2

stable in the sense that all the closed-loop signals are uniformly ultimately bounded and the output tracking error can be driven to a small region around the origin.

In summary, the main contributions of this chapter are threefold:

1. Differ from [1, 112, 113, 114], where noncontinuous sign function is applied in the designed controllers to handle disturbances, an auxiliary function is employed in this chapter to obtain a smooth control input, avoiding the chattering phenomenon in the control signal.
2. In contrast to [93, 94, 95, 97, 98, 99, 100, 101, 102, 103, 104, 105, 106, 107, 108, 109, 1], a novel approach for computing the fractional derivative of composite functions is proposed to ensure that based on the analysis of the original fractional-order systems, with an assumption that the approximation errors between the actual values of fractional derivatives of composite functions and their approximated values are globally bounded, the closed-loop system is stable in the sense that all signals are bounded with output tracking error to a given reference trajectory is asymptotically stabilized.
3. The aforementioned assumption in contribution 2 is further relaxed by a newly proposed alternative adaptive backstepping smooth control scheme.

This chapter is organized as follows. Section 2.2 formulates the problem. The main results on the asymptotic tracking controller design are presented in Section 2.3, and Section 2.4 introduces the alternative controller design with relaxed assumption. Some simulation results are given in Section 2.5 to illustrate and verify the effectiveness of the proposed schemes, and Section 2.6 concludes this chapter.

2.2 Problem Formulation

Consider the following class of Caputo-type nth-order SISO uncertain commensurate fractional-order nonlinear systems with time-varying disturbance in strict feedback form:

$$\begin{cases} {}_0\mathcal{D}_t^{\alpha} x_i(t) = x_{i+1}(t) + \phi_i^T(\overline{x}_i(t))\theta_i \\ {}_0\mathcal{D}_t^{\alpha} x_n(t) = g(x(t))u(t) + \phi_n^T(x(t))\theta_n + d(t) \\ y(t) = x_1(t) \end{cases} \tag{2.1}$$

where $i = 1, \cdots, n-1$, the fractional-orders of all the state equations are equal to $\alpha \in (0, 1]$, $u(t) \in \mathbb{R}$ and $y(t) \in \mathbb{R}$ represent the control input and system output respectively, $x(t) = [x_1(t), x_2(t), \cdots, x_n(t)]^T \in \mathbb{R}^n$ denotes the measurable state vector, $\overline{x}_i(t) = [x_1(t), x_2(t), \cdots, x_i(t)]^T \in \mathbb{R}^i$, $\phi_j(\overline{x}_j(t)) \in \mathbb{R}^p$ $(j = 1, 2, \cdots, n)$ is a vector with its elements being known nonlinear smooth functions, $\theta_j \in \mathbb{R}^p$ is an unknown constant vector, $g(x(t))$ is a known non-zero smooth nonlinear function, and $d(t)$ stands for an unknown bounded time-varying external disturbance with unknown bound D.

Assumption 2.1 *The given reference signal $r(t)$ and its αth-order Caputo-type fractional derivative ${}_0\mathcal{D}_t^{\alpha} r(t)$ are smooth and bounded.*

The control objective is to design the backstepping-based adaptive controller for uncertain fractional-order nonlinear systems (2.1) to ensure the desired reference trajectory is tracked in the presence of external disturbances while closed-loop signals remain bounded.

Remark 2.1 *Different from the control strategies in [182, 183, 184, 185] that are designed with disturbance observer or disturbance resilience, a fractional-order adaptive law, which can be found in (2.22) and (2.52), is to be derived to estimate the unknown upper bound D of the disturbance. The external time-varying disturbance $d(t)$ is allowed to be norm-bounded without any condition on its derivative in this chapter. By employing the estimated bound in designing control input u, such unknown disturbance can be compensated, which helps us to solve the above control problem.*

2.3 Asymptotic Tracking Controller Design and Stability Analysis

The control problem in this section is to design an adaptive controller for the class of systems described in (2.1) such that the following objectives are achieved: 1) The closed-loop system is globally stable in the sense that all the signals including parameter estimates are bounded; 2) The system output $y(t)$ asymptotically tracks a reference signal $r(t)$, i.e., $\lim_{t\to\infty}[y(t) - r(t)] = 0$.

To develop and analyze the proposed control scheme, the following lemma from [186] is introduced.

Lemma 2.1 ([186]) *The following inequality holds for $x(t) \in C^1([0,t], \mathbb{R})$:*

$$|x| < \epsilon + xsg(x, \epsilon) \tag{2.2}$$

where

$$sg(x, \epsilon) = \frac{x}{\sqrt{x^2 + \epsilon^2}} \tag{2.3}$$

and $\epsilon(t)$ is a differentiable function which meets $\epsilon > 0$ and $\int_0^t \epsilon(\tau)\mathrm{d}\tau < \infty$, $\forall t \geq 0$.

From equation (2.3), it can be derived that $sg(x, \epsilon)$ is differentiable with respect to time t and thus it is a smooth function for $\forall t \geq 0$.

2.3.1 Backstepping-Based Adaptive Controller Design for Asymptotic Tracking of Reference Signal

To achieve the above objectives, an adaptive controller is designed based on the backstepping design procedure. Defining virtual errors as follows: $z_1 = x_1 - r$, which is also known as tracking error, $z_i = x_i - \alpha_{i-1}$ $(i = 2, \ldots, n)$, where α_j $(j = 1, 2, \ldots, n-1)$ denotes the virtual control signal that will be designed iteratively later.

Step 1: Consider the first Lyapunov function $\overline{V}_1 = \frac{1}{2}{z_1}^2$. According to Lemma 1.2, the Caputo derivative of $\overline{V}_1$ is

$$\begin{aligned}
{}_0\mathcal{D}_t^\alpha \overline{V}_1 &= \frac{1}{2}{}_0\mathcal{D}_t^\alpha {z_1}^2 \leq z_1 {}_0\mathcal{D}_t^\alpha z_1 = z_1\left[{}_0\mathcal{D}_t^\alpha x_1 - {}_0\mathcal{D}_t^\alpha r\right] \\
&= z_1\left[x_2 + \phi_1^T(x_1)\theta_1 - {}_0\mathcal{D}_t^\alpha r\right] = z_1\left[z_2 + \alpha_1 + \phi_1^T(x_1)\theta_1 - {}_0\mathcal{D}_t^\alpha r\right] \\
&= z_1 z_2 + z_1\alpha_1 + z_1\phi_1^T(x_1)\theta_1 - z_1 {}_0\mathcal{D}_t^\alpha r.
\end{aligned} \tag{2.4}$$

Let virtual control signal α_1 be

$$\alpha_1 = -c_1 z_1 - \phi_1^T(x_1)\hat{\theta}_1 + {}_0\mathcal{D}_t^\alpha r, \tag{2.5}$$

where $\hat{\theta}_1$ stands for the estimate of θ_1 and c_1 is a positive design parameter. Defining $\tilde{\theta}_1 = \theta_1 - \hat{\theta}_1$, then we have ${}_0\mathcal{D}_t^\alpha \tilde{\theta}_1 = -{}_0\mathcal{D}_t^\alpha \hat{\theta}_1$. Therefore, inequality (2.4) becomes

$$\begin{aligned}{}_0\mathcal{D}_t^\alpha \overline{V}_1 \leq & z_1 z_2 + z_1\alpha_1 + z_1\phi_1^T(x_1)\theta_1 - z_1{}_0\mathcal{D}_t^\alpha r\\ =& z_1z_2 - c_1{z_1}^2 - z_1\phi_1^T(x_1)\hat{\theta}_1 + z_1{}_0\mathcal{D}_t^\alpha r + z_1\phi_1^T(x_1)\theta_1 - z_1{}_0\mathcal{D}_t^\alpha r\\ =& z_1z_2 - c_1{z_1}^2 + z_1\phi_1^T(x_1)\tilde{\theta}_1. \end{aligned} \tag{2.6}$$

Let $V_1 = \overline{V}_1 + \frac{1}{2}\tilde{\theta}_1^T\Gamma_1^{-1}\tilde{\theta}_1$, where Γ_1 is a positive definite matrix, then the Caputo derivative of V_1 is

$$\begin{aligned}{}_0\mathcal{D}_t^\alpha V_1 =& {}_0\mathcal{D}_t^\alpha \overline{V}_1 + \frac{1}{2}{}_0\mathcal{D}_t^\alpha \tilde{\theta}_1^T\Gamma_1^{-1}\tilde{\theta}_1 \leq {}_0\mathcal{D}_t^\alpha \overline{V}_1 + \tilde{\theta}_1^T{\Gamma_1}^{-1}{}_0\mathcal{D}_t^\alpha\tilde{\theta}_1\\ \leq& z_1z_2 - c_1{z_1}^2 + z_1\phi_1^T(x_1)\tilde{\theta}_1 + \tilde{\theta}_1^T{\Gamma_1}^{-1}{}_0\mathcal{D}_t^\alpha\tilde{\theta}_1\\ =& z_1z_2 - c_1{z_1}^2 + z_1\phi_1^T(x_1)\tilde{\theta}_1 - \tilde{\theta}_1^T{\Gamma_1}^{-1}{}_0\mathcal{D}_t^\alpha\hat{\theta}_1. \end{aligned} \tag{2.7}$$

Designing adaptive law as

$${}_0\mathcal{D}_t^\alpha \hat{\theta}_1 = z_1\Gamma_1\phi_1(x_1), \tag{2.8}$$

then inequality (2.7) becomes

$${}_0\mathcal{D}_t^\alpha V_1 \leq z_1z_2 - c_1{z_1}^2 + z_1\phi_1^T(x_1)\tilde{\theta}_1 - \tilde{\theta}_1^T{\Gamma_1}^{-1}\Gamma_1\phi_1(x_1)z_1 = z_1z_2 - c_1{z_1}^2. \tag{2.9}$$

Step i ($i = 2, \cdots, n-1$): The fractional-order derivative of z_i is

$$\begin{aligned}{}_0\mathcal{D}_t^\alpha z_i =& {}_0\mathcal{D}_t^\alpha x_i - {}_0\mathcal{D}_t^\alpha \alpha_{i-1} = x_{i+1} + \phi_i^T(\overline{x}_i)\theta_i - {}_0\mathcal{D}_t^\alpha\alpha_{i-1}\\ =& z_{i+1} + \alpha_i + \phi_i^T(\overline{x}_i(t))\theta_i - {}_0\mathcal{D}_t^\alpha\alpha_{i-1}. \end{aligned} \tag{2.10}$$

Inspired by the idea in [187] and [188], where ${}_0\mathcal{D}_x^\alpha f[u(x)]$ is approximated by $f_u'(u){}_0\mathcal{D}_x^\alpha u(x)$ with approximation error $\rho(t)$ which is bounded by an unknown bound $\bar{\rho}$, ${}_0\mathcal{D}_t^\alpha\alpha_{i-1}$ is approximated by $\sum_{j=1}^{i-1}\left(\frac{\partial\alpha_{i-1}}{\partial x_j}\right){}_0\mathcal{D}_t^\alpha x_j + \sum_{j=1}^{i-1}\left(\frac{\partial\alpha_{i-1}}{\partial\hat{\theta}_{v,j}}\right){}_0\mathcal{D}_t^\alpha\hat{\theta}_{v,j}$. Therefore by defining $\rho_{i-1}(t)$ as the approximation error with bound $\bar{\rho}_{i-1}$, (2.10) can be written as

$${}_0\mathcal{D}_t^\alpha z_i = z_{i+1} + \alpha_i + \phi_{v,i}^T\theta_{v,i} + \zeta_i + \rho_{i-1}. \tag{2.11}$$

The functions and parameters in (2.11) are given below

$$\begin{aligned}\phi_{v,i} =& \left[\phi_i^T(\overline{x}_i), \left(-\frac{\partial\alpha_{i-1}}{\partial x_{i-1}}\right)\phi_{i-1}^T(\overline{x}_{i-1}), \cdots, \left(-\frac{\partial\alpha_{i-1}}{\partial x_1}\right)\phi_1^T(\overline{x}_1)\right]^T,\\ \theta_{v,i} =& \left[\theta_i^T, \theta_{i-1}^T, \cdots, \theta_1^T\right]^T,\\ \zeta_i =& \sum_{j=2}^{i}\left(-\frac{\partial\alpha_{i-1}}{\partial x_{j-1}}\right)x_j + \sum_{j=1}^{i-1}\left(-\frac{\partial\alpha_{i-1}}{\partial\hat{\theta}_{v,j}}\right){}_0\mathcal{D}_t^\alpha\hat{\theta}_{v,j}, \end{aligned}$$

where $\hat{\theta}_{v,j}$ is the estimate of $\theta_{v,j}$, $j = 1, 2, \cdots, i-1$.

Choose virtual control signal α_i as

$$\alpha_i = -z_{i-1} - c_iz_i - \phi_{v,i}^T\hat{\theta}_{v,i} - \zeta_i - \hat{\rho}_{i-1}, \tag{2.12}$$

where c_i denotes a positive design parameter, $\hat{\theta}_{v,i}$ means the estimate of $\theta_{v,i}$, and $\hat{\rho}_{i-1}$ is the estimation of $\bar{\rho}_{i-1}$ with their adaptive law, respectively, designed as

$$\begin{cases}{}_0\mathcal{D}_t^\alpha\hat{\theta}_{v,i} = z_i\Gamma_i\phi_{v,i},\\ {}_0\mathcal{D}_t^\alpha\hat{\rho}_{i-1} = z_i\lambda_{i-1},\end{cases} \tag{2.13}$$

where Γ_i is positive definite and $\lambda_{i-1} > 0$. Let $V_i = V_{i-1} + \frac{1}{2}{z_i}^2 + \frac{1}{2}\tilde{\theta}_{v,i}^T \Gamma_i^{-1}\tilde{\theta}_{v,i} + \frac{1}{2\lambda_{i-1}}\tilde{\rho}_{i-1}^2$, where $\tilde{\theta}_{v,i} = \theta_{v,i} - \hat{\theta}_{v,i}$ and $\tilde{\rho}_{i-1} = \bar{\rho}_{i-1} - \hat{\rho}_{i-1}$, then

$$\begin{aligned} {}_0\mathcal{D}_t^\alpha V_i \leq & {}_0\mathcal{D}_t^\alpha V_{i-1} + z_i {}_0\mathcal{D}_t^\alpha z_i + \tilde{\theta}_{v,i}^T \Gamma_i^{-1} {}_0\mathcal{D}_t^\alpha \tilde{\theta}_{v,i} + \frac{1}{\lambda_{i-1}} \tilde{\rho}_{i-1} {}_0\mathcal{D}_t^\alpha \tilde{\rho}_{i-1} \\ = & {}_0\mathcal{D}_t^\alpha V_{i-1} + z_i {}_0\mathcal{D}_t^\alpha z_i - \tilde{\theta}_{v,i}^T \Gamma_i^{-1} {}_0\mathcal{D}_t^\alpha \hat{\theta}_{v,i} - \frac{1}{\lambda_{i-1}} \tilde{\rho}_{i-1} {}_0\mathcal{D}_t^\alpha \hat{\rho}_{i-1}. \end{aligned} \tag{2.14}$$

Since ${}_0\mathcal{D}_t^\alpha V_{i-1} \leq z_{i-1}z_i - \sum_{j=1}^{i-1} c_j {z_j}^2$, thus we get

$${}_0\mathcal{D}_t^\alpha V_i \leq z_i z_{i+1} - \sum_{j=1}^{i} c_j {z_j}^2. \tag{2.15}$$

Step n: The fractional-order derivative of z_n is

$${}_0\mathcal{D}_t^\alpha z_n = g(x)u + d + \phi_{v,n}^T \theta_{v,n} + \zeta_n + \rho_{n-1}, \tag{2.16}$$

where

$$\begin{aligned} \phi_{v,n} = & \left[\phi_n^T(x), \left(-\frac{\partial \alpha_{n-1}}{\partial x_{n-1}} \right) \phi_{n-1}^T(\overline{x}_{n-1}), \cdots, \left(-\frac{\partial \alpha_{n-1}}{\partial x_1} \right) \phi_1^T(\overline{x}_1) \right]^T, \\ \theta_{v,n} = & \left[\theta_n^T, \theta_{n-1}^T, \cdots, \theta_1^T \right]^T, \\ \zeta_n = & \sum_{j=2}^{n} \left(-\frac{\partial \alpha_{n-1}}{\partial x_{j-1}} \right) x_j + \sum_{j=1}^{n-1} \left(-\frac{\partial \alpha_{n-1}}{\partial \hat{\theta}_{v,j}} \right) {}_0\mathcal{D}_t^\alpha \hat{\theta}_{v,j}, \\ \rho_{n-1} = & \sum_{j=1}^{n-1} \left(\frac{\partial \alpha_{n-1}}{\partial x_j} \right) {}_0\mathcal{D}_t^\alpha x_j + \sum_{j=1}^{n-1} \left(\frac{\partial \alpha_{n-1}}{\partial \hat{\theta}_{v,j}} \right) {}_0\mathcal{D}_t^\alpha \hat{\theta}_{v,j} - {}_0\mathcal{D}_t^\alpha \alpha_{n-1}, \\ \bar{\rho}_{n-1} \geq & \left| \rho_{n-1} \right|. \end{aligned}$$

Considering the Lyapunov function $\overline{V}_n = V_{n-1} + \frac{1}{2}{z_n}^2$, then the Caputo derivative of $\overline{V}_n$ is

$$\begin{aligned} {}_0\mathcal{D}_t^\alpha \overline{V}_n \leq & {}_0\mathcal{D}_t^\alpha V_{n-1} + z_n {}_0\mathcal{D}_t^\alpha z_n \\ \leq & z_{n-1} z_n - \sum_{j=1}^{n-1} c_j {z_j}^2 + z_n \left[g(x)u + d + \phi_{v,n}^T \theta_{v,n} + \zeta_n + \bar{\rho}_{n-1} \right] \\ \leq & z_{n-1} z_n - \sum_{j=1}^{n-1} c_j {z_j}^2 + z_n \left[g(x)u + \phi_{v,n}^T \theta_{v,n} + \zeta_n + \bar{\rho}_{n-1} \right] + \left| z_n \right| D, \end{aligned} \tag{2.17}$$

where D is an unknown bound of $d(t)$. Considering Lemma 2.1, since $|z_n| < \epsilon + z_n sg(z_n, \epsilon)$, hence $|z_n| D \leq \epsilon D + z_n sg(z_n, \epsilon) D$. Hence (2.17) can be written as

$$\begin{aligned} {}_0\mathcal{D}_t^\alpha \overline{V}_n \leq & z_{n-1} z_n - \sum_{j=1}^{n-1} c_j {z_j}^2 + z_n \left[g(x)u + \phi_{v,n}^T \theta_{v,n} + \zeta_n + \bar{\rho}_{n-1} \right] \\ & + \epsilon D + z_n sg(z_n, \epsilon) D. \end{aligned} \tag{2.18}$$

Let $V_n = V_{n-1} + \frac{1}{2}{z_n}^2 + \frac{1}{2}\tilde{\theta}_{v,n}^T \Gamma_n^{-1}\tilde{\theta}_{v,n} + \frac{1}{2\lambda_{n-1}}\tilde{\rho}_{n-1}^2 + \frac{1}{2\eta}\tilde{D}^2$, where Γ_n is positive definite, $\eta > 0$, $\lambda_{n-1} > 0$, $\tilde{\theta}_{v,n} = \theta_{v,n} - \hat{\theta}_{v,n}$, $\tilde{\rho}_{n-1} = \bar{\rho}_{n-1} - \hat{\rho}_{n-1}$, $\tilde{D} = D - \hat{D}$, and $\hat{\theta}_{v,n}$, $\hat{\rho}_{n-1}$ and $\hat{D}$ represent the estimates of $\theta_{v,n}$, $\bar{\rho}_{n-1}$ and D, respectively. Then

$$
\begin{aligned}
{}_0\mathcal{D}_t^\alpha V_n \leq & {}_0\mathcal{D}_t^\alpha V_{n-1} + z_n {}_0\mathcal{D}_t^\alpha z_n + \tilde{\theta}_{v,n}^T \Gamma_n^{-1} {}_0\mathcal{D}_t^\alpha \tilde{\theta}_{v,n} + \frac{1}{\lambda_{n-1}}\tilde{\rho}_{n-1} {}_0\mathcal{D}_t^\alpha \tilde{\rho}_{n-1} + \frac{1}{\eta}\tilde{D} {}_0\mathcal{D}_t^\alpha \tilde{D} \\
= & {}_0\mathcal{D}_t^\alpha V_{n-1} + z_n {}_0\mathcal{D}_t^\alpha z_n - \tilde{\theta}_{v,n}^T \Gamma_n^{-1} {}_0\mathcal{D}_t^\alpha \hat{\theta}_{v,n} - \frac{1}{\lambda_{n-1}}\tilde{\rho}_{n-1} {}_0\mathcal{D}_t^\alpha \hat{\rho}_{n-1} - \frac{1}{\eta}\tilde{D} {}_0\mathcal{D}_t^\alpha \hat{D}. \quad (2.19)
\end{aligned}
$$

According to (2.17) and (2.18), we then have

$$
\begin{aligned}
{}_0\mathcal{D}_t^\alpha V_n \leq & z_{n-1}z_n - \sum_{j=1}^{n-1} c_j {z_j}^2 + z_n\big[g(x)u + \phi_{v,n}^T\theta_{v,n} + \zeta_n + \bar{\rho}_{n-1}\big] + \epsilon D \\
& + z_n sg(z_n, \epsilon)D - \tilde{\theta}_{v,n}^T \Gamma_n^{-1} {}_0\mathcal{D}_t^\alpha \hat{\theta}_{v,n} - \frac{1}{\lambda_{n-1}}\tilde{\rho}_{n-1} {}_0\mathcal{D}_t^\alpha \hat{\rho}_{n-1} - \frac{\tilde{D} {}_0\mathcal{D}_t^\alpha \hat{D}}{\eta}. \quad (2.20)
\end{aligned}
$$

Finally the real control input u is designed as

$$
u = \frac{1}{g(x)}\Big[- z_{n-1} - c_n z_n - \phi_{v,n}^T\hat{\theta}_{v,n} - sg(z_n, \epsilon)\hat{D} - \zeta_n - \hat{\rho}_{n-1}\Big], \quad (2.21)
$$

where function $\epsilon(t) = e^{-at}$ with a being a positive constant and c_n is a positive design parameter.

Also we have the following adaptive laws designed

$$
\begin{cases}
{}_0\mathcal{D}_t^\alpha \hat{\theta}_{v,n} = z_n \Gamma_n \phi_{v,n}, \\
{}_0\mathcal{D}_t^\alpha \hat{\rho}_{n-1} = z_n \lambda_{n-1}, \\
{}_0\mathcal{D}_t^\alpha \hat{D} = \eta z_n sg(z_n, \epsilon).
\end{cases} \quad (2.22)
$$

Hence by substituting (2.21) and (2.22) into (2.20), we have

$$
{}_0\mathcal{D}_t^\alpha V_n \leq -\sum_{j=1}^{n} c_j {z_j}^2 + \epsilon D. \quad (2.23)
$$

Remark 2.2 *Function $sg(z_n, \epsilon)$ in the control law (2.21) is engaged to compensate for the effects of the unknown external time-varying disturbance. Due to the fact that $z_n d \leq |z_n|D$ shown in (2.17), such compensation is achieved by handling its bound D. Furthermore, to employ $sg(z_n, \epsilon)$ in the control input for the compensation, we choose function $\epsilon(t) = e^{-at} > 0$ where $a > 0$. Since this function satisfies $\int_0^t \epsilon(\tau)d\tau < \infty$, $\forall t \geq 0$, it, together with other parts of the proposed adaptive controller including the estimated $\hat{D}$ obtained from the adaptive law in (2.22), ensures that perfect asymptotic output tracking is achieved in this section, which is proved in the next subsection. Also since $sg(z_n, \epsilon)$ is differentiable, the obtained control signal in (2.21) is smooth.*

Remark 2.3 *Different from some existing control schemes such as that in [185], we do not estimate the actual disturbance for achieving the control objectives. Instead, we estimate the upper bound D of the disturbance. In addition, the adaptive laws in (2.22), including the law for updating $\hat{D}$, and control law (2.21) are derived from (2.20) in such a way that (2.23) is obtained. Our main results in the next subsection are proved based on (2.23). As seen from the proof, $\hat{D}$ is not required to converge to the true unknown bound D and, in fact, such a convergence is not one of the control objectives.*

2.3.2 Stability Analysis

For the adaptive controller design above, we have the following result.

Theorem 2.1 *Consider the closed-loop system consisting of fractional system (2.1) and adaptive controller with control law (2.21) and adaptive laws (2.8), (2.13) and (2.22). The system is globally stable in the sense that all signals in the closed-loop system are uniformly bounded and also, asymptotic tracking is achieved, i.e.,* $\lim_{t\to\infty}[y(t) - r(t)] = 0$.

Proof: Taking fractional integration of both sides of inequality (2.23) gives

$$V_n(t) \le {}_0I_t^\alpha\Big(-\sum_{j=1}^{n} c_j{z_j}^2\Big) + {}_0I_t^\alpha(\epsilon D) + V_n(0) \le {}_0I_t^\alpha(\epsilon D) + V_n(0). \tag{2.24}$$

In (2.24), the first term on the right-hand side can be computed as follows:

$$\begin{aligned} {}_0I_t^\alpha(\epsilon D) &= \frac{D}{\Gamma(\alpha)}\int_0^t \frac{\epsilon(\tau)}{(t-\tau)^{1-\alpha}}\mathrm{d}\tau = \frac{D}{\Gamma(\alpha)}\int_0^t \frac{e^{-a\tau}}{(t-\tau)^{1-\alpha}}\mathrm{d}\tau \\ &= Dt^\alpha \mathrm{E}_{1,\,\alpha+1}(-at). \end{aligned} \tag{2.25}$$

According to Lemma 1.1, since $|\arg(-at)| = \pi$ and $|-at| \to \infty$ for $t \to \infty$, by choosing integer $n = 1$, then we have

$$\mathrm{E}_{1,\,\alpha+1}(-at) = -\frac{(-at)^{-1}}{\Gamma(\alpha+1-1)} + \mathrm{o}(|-at|^{-2}) = \frac{a^{-1}}{\Gamma(\alpha)t} + \mathrm{o}\Big(\frac{1}{|-at|^2}\Big). \tag{2.26}$$

Hence following (2.26),

$$t^\alpha \mathrm{E}_{1,\,\alpha+1}(-at) = \frac{a^{-1}}{\Gamma(\alpha)t^{1-\alpha}} + t^\alpha \mathrm{o}\Big(\frac{1}{|-at|^2}\Big). \tag{2.27}$$

As $t \to \infty$, the right-hand side of (2.27) tends to 0. Therefore, we get $\lim_{t\to\infty} t^\alpha \mathrm{E}_{1,\,\alpha+1}(-at) = 0$, which implies $\lim_{t\to\infty} {}_0I_t^\alpha(\epsilon D)$ is also 0. From Lemma 1.6, we get $\lim_{t\to\infty}(\epsilon D) = 0$. Finally, we can come to the conclusion that $V_n(t)$ is bounded from (2.24), thus every signal in $V_n(t)$ is bounded. Then all the virtual control signals $\alpha_j\,(j = 1, 2, \ldots, n-1)$ are bounded in accordance with (2.5) and (2.12) and further $x_i\,(i = 1, 2, \ldots, n)$ are also bounded. Besides, from (2.21) it can be noticed that u is bounded. Therefore, from system (2.1) it can be shown that the derivatives of x_i exist and are bounded, which reveals that x_i are uniformly continuous.

Since r is uniformly continuous according to Assumption 2.1, hence $z_1 = x_1 - r$ is also uniformly continuous. As a result, α_1 is uniformly continuous from (2.5) and (2.8). Subsequently, as $z_i = x_i - \alpha_{i-1}\,(i = 2, \ldots, n)$, z_i can be proved recursively that are uniformly continuous. From (2.24) we know that ${}_0I_t^\alpha(\sum_{j=1}^n c_j{z_j}^2)$ is bounded. According to Lemma 1.7, we have $\lim_{t\to\infty} z_j = 0$, $j = 1, 2, \cdots, n$. Therefore the output $y(t)$ tracks reference signal $r(t)$ asymptotically. ■

Remark 2.4 *If the control objective is to globally asymptotically stabilize the system, it could be ensured with the same procedure by treating* $r(t) = 0$. *In addition, if* $\alpha = 1$, *then the corresponding results in this chapter become those for the integer-order case.*

2.4 Controller Design and Stability Analysis under Relaxed Assumption

In Section 2.3, a state feedback smooth adaptive backstepping control scheme is proposed to achieve asymptotic output tracking for fractional systems (2.1). Nevertheless, the conclusion is established under the assumption that the approximation errors between the true values of fractional derivatives of virtual control signals and their approximated values are globally bounded. The main purpose of this section is to relax such an assumption. An adaptive controller for systems (2.1) will be designed in this section to achieve: 1) The corresponding closed-loop system is stable in the sense that all the signals including parameter estimates are uniformly ultimately bounded; 2) The system output $y(t)$ can track a reference signal $r(t)$ well.

2.4.1 Design of Adaptive Backstepping Controller with Relaxed System Assumption

Similar to the previous section, this subsection designs an adaptive controller based on the backstepping design procedure. Defining tracking error and virtual errors respectively as $z_1 = x_1 - r$ and $z_i = x_i - \alpha_{i-1}\,(i = 2, \ldots, n)$ with $\alpha_j\,(j = 1, 2, \ldots, n-1)$ signifying the virtual control signal to be designed.

Step 1: Construct the first Lyapunov function $\overline{V}_1 = \frac{1}{2}{z_1}^2$. In accordance with Lemma 1.2, the fractional derivative of $\overline{V}_1$ is

$$ {}_0\mathcal{D}_t^{\alpha}\overline{V}_1 \leq z_1 {}_0\mathcal{D}_t^{\alpha} z_1 = z_1 z_2 + z_1\alpha_1 + z_1\phi_1^T(x_1)\theta_1 - z_1 {}_0\mathcal{D}_t^{\alpha} r. \tag{2.28} $$

Let virtual control signal α_1 be designed as

$$ \alpha_1 = -c_1 z_1 - \phi_1^T(x_1)\hat{\theta}_1 + {}_0\mathcal{D}_t^{\alpha} r, \tag{2.29} $$

where $\hat{\theta}_1$ denotes the estimate of θ_1 and c_1 is a positive design parameter. Defining $\tilde{\theta}_1 = \hat{\theta}_1 - \theta_1$, then inequality (2.28) becomes

$$ {}_0\mathcal{D}_t^{\alpha}\overline{V}_1 \leq z_1 z_2 - c_1 {z_1}^2 - z_1\phi_1^T(x_1)\tilde{\theta}_1. \tag{2.30} $$

Let $V_1 = \overline{V}_1 + \frac{1}{2}\tilde{\theta}_1^T{\Gamma_1}^{-1}\tilde{\theta}_1$, where Γ_1 is a positive definite matrix, then the fractional derivative of V_1 is

$$ {}_0\mathcal{D}_t^{\alpha} V_1 \leq z_1 z_2 - c_1 {z_1}^2 - z_1\phi_1^T(x_1)\tilde{\theta}_1 + \tilde{\theta}_1^T\Gamma_1^{-1} {}_0\mathcal{D}_t^{\alpha}\hat{\theta}_1. \tag{2.31} $$

Design an adaptive law to estimate θ_1 as

$$ {}_0\mathcal{D}_t^{\alpha}\hat{\theta}_1 = z_1\Gamma_1\phi_1(x_1) - \nu\Gamma_1\hat{\theta}_1, \tag{2.32} $$

where $\nu > 0$. Then inequality (2.31) becomes

$$ {}_0\mathcal{D}_t^{\alpha} V_1 \leq z_1 z_2 - c_1 {z_1}^2 - \nu\tilde{\theta}_1^T\hat{\theta}_1. \tag{2.33} $$

Step 2: The fractional-order derivative of z_2 is

$$ {}_0\mathcal{D}_t^{\alpha} z_2 = {}_0\mathcal{D}_t^{\alpha} x_2 - {}_0\mathcal{D}_t^{\alpha}\alpha_1 = z_3 + \alpha_2 + \phi_2^T(\overline{x}_2)\theta_2 - {}_0\mathcal{D}_t^{\alpha}\alpha_1. \tag{2.34} $$

Let ${}_0\mathcal{D}_t^{\alpha}\alpha_1$ be computed as follows

$$ {}_0\mathcal{D}_t^{\alpha}\alpha_1 = \frac{\partial\alpha_1}{\partial x_1} {}_0\mathcal{D}_t^{\alpha} x_1 + \frac{\partial\alpha_1}{\partial\hat{\theta}_1} {}_0\mathcal{D}_t^{\alpha}\hat{\theta}_1 - \rho_1, \tag{2.35} $$

where ρ_1 denotes the unknown approximation error function.

Remark 2.5 *The fractional derivative of virtual control signal α_1 in the form of (2.35) is different from that in [109, 1] where fuzzy logic is used to approximate the fractional derivative of α_1 without considering the information of system states and α_1. Note that in Section 2.3, the error function ρ_i $(i = 1, 2, \cdots, n-1)$ is assumed to be globally bounded by an unknown constant and an adaptive law is developed to estimate the bound for controller design. In this section, we do not impose such an assumption. Instead, we use approximated fractional derivatives, such as the first two terms on the right-hand side of (2.35), of virtual controls in design and will take the resulting approximation errors including ρ_1 into account in analysis, as seen in the subsequent subsection.*

Let $\overline{V}_2 = V_1 + \frac{1}{2}{z_2}^2$. Then the derivative of $\overline{V}_2$ satisfies

$$ {}_0\mathcal{D}_t^\alpha \overline{V}_2 \le {}_0\mathcal{D}_t^\alpha V_1 + z_2 {}_0\mathcal{D}_t^\alpha z_2. \tag{2.36}$$

Substituting (2.34) and (2.35) into (2.36), we get

$$\begin{aligned} {}_0\mathcal{D}_t^\alpha \overline{V}_2 \le & z_1 z_2 - c_1 {z_1}^2 - \nu \tilde{\theta}_1^T \hat{\theta}_1 + z_2 z_3 + z_2 \alpha_2 + z_2 \phi_2^T(\overline{x}_2)\theta_2 + z_2(-\frac{\partial \alpha_1}{\partial x_1})x_2 \\ & + z_2(-\frac{\partial \alpha_1}{\partial x_1})\phi_1^T(x_1)\theta_1 + z_2(-\frac{\partial \alpha_1}{\partial \hat{\theta}_1}){}_0\mathcal{D}_t^\alpha \hat{\theta}_1 + z_2\rho_1 \\ = & z_1 z_2 - c_1 {z_1}^2 - \nu \tilde{\theta}_1^T \hat{\theta}_1 + z_2 z_3 + z_2 \alpha_2 + z_2 \phi_{v,2}^T \theta_{v,2} + z_2 \zeta_2 + z_2 \rho_1, \end{aligned} \tag{2.37}$$

in which $\phi_{v,2} = [\phi_2^T(\overline{x}_2), (-\frac{\partial \alpha_1}{\partial x_1})\phi_1^T(\overline{x}_1)]^T$, $\theta_{v,2} = [\theta_2^T, \theta_1^T]^T$, and $\zeta_2 = (-\frac{\partial \alpha_1}{\partial x_1})x_2 + (-\frac{\partial \alpha_1}{\partial \hat{\theta}_1}){}_0\mathcal{D}_t^\alpha \hat{\theta}_1$. Design virtual control signal α_2 as

$$\alpha_2 = -z_1 - c_2 z_2 - \phi_{v,2}^T \hat{\theta}_{v,2} - \zeta_2, \tag{2.38}$$

where $c_2 > 0$ and $\hat{\theta}_{v,2}$ is the estimate of $\theta_{v,2}$, then (2.37) becomes

$$ {}_0\mathcal{D}_t^\alpha \overline{V}_2 \le - c_1 {z_1}^2 - c_2 {z_2}^2 + z_2 z_3 - \nu \tilde{\theta}_1^T \hat{\theta}_1 - z_2 \phi_{v,2}^T \tilde{\theta}_{v,2} + z_2 \rho_1 \tag{2.39}$$

with $\tilde{\theta}_{v,2} = \hat{\theta}_{v,2} - \theta_{v,2}$.

Defining $V_2 = \overline{V}_2 + \frac{1}{2}\tilde{\theta}_{v,2}^T \Gamma_2^{-1} \tilde{\theta}_{v,2}$ where Γ_2 is positive definite, we can obtain

$$\begin{aligned} {}_0\mathcal{D}_t^\alpha V_2 \le & - c_1 {z_1}^2 - c_2 {z_2}^2 + z_2 z_3 - \nu \tilde{\theta}_1^T \hat{\theta}_1 - z_2 \phi_{v,2}^T \tilde{\theta}_{v,2} + z_2 \rho_1 \\ & + \tilde{\theta}_{v,2}^T \Gamma_2^{-1} {}_0\mathcal{D}_t^\alpha \hat{\theta}_{v,2}. \end{aligned} \tag{2.40}$$

Designing adaptive law as

$$ {}_0\mathcal{D}_t^\alpha \hat{\theta}_{v,2} = z_2 \Gamma_2 \phi_{v,2} - \nu \Gamma_2 \hat{\theta}_{v,2}, \tag{2.41}$$

then

$$ {}_0\mathcal{D}_t^\alpha V_2 \le - c_1 {z_1}^2 - c_2 {z_2}^2 + z_2 z_3 - \nu \tilde{\theta}_1^T \hat{\theta}_1 - \nu \tilde{\theta}_{v,2}^T \hat{\theta}_{v,2} + z_2 \rho_1. \tag{2.42}$$

Step i $(i = 3, \cdots, n-1)$*:* The fractional derivative of z_i is

$$ {}_0\mathcal{D}_t^\alpha z_i = {}_0\mathcal{D}_t^\alpha x_i - {}_0\mathcal{D}_t^\alpha \alpha_{i-1} = z_{i+1} + \alpha_i + \phi_i^T(\overline{x}_i(t))\theta_i - {}_0\mathcal{D}_t^\alpha \alpha_{i-1}. \tag{2.43}$$

Let ${}_0\mathcal{D}_t^\alpha \alpha_{i-1}$ be computed with an unknown approximation error function ρ_{i-1} as

$$\begin{aligned} {}_0\mathcal{D}_t^\alpha \alpha_{i-1} = & \sum_{j=1}^{i-1} \frac{\partial \alpha_{i-1}}{\partial x_j} {}_0\mathcal{D}_t^\alpha x_j + \sum_{j=1}^{i-1} \frac{\partial \alpha_{i-1}}{\partial \hat{\theta}_{v,j}} {}_0\mathcal{D}_t^\alpha \hat{\theta}_{v,j} - \rho_{i-1} \\ = & \sum_{j=2}^{i} \frac{\partial \alpha_{i-1}}{\partial x_{j-1}} x_j + \sum_{j=1}^{i-1} \frac{\partial \alpha_{i-1}}{\partial x_j} \phi_j^T(\overline{x}_j)\theta_j + \sum_{j=1}^{i-1} \frac{\partial \alpha_{i-1}}{\partial \hat{\theta}_{v,j}} {}_0\mathcal{D}_t^\alpha \hat{\theta}_{v,j} - \rho_{i-1}, \end{aligned} \tag{2.44}$$

then (2.43) can be expressed as

$$ {}_0\mathcal{D}_t^\alpha z_i = z_{i+1} + \alpha_i + \phi_{v,i}^T\theta_{v,i} + \zeta_i + \rho_{i-1}. \tag{2.45}$$

Choosing virtual control signal α_i and adaptive law as

$$\alpha_i = -z_{i-1} - c_i z_i - \phi_{v,i}^T\hat{\theta}_{v,i} - \zeta_i, \tag{2.46}$$

$$ {}_0\mathcal{D}_t^\alpha \hat{\theta}_{v,i} = z_i\Gamma_i\phi_{v,i} - \nu\Gamma_i\hat{\theta}_{v,i}, \tag{2.47}$$

where Γ_i is positive definite and $c_i > 0$. Let $V_i = V_{i-1} + \frac{1}{2}{z_i}^2 + \frac{1}{2}\tilde{\theta}_{v,i}^T\Gamma_i^{-1}\tilde{\theta}_{v,i}$, where $\tilde{\theta}_{v,i} = \hat{\theta}_{v,i} - \theta_{v,i}$, then we get

$$ {}_0\mathcal{D}_t^\alpha V_i \le z_i z_{i+1} - \sum_{j=1}^{i} c_j {z_j}^2 - \nu\sum_{j=1}^{i}\tilde{\theta}_{v,j}^T\hat{\theta}_{v,j} + \sum_{j=1}^{i-1} z_{j+1}\rho_j. \tag{2.48}$$

Step n: The fractional derivative of z_n is

$$ {}_0\mathcal{D}_t^\alpha z_n = g(x)u + d + \phi_{v,n}^T\theta_{v,n} + \zeta_n + \rho_{n-1}. \tag{2.49}$$

Constructing the Lyapunov function $\overline{V}_n = V_{n-1} + \frac{1}{2}{z_n}^2$, then the fractional derivative of $\overline{V}_n$ is

$$ {}_0\mathcal{D}_t^\alpha \overline{V}_n \le {}_0\mathcal{D}_t^\alpha V_{n-1} + z_n[g(x)u + \phi_{v,n}^T\theta_{v,n} + \zeta_n + \rho_{n-1}] + |z_n|\,D, \tag{2.50}$$

where D is an unknown bound of $d(t)$. In the light of Lemma 2.1, since $|z_n| < \epsilon + z_n sg(z_n,\epsilon)$, hence $|z_n|\,D \le \epsilon D + z_n sg(z_n,\epsilon)D$.

Let the control input u be designed as

$$u = \frac{1}{g(x)}[-z_{n-1} - c_n z_n - \phi_{v,n}^T\hat{\theta}_{v,n} - sg(z_n,\epsilon)\hat{D} - \zeta_n], \tag{2.51}$$

where function $\epsilon(t) = e^{-at}$ $(a > 0)$, $c_n > 0$, $\hat{\theta}_{v,n}$ and $\hat{D}$ stand for the estimates of $\theta_{v,n}$ and D respectively. The adaptive laws are designed as

$$\begin{cases} {}_0\mathcal{D}_t^\alpha \hat{\theta}_{v,n} = z_n\Gamma_n\phi_{v,n} - \nu\Gamma_n\hat{\theta}_{v,n}, \\ {}_0\mathcal{D}_t^\alpha \hat{D} = \eta z_n sg(z_n,\epsilon) - \iota\hat{D}, \end{cases} \tag{2.52}$$

where Γ_n is positive definite, $\eta > 0$, and $\iota > 0$. Let $V_n = \overline{V}_n + \frac{1}{2}\tilde{\theta}_{v,n}^T\Gamma_n^{-1}\tilde{\theta}_{v,n} + \frac{1}{2\eta}\tilde{D}^2$, $\tilde{\theta}_{v,n} = \hat{\theta}_{v,n} - \theta_{v,n}$ and $\tilde{D} = \hat{D} - D$. Then combining (2.50), (2.51) and (2.52), we have

$$ {}_0\mathcal{D}_t^\alpha V_n \le -\sum_{j=1}^{n} c_j {z_j}^2 - \nu\sum_{j=1}^{n}\tilde{\theta}_{v,j}^T\hat{\theta}_{v,j} + \sum_{j=1}^{n-1} z_{j+1}\rho_j + \epsilon D - \frac{\iota}{\eta}\tilde{D}\hat{D}. \tag{2.53}$$

2.4.2 Stability Analysis

The following theorem presents the main result of this section.

Theorem 2.2 *Consider the closed-loop fractional-order system comprising fractional system (2.1) and adaptive controller with control law (2.51), virtual control laws (2.29), (2.38) and (2.46) and adaptive laws (2.32), (2.41), (2.47) and (2.52). Given any positive numbers $B_0 > 0$ and $p > 0$, for all initial conditions satisfying $\sum_{j=1}^{i-1}({}_0\mathcal{D}_t^\alpha\rho_j)^2 \le B_0$ and $(\sum_{j=1}^{i}{z_j}^2 + \sum_{j=1}^{i}\tilde{\theta}_{v,j}^T\Gamma_j^{-1}\tilde{\theta}_{v,j} + \sum_{j=2}^{i}\rho_{j-1}^2 + \frac{1}{\eta}\tilde{D}^2) \le 2p$, $i = 1, 2, \cdots, n$, the system is stable in the sense that all signals in the closed-loop system and the tracking error are ultimately uniformly bounded.*

Proof: Since $\tilde{\theta}_{v,i} = \hat{\theta}_{v,i} - \theta_{v,i}$, thus according to Young's inequality,

$$
\begin{aligned}
-\tilde{\theta}_{v,i}^T \hat{\theta}_{v,i} &= -\tilde{\theta}_{v,i}^T(\theta_{v,i} + \tilde{\theta}_{v,i}) = -\|\tilde{\theta}_{v,i}\|^2 - \tilde{\theta}_{v,i}^T \theta_{v,i} \\
&\leq -\|\tilde{\theta}_{v,i}\|^2 + \frac{1}{2}\|\tilde{\theta}_{v,i}\|^2 + \frac{1}{2}\|\theta_{v,i}\|^2 = -\frac{1}{2}\|\tilde{\theta}_{v,i}\|^2 + \frac{1}{2}\|\theta_{v,i}\|^2. \qquad (2.54)
\end{aligned}
$$

Similarly, $-\tilde{D}\hat{D} \leq -\frac{1}{2}\tilde{D}^2 + \frac{1}{2}D^2$. Therefore, (2.53) becomes

$$
\begin{aligned}
{}_0\mathcal{D}_t^\alpha V_n \leq & -\sum_{j=1}^n c_j {z_j}^2 - \frac{\nu}{2}\sum_{j=1}^n \|\tilde{\theta}_{v,j}\|^2 + \frac{\nu}{2}\sum_{j=1}^n \|\theta_{v,j}\|^2 + \sum_{j=1}^{n-1} z_{j+1}\rho_j + \epsilon D \\
& - \frac{\iota}{2\eta}\tilde{D}^2 + \frac{\iota}{2\eta}D^2. \qquad (2.55)
\end{aligned}
$$

Defining $V = V_n + \frac{1}{2}\sum_{j=1}^{n-1}\rho_j^2$, then the fractional derivative of V is

$$
{}_0\mathcal{D}_t^\alpha V \leq {}_0\mathcal{D}_t^\alpha V_n + \sum_{j=1}^{n-1} \rho_j {}_0\mathcal{D}_t^\alpha \rho_j. \qquad (2.56)
$$

Since given any $B_0 > 0$ and $p > 0$, sets $\Omega_i := \{(\sum_{j=1}^i {z_j}^2 + \sum_{j=1}^i \tilde{\theta}_{v,j}^T \Gamma_j^{-1}\tilde{\theta}_{v,j} + \sum_{j=2}^i \rho_{j-1}^2 + \frac{1}{\eta}\tilde{D}^2) \leq 2p\}$, $i = 1, 2, \cdots, n$ and $\Omega_0 := \{\sum_{j=1}^{i-1}({}_0\mathcal{D}_t^\alpha \rho_j)^2 \leq B_0\}$, are compact in $\mathbb{R}^{\sum_{j=1}^i N_j + 2i}$ and $\mathbb{R}^{i-1}$ respectively, where N_j represents the dimension of $\tilde{\theta}_{v,j}$. Thus, $\Omega_0 \times \Omega_i$ is compact in $\mathbb{R}^{\sum_{j=1}^i N_j + 3i - 1}$. Therefore, $|\rho_m|$ and $|{}_0\mathcal{D}_t^\alpha \rho_m|$ have maximum P_m and M_m on $\Omega_0 \times \Omega_m$ accordingly ($m = 1, 2, \cdots, n-1$). Consequently, by substituting (2.55) to (2.56), we have

$$
\begin{aligned}
{}_0\mathcal{D}_t^\alpha V \leq & -\sum_{j=1}^n c_j {z_j}^2 - \frac{\nu}{2}\sum_{j=1}^n \|\tilde{\theta}_{v,j}\|^2 + \frac{\nu}{2}\sum_{j=1}^n \|\theta_{v,j}\|^2 + \epsilon D + \sum_{j=1}^{n-1} z_{j+1}\rho_j \\
& - \frac{\iota}{2\eta}\tilde{D}^2 + \frac{\iota}{2\eta}D^2 + \sum_{j=1}^{n-1} |\rho_j {}_0\mathcal{D}_t^\alpha \rho_j| \\
\leq & -\sum_{j=1}^n c_j {z_j}^2 - \frac{\nu}{2}\sum_{j=1}^n \frac{1}{\lambda_{max}(\Gamma_j^{-1})}\tilde{\theta}_{v,j}^T \Gamma_j^{-1}\tilde{\theta}_{v,j} + \epsilon D + \frac{\nu}{2}\sum_{j=1}^n \|\theta_{v,j}\|^2 \\
& + \frac{1}{2\sigma}\sum_{j=1}^{n-1} {z_{j+1}}^2 + \frac{\sigma}{2}\sum_{j=1}^{n-1} {\rho_j}^2 - \frac{\iota\tilde{D}^2}{2\eta} + \frac{\iota D^2}{2\eta} + \sum_{j=1}^{n-1} |\rho_j {}_0\mathcal{D}_t^\alpha \rho_j|, \qquad (2.57)
\end{aligned}
$$

where $\sigma > 0$. Choose $c_1 = \beta_0$ and $c_i = \beta_0 + \frac{1}{2\sigma}$, $i = 2, \cdots, n$,where β_0 is a positive constant. Since for any positive number χ, we get $|\rho_j {}_0\mathcal{D}_t^\alpha \rho_j| \leq [{\rho_j}^2({}_0\mathcal{D}_t^\alpha \rho_j)^2]/(2\chi) + (\chi/2)$, hence (2.57) becomes

$$
\begin{aligned}
{}_0\mathcal{D}_t^\alpha V \leq & \sum_{j=1}^n \left(-\beta_0 {z_j}^2 - \frac{\nu}{2\lambda_{max}(\Gamma_j^{-1})}\tilde{\theta}_{v,j}^T \Gamma_j^{-1}\tilde{\theta}_{v,j} \right) - \frac{\iota\tilde{D}^2}{2\eta} + \epsilon D + \frac{\nu}{2}\sum_{j=1}^n \|\theta_{v,j}\|^2 \\
& + \frac{\iota D^2}{2\eta} + \frac{n-1}{2}\chi + \sum_{j=1}^{n-1} \left(\frac{\sigma}{2}{\rho_j}^2 + \frac{{\rho_j}^2({}_0\mathcal{D}_t^\alpha \rho_j)^2 M_j^2}{2\chi M_j^2} + \frac{M_j^2}{\chi}{\rho_j}^2 - \frac{M_j^2}{2\chi}{\rho_j}^2 \right). \qquad (2.58)
\end{aligned}
$$

Choose $\iota = 2\beta_0$, then we have

$$\begin{aligned}
{}_0\mathcal{D}_t^\alpha V \leq & \sum_{j=1}^{n}\Big(-\beta_0 {z_j}^2 - \frac{\nu}{2\lambda_{max}(\Gamma_j^{-1})}\tilde{\theta}_{v,j}^T\Gamma_j^{-1}\tilde{\theta}_{v,j}\Big) + \epsilon D - \frac{\beta_0\tilde{D}^2}{\eta} + \frac{\nu}{2}\sum_{j=1}^{n}\|\theta_{v,j}\|^2 \\
& + \frac{\beta_0 D^2}{\eta} + \frac{n-1}{2}\chi + \sum_{j=1}^{n-1}\Big(\frac{\sigma}{2}{\rho_j}^2 + \frac{M_j^2}{\chi}{\rho_j}^2 - \Big[1 - \frac{({}_0\mathcal{D}_t^\alpha \rho_j)^2}{M_j^2}\Big]\frac{M_j^2}{2\chi}{\rho_j}^2\Big) \\
\leq & \sum_{j=1}^{n}\Big(-\beta_0 {z_j}^2 - \frac{\nu}{2\lambda_{max}(\Gamma_j^{-1})}\tilde{\theta}_{v,j}^T\Gamma_j^{-1}\tilde{\theta}_{v,j}\Big) - \frac{\beta_0\tilde{D}^2}{\eta} + \epsilon D + \frac{\nu}{2}\sum_{j=1}^{n}\|\theta_{v,j}\|^2 \\
& + \frac{\beta_0 D^2}{\eta} + \frac{n-1}{2}\chi + \sum_{j=1}^{n-1}\Big(\frac{\sigma}{2}{\rho_j}^2 + \frac{M_j^2}{\chi}{\rho_j}^2 - \beta_0{\rho_j}^2 + \beta_0{\rho_j}^2\Big).
\end{aligned} \tag{2.59}$$

Let $\frac{\nu}{2\lambda_{max}(\Gamma_j^{-1})} = \beta_0$, then

$$\begin{aligned}
{}_0\mathcal{D}_t^\alpha V \leq & -2\beta_0 V + \epsilon D + \frac{\nu}{2}\sum_{j=1}^{n}\|\theta_{v,j}\|^2 + \frac{\beta_0 D^2}{\eta} + \frac{n-1}{2}\chi \\
& + \sum_{j=1}^{n-1}\Big(\frac{\sigma}{2} + \frac{M_j^2}{\chi} + \beta_0\Big)P_j^2.
\end{aligned} \tag{2.60}$$

Design η, σ and χ to satisfy

$$\beta_0 > \frac{1}{2p}\Big[\epsilon D + \frac{\nu}{2}\sum_{j=1}^{n}\|\theta_{v,j}\|^2 + \frac{\beta_0 D^2}{\eta} + \frac{n-1}{2}\chi + \sum_{j=1}^{n-1}\Big(\frac{\sigma}{2} + \frac{M_j^2}{\chi} + \beta_0\Big)P_j^2\Big], \tag{2.61}$$

then ${}_0\mathcal{D}_t^\alpha V \leq 0$ on $V \leq p$. Define ψ as a time-varying function satisfying that

$${}_0\mathcal{D}_t^\alpha V + \psi = 0. \tag{2.62}$$

Clearly $\psi \geq 0$. By taking Laplace transform to (2.62), we get

$$V(s) = s^{-1}V(0) - s^{-\alpha}\Psi(s), \tag{2.63}$$

where $V(s)$ and $\Psi(s)$ are the Laplace transform of $V(t)$ and ψ, respectively. Applying inverse Laplace transform to (2.63) yields

$$V(t) = V(0) - \frac{t^{\alpha-1}}{\Gamma(\alpha)} * \psi \tag{2.64}$$

in which $*$ represents the convolution operator. Since $t^{\alpha-1}$ and $\Gamma(\alpha)$ are non-negative, thus $V(t) \leq V(0) \leq p$. Hence $V \leq p$ is an invariant set, which validates (2.60) holds $\forall t \geq 0$ with $V(0) \leq p$. Thus all signals in $V(t)$ are uniformly ultimately bounded. ■

Remark 2.6 *The arbitrary positive numbers B_0 and p determine the restriction on the initial values in the systems, i.e., all relevant variables remain in the compact set $\Omega_0 \times \Omega_i$ $(i = 1, \cdots, n)$, as mentioned in the statement under (2.56). The actual values of these two numbers are not necessary to be derived for controller design yet the trade-off between a smaller bound of tracking error and a more relaxed restriction on the initial compact set exists since under larger B_0 and p, larger β_0, Γ_i, and η and smaller σ are required for achieving smaller tracking error bound, leading to a higher control force.*

2.5 Simulation Studies

In this section, second-order and third-order fractional nonlinear systems are respectively presented as examples for demonstrating the proposed control methods in Section 2.3 and Section 2.4.

2.5.1 Example for Asymptotic Case

The system to be controlled is given as follows

$$\begin{cases} {}_0\mathcal{D}_t^\alpha x_1(t) = x_2(t) + \phi_1(x_1(t))\theta_1 \\ {}_0\mathcal{D}_t^\alpha x_2(t) = u(t) + \phi_2(x_1(t), x_2(t))\theta_2 + d(t) \\ y(t) = x_1(t) \end{cases} \tag{2.65}$$

where $\alpha = 0.95$, $\phi_1(x_1) = -0.4x_1^2$, $\phi_2(x_1, x_2) = -0.1x_2 + \frac{x_2 - 0.5x_1^2}{1+x_1^4}$. Here $\theta_1 = \theta_2 = 1$ and external time-varying disturbance $d(t) = \sin(t) + \cos(t) + 2\mathrm{U}(t-15)$ where $\mathrm{U}(t)$ is the unit step function is used for simulation purposes, but unknown to the designer. The term $2\mathrm{U}(t-15)$ implies that disturbance $d(t)$ has a step jump at $t = 15$ s and thus it does not have bounded derivative.

Firstly, the system is simulated when no control is applied, i.e. $u(t) = 0$, and the result is given in Fig. 2.1. As observed from the figure, in this case, the system (2.65) is unstable in the presence of disturbance $d(t)$. To stabilize the system, we apply our proposed adaptive control scheme presented in Section 2.3. The designed control parameters are selected as $c_1 = 30$, $c_2 = 1$, $\Gamma_1 = \Gamma_2 = 2I$, $\eta = 4$, and $\lambda_1 = 2$. The responses in this case are shown in Fig. 2.2 to Fig. 2.4. It could be found from Fig. 2.2 that the system output y and the virtual error z_2 are driven to 0 eventually. Fig. 2.3 and Fig. 2.4 show the estimations of uncertain parameters $\hat{\theta}_{v,1}$ and $\hat{\theta}_{v,2}$ and bound of external disturbance $\hat{D}$, as well as the bound of $\rho_1(t)$ accordingly. The control signal u is shown in Fig. 2.5.

To better illustrate the effectiveness of the proposed control algorithm, a comparative simulation study between the schemes in [1] and Section 2.3 is conducted under the same control objective that the output signal tracks a reference signal $r(t) = \sin(0.5t)$ while ensuring all the signals bounded. In [1], fuzzy logic systems are employed to approximate unknown compounded nonlinear functions in the systems and also the fractional-order derivatives of the virtual control functions. The sign function is used to compensate for the effects caused

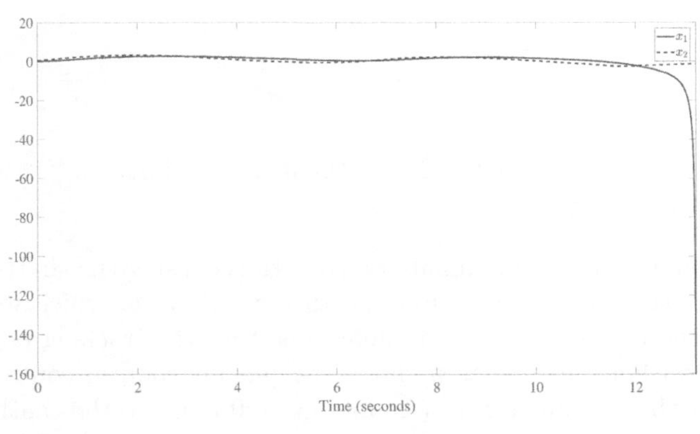

FIGURE 2.1: State variables of uncontrolled system (2.65).

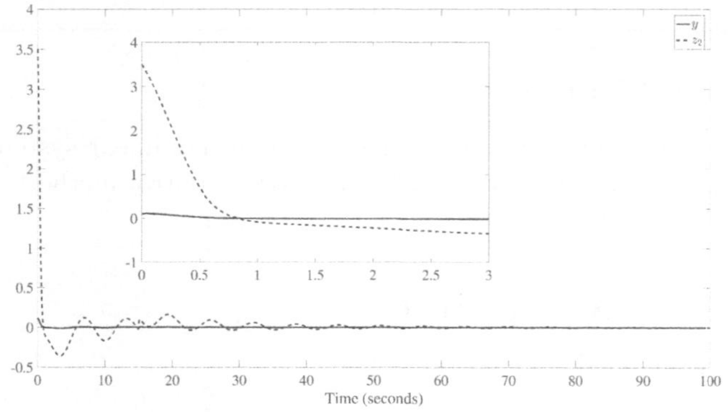

FIGURE 2.2: System output $y = x_1 = z_1$ and virtual error z_2.

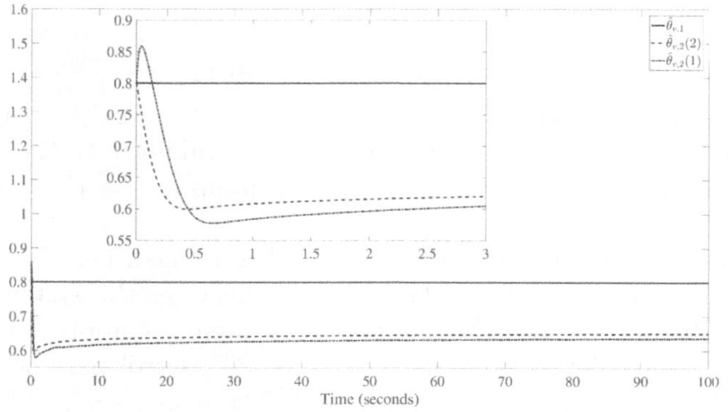

FIGURE 2.3: Estimates $\hat{\theta}_{v,1}$ and $\hat{\theta}_{v,2}$.

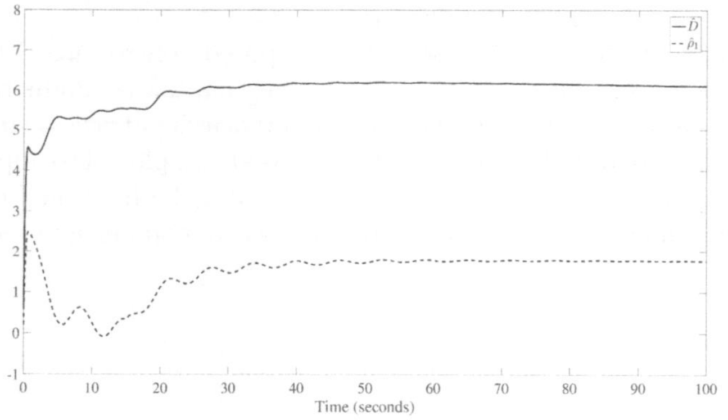

FIGURE 2.4: Estimates $\hat{D}$ and $\hat{\rho}_1$.

by system uncertainties and approximated errors, thus causing the chattering phenomenon. The comparison results are given in Fig. 2.6 to Fig. 2.7, from which we can observe that the proposed method in Section 2.3 can guarantee that the tracking error z_1 converges to 0 asymptotically without leading to a chattering phenomenon in control signal u. On the contrary, not only the tracking error z_1 cannot go to 0 but also the chattering phenomenon in the control signal u is caused by using the control signal in [1], which is also consistent

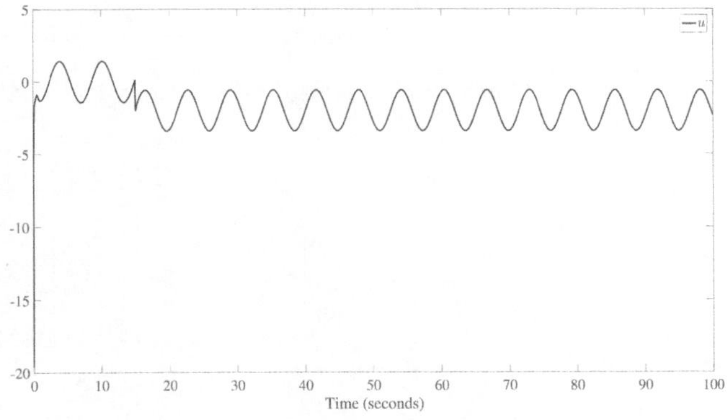

FIGURE 2.5: Control input u.

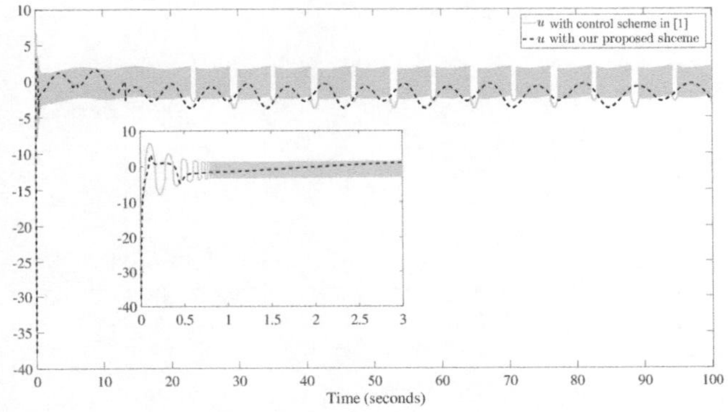

FIGURE 2.6: Control input u with control scheme involving sign(·) in [1] and with scheme in Section 2.3.

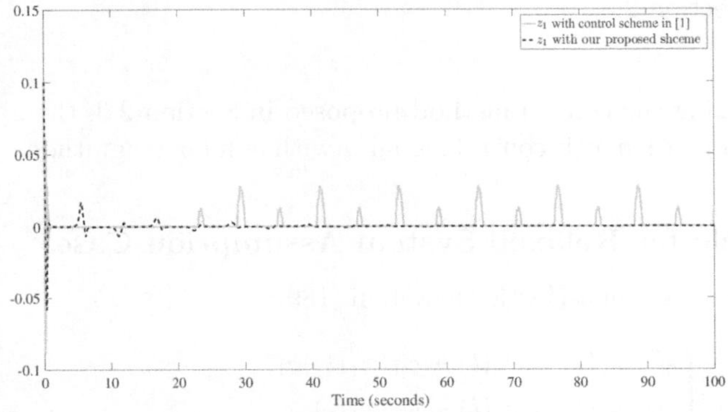

FIGURE 2.7: Output tracking error z_1 with control scheme involving sign(·) in [1] and with scheme in Section 2.3.

with its theoretical results established. Furthermore, it is proposed in [1] that the chattering phenomenon can be avoided by replacing sign(·) with arctan(10·) in the controller design, without theoretical analysis on stability and tracking performance provided. Comparison simulations are also carried out with such a replacement and the results can be found in Fig. 2.8 to Fig. 2.9. It can be seen from these figures that although the control signal in [1] becomes smooth, the tracking error z_1 can only be driven to a small region near 0.

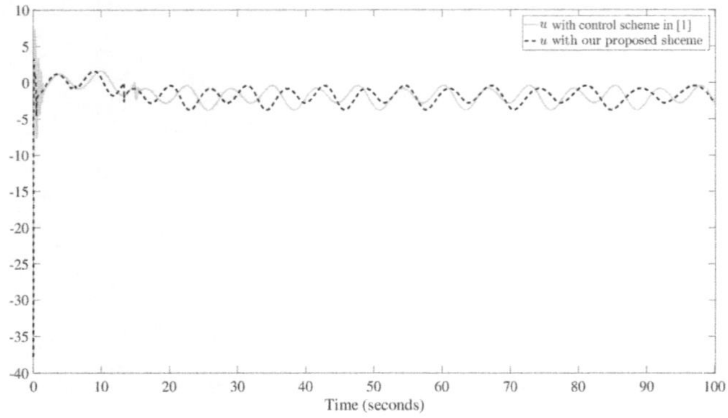

FIGURE 2.8: Control input u with control scheme involving arctan(10·) in [1] and with scheme in Section 2.3.

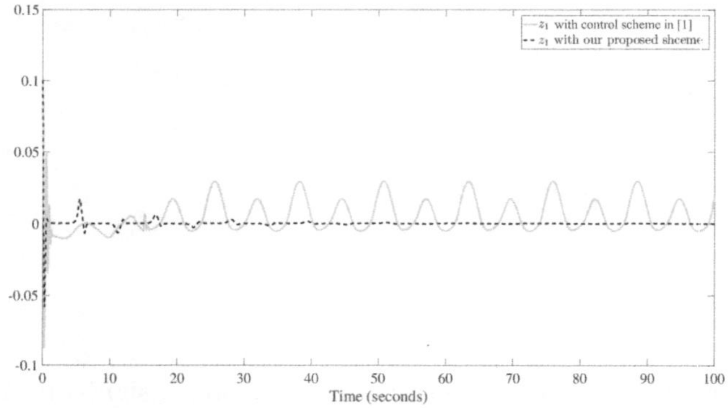

FIGURE 2.9: Output tracking error z_1 with control scheme involving arctan(10·) in [1] and with scheme in Section 2.3.

However, by utilizing the control method proposed in Section 2.3, the z_1 can be asymptotically stabilized under smooth control signal u with similar magnitude.

2.5.2 Example for Relaxed System Assumption Case

Consider the following Chua-Hartley's system [189]

$$\begin{cases} {}_0\mathcal{D}_t^\alpha x_1(t) = x_2(t) + \phi_1(x_1(t))\theta_1 \\ {}_0\mathcal{D}_t^\alpha x_2(t) = x_3(t) + \phi_2(x_1(t), x_2(t))\theta_2 \\ {}_0\mathcal{D}_t^\alpha x_3(t) = u(t) + \phi_3(x_1(t), x_2(t), x_3(t))\theta_3 + d(t) \\ y(t) = x_1(t) \end{cases} \tag{2.66}$$

with $\alpha = 0.98$, $\phi_1(x_1) = \frac{10}{7}(x_1 - x_1^3)$, $\phi_2(x_1, x_2) = 10x_1 - x_2$, $\phi_3(x_1, x_2, x_3) = -\frac{100}{7}x_2$, $\theta_1 = \theta_2 = \theta_3 = 1$, and $d(t) = 0.5\sin(2t) + 3\mathrm{U}(t-10)$. The designed control parameters are chosen as $c_1 = 1$, $c_2 = c_3 = 2$, $\Gamma_1 = \Gamma_2 = \Gamma_3 = 2I$, $\nu = 1$, $\iota = 2$, $\eta = 100$. We simulate the system without control effort and with the proposed control scheme under identical initial condition $x(0) = [0.8, -2, 1]^T$ to make a comparison. The behaviour of state variables in (2.66) are shown in Fig. 2.10 and Fig. 2.11, respectively. As observed from Fig. 2.10, in which the green dot represents the initial state, the uncontrolled system exhibits chaos.

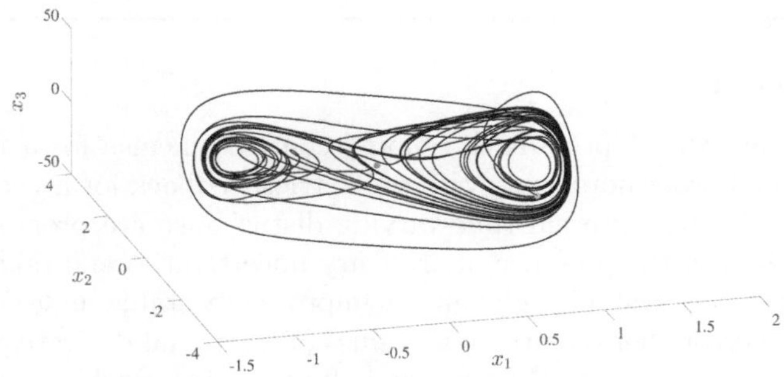

FIGURE 2.10: States of uncontrolled system (2.66) with *green dot* and *red dot* standing for initial condition and the origin respectively.

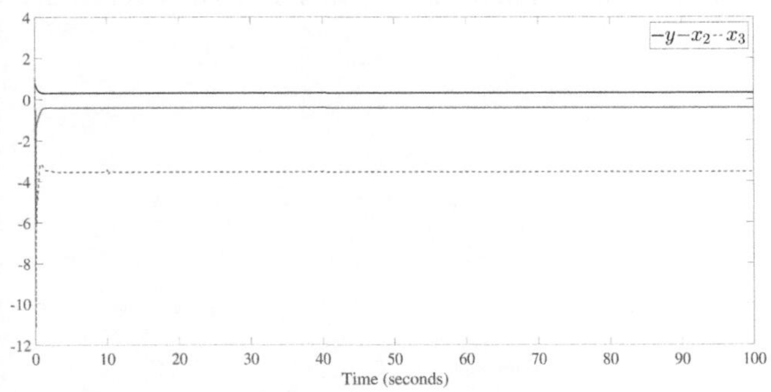

FIGURE 2.11: States of (2.66) under the proposed control scheme in Section 2.4.

From Fig. 2.11, it is found that the investigated control law in Section 2.4 eliminates the chaotic behaviors of the system without control effort. Besides, it can be seen from Fig. 2.11 that the output y is finally driven to a small region around 0 by the control input u displayed in Fig. 2.12, which implies that the closed-loop system is stabilized.

All these simulation results illustrate the effectiveness of the proposed schemes in Section 2.3 and Section 2.4 accordingly and verify our established theoretical results.

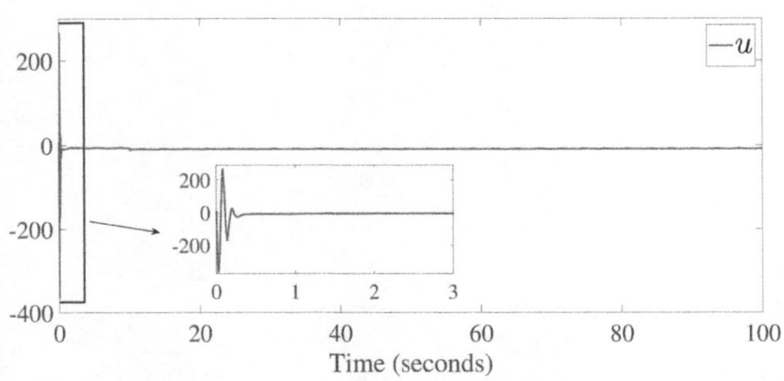

FIGURE 2.12: Control input u of (2.66).

2.6 Conclusions

In this chapter, smooth adaptive backstepping control schemes for a class of SISO commensurate fractional-order nonlinear systems in strict feedback form with uncertain system parameters and unknown external time-varying disturbance are proposed. It is proved in Section 2.3 that even in the presence of arbitrary uncertainties and bounded disturbances, the resulting closed-loop system is globally asymptotically stable under an assumption that the approximation errors between the true values of fractional derivatives of virtual control signals and their approximated values are globally bounded. Such an assumption has been relaxed in Section 2.4, where the obtained closed-loop fractional-order system under the investigated control law is ensured to be stable in the sense that all the closed-loop signals are uniformly ultimately bounded and the output tracking error can be driven to a small region around the origin. The simulation results verify the effectiveness of both proposed control schemes.

3

Adaptive Backstepping Controller Design for Commensurate Nonlinear Fractional-Order Uncertain Systems with Quantized Input

This chapter explores the development of backstepping-based adaptive control for a class of nonlinear fractional-order systems that encounter systematic uncertainties, time-varying external disturbances, and quantized control input. A novel analog chain rule computation method is proposed to overcome the difficulty in handling the fractional derivatives of composite functions (i.e. virtual control signals) encountered in intermediate recursive steps of the backstepping procedure. A quantified adaptive controller is strictly crafted to guarantee the stability of the system, with all closed-loop signals staying within bounds and the system output tracking error converging to an adjustable residual set asymptotically, leveraging the direct fractional Lyapunov method and such an approach. The simulated studies confirm the efficacy of the investigated control strategy.

3.1 Introduction

In Section 2.3, an adaptive backstepping control strategy is investigated for uncertain nonlinear fractional-order systems to realize perfect asymptotic output tracking to a predefined reference signal. The proposed adaptive controller yields global results, provided that the discrepancies between the actual fractional derivatives of virtual controllers and their approximations are assumed to be bounded globally. Despite this, the validity of such an assumption is still to be determined. Apparently, how to rigorously handle the fractional derivatives of virtual controllers for fractional-order nonlinear systems is a challenging issue and still remains open.

On the other hand, due to the influence of limited data-transmitted bandwidth in digital control systems, developing quantized control strategies is of great importance theoretically and practically. During the past few years, lots of attention has been drawn to the stabilization problem for uncertain integer-order linear or nonlinear systems with adaptive quantized control inputs, see for examples [115, 116, 117, 118] and references wherein. Despite a wide deal of research into adaptive quantized control for integer-order nonlinear systems, there are limited findings related to fractional nonlinear systems, as discussed in the fourth paragraph of Subsection 1.3.1. Thus designing appropriate adaptive quantized control input for nonlinear fractional systems without constraints on the nonlinear functions is meaningful yet challenging.

Inspired by the discussion above, this chapter considers the adaptive backstepping control for a class of nonlinear fractional-order systems with systematic uncertainties,

DOI: 10.1201/9781003620372-3

time-varying external disturbances, and quantized control input. The contributions of this chapter are summarized as follows.

1. In order to overcome the shortage of computing the fractional derivatives of virtual control signals, a novel analog chain rule computation approach is proposed to compute the fractional derivative of a composite function with an approximation error. Sufficient conditions are set up to guarantee the boundedness of the error.
2. Different from the backstepping control methods based on neural networks or fuzzy logic systems [104, 106, 1, 108], where the selection of approximation functions and parameters are heavily experience-dependent and have barely any relation with the considered fractional-order systems, system states are considered in computing fractional derivatives of virtual control laws with the help of our proposed calculation approach in this chapter.
3. Compared with the results obtained in Section 2.3, rigorously adaptive controller design and stability analysis are guaranteed under such an analog chain rule computation method. Additionally, an adaptive quantized control law is investigated for a more general class of fractional-order nonlinear systems that the global Lipschitz and boundedness conditions on the nonlinear functions are no longer required.
4. Apart from showing the stability of the corresponding closed-loop system with the direct fractional-order Lyapunov method, the system output tracking error to a predefined signal can be made to converge to a tunable residual asymptotically with the theoretical results established in this chapter.

The rest of this chapter is organized as follows. The statement of the considered problem is presented in Section 3.2. In Section 3.3, a novel method for calculating the fractional derivatives of the composite function is introduced. Section 3.4 contains a detailed description of the designed adaptive backstepping controller. The results of the simulation outlined in Section 3.5 demonstrate the efficacy of the investigated control strategy and the chapter is concluded in Section 3.6.

3.2 Problem Formulation

3.2.1 Problem Statement

In this chapter, the following class of uncertain fractional nonlinear systems is studied

$$\begin{cases} {}_0\mathcal{D}_t^\alpha x_i(t) = x_{i+1}(t) + \psi_i^T(\overline{x}_i(t))\theta_i \\ {}_0\mathcal{D}_t^\alpha x_n(t) = q(u(t)) + \psi_n^T(\overline{x}_n(t))\theta_n + d(t) \\ y(t) = x_1(t), \end{cases} \tag{3.1}$$

in which $i = 1, \cdots, n-1$, $\alpha \in (0,1]$, $x_j(t) \in \mathbb{R}(j = 1, \cdots, n)$, $u(t) \in \mathbb{R}$, and $y(t) \in \mathbb{R}$ are the states, control input, and output of the system respectively, $\overline{x}_j(t) = [x_1(t), x_2(t), \cdots, x_j(t)]^T \in \mathbb{R}^j$, $\psi_j(\overline{x}_j(t)) \in \mathbb{R}^p$ are nonlinear known smooth functions, $\theta_j \in \mathbb{R}^p$ represent unknown constant vectors, $q(u(t))$ indicates the quantized control input and $d(t)$ is the unknown disturbance bounded by an unknown constant $D \in \mathbb{R}_{\geq 0}$.

The aim of the control is to develop a backstepping-based adaptive control law for the fractional systems outlined in (3.1) such that the system output y can accurately follow the reference signal r while ensuring all closed-loop signals remain bounded.

Remark 3.1 *The proposed synchronization scheme for fractional nonlinear systems in [121] requires the contained nonlinearities to be globally Lipschitz continuous and bounded. This chapter offers more lenient conditions compared with [121], and a broad range of fractional-order systems can be taken into account provided conditions in (3.7) and (3.8) being satisfied.*

Assumption 3.1 *The reference signal r and its αith-order Caputo derivative ${}_0\mathcal{D}_t^{\alpha i}r$, $i = 1, \cdots, n$ are smooth, known, and bounded.*

3.2.2 Quantizer

This chapter takes into account the hysteresis quantizer described in [117].

$$q(u) = \begin{cases} u_i \text{sign}(u), & \frac{u_i}{1+\delta} < |u| \le u_i,\ u^{(1)} < 0, \text{or} \\ & u_i < |u| \le \frac{u_i}{1-\delta},\ u^{(1)} > 0 \\ u_i(1+\delta)\text{sign}(u), & u_i < |u| \le \frac{u_i}{1-\delta},\ u^{(1)} < 0, \text{or} \\ & \frac{u_i}{1-\delta} < |u| \le \frac{u_i(1+\delta)}{1-\delta},\ u^{(1)} > 0 \\ 0, & 0 \le |u| < \frac{u_{\min}}{1+\delta},\ u^{(1)} < 0, \text{or} \\ & \frac{u_{\min}}{1+\delta} \le u \le u_{\min},\ u^{(1)} > 0 \\ q(u(t^-)), & u^{(1)} = 0 \end{cases} \tag{3.2}$$

where $\delta = \frac{1-\sigma}{1+\sigma}$, $0 < \sigma < 1$, $u_i = \sigma^{1-i}u_{\min}$ $(i = 1, 2, \cdots)$ and $u_{\min} > 0$, and $q(u)$ takes values in the set $U = \{0,\ \pm u_i,\ \pm u_i(1+\delta)\}$. In accordance with [117] that in comparison with the logarithmic quantizer [115], the hysteresis quantizer possesses additional quantization levels and can be applied to prevent serious chattering.

Lemma 3.1 **([117])** *The quantization error satisfies*

$$|q(u) - u| \le \delta|u| + u_{\min}, \tag{3.3}$$

in which $u_{\min} > 0$ and $0 < \delta < 1$ are quantization parameters.

3.3 Analog Chain Rule Computation of Fractional Derivative of Composite Function

For a composite function $g(\xi(t))$, $g^{(p)}(\xi(t))$ $(p \in \mathbb{N})$ means the pth-order partial derivative of $g(t)$ with respect to $\xi(t)$ and a function $\Psi(g,\xi)_k$ $(k \in \mathbb{N}_{\ge 2})$ is defined as

$$\Psi(g,\xi)_k = \sum_{m=2}^{k} g^{(m)}(\xi(t))B_{k,m}(\xi^{(1)}(t), \cdots, \xi^{(k-m+1)}(t)), \tag{3.4}$$

in which $B_{k,m}(\cdot)$ denotes the partial Bell polynomial [190] in variables $\xi^{(1)}(t), \cdots, \xi^{(k-m+1)}(t)$ shown as

$$B_{k,m}(\xi^{(1)}(t), \cdots, \xi^{(k-m+1)}(t)) = \sum \frac{k!}{a_1! \cdots a_{k-m+1}!} \left(\frac{\xi^{(1)}(t)}{1!}\right)^{a_1} \left(\frac{\xi^{(2)}(t)}{2!}\right)^{a_2} \cdots \left(\frac{\xi^{(k-m+1)}(t)}{(k-m+1)!}\right)^{a_{k-m+1}}, \tag{3.5}$$

where the definition of a_j $(j = 1, \cdots, k-m+1)$ can be found in (1.38) in Property 1.3.

As stated in Subsection 1.3.1, currently no result has been reported to compute the fractional derivative of a composite function succinctly. An analog chain rule computation method is now introduced to compute such a fractional derivative.

Analog Chain Rule Computation: For a continuous and differentiable composite function $g(\xi(t))$, its fractional derivative ${}_{t_0}\mathcal{D}_t^{\alpha} g(\xi(t))$ $(0 < \alpha < 1$ and $0 \le t_0 < t)$ is computed as $g^{(1)}(\xi(t)) {}_{t_0}\mathcal{D}_t^{\alpha} \xi(t)$ with an approximation error $\rho(t)$ defined as

$$\rho(t) = g^{(1)}(\xi(t)) {}_{t_0}\mathcal{D}_t^{\alpha} \xi(t) - {}_{t_0}\mathcal{D}_t^{\alpha} g(\xi(t)). \tag{3.6}$$

Unlike Section 2.3 where $\rho(t)$ is assumed to be globally bounded without giving any validation, sufficient conditions are now established in the following lemma to guarantee the boundedness of $\rho(t)$.

Lemma 3.2 *For a continuous and differentiable composite function $g(\xi(t))$, if there exists a constant $M \in \mathbb{R}^+$ such that the following two conditions are satisfied for $\forall t > t_0$:*

$$(i) \quad \Psi(g, \xi)_{k+1} = O(\Psi(g, \xi)_k); \tag{3.7}$$

$$(ii) \quad \left| \frac{g(\xi(t)) - g^{(1)}(\xi(t))\xi(t)}{\Gamma(1-\alpha)\left[(t-t_0)^{\alpha} + \epsilon\right]} \right| \le M; \tag{3.8}$$

where $\epsilon > 0$ is a small constant, then the approximation error $\rho(t)$ in (3.6) is ensured to remain bounded.

Proof: According to Property 1.3 and the definition of partial Bell polynomial in variables $\xi^{(1)}(t), \cdots, \xi^{(k-m+1)}(t)$ given above, ${}_{t_0}\mathcal{D}_t^{\alpha} g(\xi(t))$ and ${}_{t_0}\mathcal{D}_t^{\alpha} \xi(t)$ can be respectively calculated as

$$\begin{aligned}
{}_{t_0}\mathcal{D}_t^{\alpha} g(\xi(t)) =& \frac{\alpha(t-t_0)^{1-\alpha}}{\Gamma(2-\alpha)} g^{(1)}(\xi(t))\xi^{(1)}(t) + \frac{(t-t_0)^{-\alpha}}{\Gamma(1-\alpha)} g(\xi(t)) \\
&+ \sum_{k=2}^{\infty} \binom{\alpha}{k} \frac{(t-t_0)^{k-\alpha}}{\Gamma(k-\alpha+1)} \left[\Psi(g,\xi)_k + g^{(1)}(\xi(t))\xi^{(k)}(t)\right] \\
=& \frac{\alpha(t-t_0)^{1-\alpha}}{\Gamma(2-\alpha)} g^{(1)}(\xi(t))\xi^{(1)}(t) + \frac{(t-t_0)^{-\alpha}}{\Gamma(1-\alpha)} g(\xi(t)) \\
&+ \sum_{k=2}^{\infty} \frac{\Gamma(\alpha+1)\sin\left[\pi(k-\alpha)\right]}{\Gamma(k+1)\pi(k-\alpha)} (t-t_0)^{k-\alpha} \Big[\Psi(g,\xi)_k \\
&+ g^{(1)}(\xi(t))\xi^{(k)}(t)\Big]
\end{aligned} \tag{3.9}$$

and

$$
\begin{aligned}
{}_{t_0}\mathcal{D}_t^\alpha \xi(t) = & \frac{\alpha(t-t_0)^{1-\alpha}}{\Gamma(2-\alpha)}\xi^{(1)}(t) + \frac{(t-t_0)^{-\alpha}}{\Gamma(1-\alpha)}\xi(t) + \\
& \sum_{k=2}^{\infty} \frac{\Gamma(\alpha+1)\sin[\pi(k-\alpha)]}{\Gamma(k+1)\pi(k-\alpha)}(t-t_0)^{k-\alpha}\xi^{(k)}(t).
\end{aligned} \tag{3.10}
$$

Hence the approximation error $\rho(t)$ is computed as

$$
\begin{aligned}
\rho(t) = & -\frac{(t-t_0)^{-\alpha}}{\Gamma(1-\alpha)}\left[g(\xi(t)) - g^{(1)}(\xi(t))\xi(t)\right] \\
& -\sum_{k=2}^{\infty} \frac{\Gamma(\alpha+1)\sin[\pi(k-\alpha)]}{\Gamma(k+1)\pi(k-\alpha)}(t-t_0)^{k-\alpha}\Big[\Psi(g,\xi)_k \\
& + g^{(1)}(\xi(t))\xi^{(k)}(t)\Big] + \sum_{k=2}^{\infty} \frac{\Gamma(\alpha+1)\sin[\pi(k-\alpha)]}{\Gamma(k+1)\pi(k-\alpha)} \\
& (t-t_0)^{k-\alpha} g^{(1)}(\xi(t))\xi^{(k)}(t) \\
= & -\frac{(t-t_0)^{-\alpha}}{\Gamma(1-\alpha)}\left[g(\xi(t)) - g^{(1)}(\xi(t))\xi(t)\right] - \sum_{k=2}^{\infty} S_k
\end{aligned} \tag{3.11}
$$

where

$$
S_k = \frac{\sin[\pi(k-\alpha)]\Gamma(\alpha+1)}{\pi(k-\alpha)\Gamma(k+1)}(t-t_0)^{k-\alpha}\Psi(g,\xi)_k. \tag{3.12}
$$

If condition (i) in Lemma 3.2 is satisfied, we have

$$
\lim_{k\to\infty} \frac{\Psi(g,\xi)_{k+1}}{\Psi(g,\xi)_k} = c,\ c \in \mathbb{R} \text{ and } c \neq 0, \infty. \tag{3.13}
$$

Furthermore, with the relationship given below

$$
\begin{aligned}
\sin[\pi(k+1-\alpha)] &= \sin[\pi(k-\alpha)]\cos\pi + \cos[\pi(k-\alpha)]\sin\pi \\
&= -\sin[\pi(k-\alpha)],
\end{aligned} \tag{3.14}
$$

we get

$$
\begin{aligned}
\lim_{k\to\infty}\left|\frac{S_{k+1}}{S_k}\right| &= \lim_{k\to\infty}\left|\frac{\frac{\Gamma(\alpha+1)\sin[\pi(k+1-\alpha)](t-t_0)^{k+1-\alpha}\Psi(g,\xi)_{k+1}}{\Gamma(k+2)\pi(k+1-\alpha)}}{\frac{\Gamma(\alpha+1)\sin[\pi(k-\alpha)](t-t_0)^{k-\alpha}\Psi(g,\xi)_k}{\Gamma(k+1)\pi(k-\alpha)}}\right| \\
&= \lim_{k\to\infty}\left|\frac{\sin[\pi(k+1-\alpha)](t-t_0)(k-\alpha)\Psi(g,\xi)_{k+1}}{\sin[\pi(k-\alpha)](k+1)(k+1-\alpha)\Psi(g,\xi)_k}\right| \\
&= \lim_{k\to\infty}\left|\frac{-(t-t_0)(k-\alpha)\Psi(g,\xi)_{k+1}}{(k+1)(k+1-\alpha)\Psi(g,\xi)_k}\right| = 0,
\end{aligned} \tag{3.15}
$$

which implies series $\sum_{k=2}^{\infty} S_k$ is convergent. Additionally, from (3.11) it yields

$$
\begin{aligned}
|\rho(t)| \leq & \frac{\left|g(\xi(t)) - g^{(1)}(\xi(t))\xi(t)\right|}{\Gamma(1-\alpha)(t-t_0)^{\alpha}} + \left|\sum_{k=2}^{\infty} S_k\right| \\
\leq & \frac{\left|g(\xi(t)) - g^{(1)}(\xi(t))\xi(t)\right|}{\Gamma(1-\alpha)(t-t_0)^{\alpha}} + \left|\sum_{k=2}^{\infty} S_k\right| + \frac{\left|g(\xi(t)) - g^{(1)}(\xi(t))\xi(t)\right|}{\Gamma(1-\alpha)\epsilon} \\
= & \left|\frac{g(\xi(t)) - g^{(1)}(\xi(t))\xi(t)}{\Gamma(1-\alpha)\left[(t-t_0)^{\alpha}+\epsilon\right]}\right| + \left|\sum_{k=2}^{\infty} S_k\right|.
\end{aligned}
\tag{3.16}
$$

Combining (3.15) and (3.16) with condition (ii) in Lemma 3.2, it can be obtained that $\rho(t)$ is bounded for $\forall t > t_0$. ■

Remark 3.2 *This chapter has offered Lemma 3.2 as one of its noteworthy contributions. As shown in Section 3.4, such a lemma helps us to overcome the difficulties in designing the adaptive backstepping controller for the class of fractional nonlinear systems, provided that conditions in (3.7) and (3.8) are fulfilled, and show rigorous stability analysis for the associated closed-loop system. The examination of such two conditions will be illustrated in our simulation studies.*

3.4 Adaptive Controller Design and Stability Analysis

3.4.1 Adaptive Controller Design

A backstepping-based adaptive controller will be constructed for the class of fractional nonlinear uncertain systems (3.1) in this section. Firstly, virtual errors are defined as

$$e_1 = x_1 - r, \tag{3.17}$$

which is also treated as the output tracking error, and

$$e_i = x_i - \beta_{i-1}, \tag{3.18}$$

where $\beta_{i-1}\,(i = 2, \ldots, n)$ are virtual control signals. The design process is summarized in the following steps.

Step 1: Consider the Lyapunov function $\overline{W}_1 = \frac{1}{2}{e_1}^2$, then according to Lemma 1.2 and Property 1.2, the fractional derivative of $\overline{W}_1$ satisfies

$${}_0\mathcal{D}_t^{\alpha}\overline{W}_1 \leq e_1 {}_0\mathcal{D}_t^{\alpha} e_1 = e_1[e_2 + \beta_1 + \psi_1^T(x_1)\theta_1 - {}_0\mathcal{D}_t^{\alpha} r]. \tag{3.19}$$

The virtual control signal β_1 is designed as

$$\beta_1 = -c_1 e_1 - \psi_1^T(x_1)\hat{\theta}_1 + {}_0\mathcal{D}_t^{\alpha} r, \tag{3.20}$$

where $\hat{\theta}_1$ denotes the estimate of θ_1 and $c_1 > 0$ is a design parameter. Defining $\tilde{\theta}_1 = \theta_1 - \hat{\theta}_1$, then inequality (3.19) becomes

$${}_0\mathcal{D}_t^{\alpha}\overline{W}_1 \leq e_1 e_2 - c_1 {e_1}^2 + e_1\psi_1^T(x_1)\tilde{\theta}_1. \tag{3.21}$$

Let $W_1 = \overline{W}_1 + \frac{1}{2}\tilde{\theta}_1^T\Gamma_1^{-1}\tilde{\theta}_1$, where matrix Γ_1 is positive definite. The fractional-order derivative of W_1 is

$${}_0\mathcal{D}_t^{\alpha} W_1 \leq e_1 e_2 - c_1 {e_1}^2 + e_1\psi_1^T(x_1)\tilde{\theta}_1 - \tilde{\theta}_1^T\Gamma_1^{-1}{}_0\mathcal{D}_t^{\alpha}\hat{\theta}_1. \tag{3.22}$$

Design the adaptive law as

$$ {}_0\mathcal{D}_t^\alpha \hat{\theta}_1 = e_1 \Gamma_1 \psi_1(x_1) - \iota_1 \Gamma_1 \hat{\theta}_1, \tag{3.23} $$

where $\frac{1}{2} < \iota_1 < 1$ is a design parameter. Then inequality (3.22) becomes

$$ {}_0\mathcal{D}_t^\alpha W_1 \leq e_1 e_2 - c_1 {e_1}^2 + \iota_1 \tilde{\theta}_1^T \hat{\theta}_1. \tag{3.24} $$

Step 2: According to (3.1) and (3.18), the fractional derivative of e_2 is computed as

$$ {}_0\mathcal{D}_t^\alpha e_2 = e_3 + \beta_2 + \psi_2^T \theta_2 - {}_0\mathcal{D}_t^\alpha \beta_1. \tag{3.25} $$

In light of our proposed analog chain rule, ${}_0\mathcal{D}_t^\alpha \beta_1$ can be computed as

$$ \begin{aligned} {}_0\mathcal{D}_t^\alpha \beta_1 = &\beta_1^{(1)}(x_1) {}_0\mathcal{D}_t^\alpha x_1 + \beta_1^{(1)}(\hat{\theta}_1) {}_0\mathcal{D}_t^\alpha \hat{\theta}_1 + \beta_1^{(1)}(r) {}_0\mathcal{D}_t^\alpha r \\ &+ \beta_1^{(1)}({}_0\mathcal{D}_t^\alpha r) {}_0\mathcal{D}_t^{2\alpha} r - \rho_1 \end{aligned} \tag{3.26} $$

with $\rho_1(t)$ being the approximation error which is upper bounded by an unknown parameter $\bar{\rho}_1$ if system (3.1) satisfies conditions in (3.7) and (3.8) with function $g(\xi)$ being $\beta_1(x_1, \hat{\theta}_1, r, {}_0\mathcal{D}_t^\alpha r)$ here.

Thus by combining (3.26) with (3.1), (3.25) can be written as

$$ \begin{aligned} {}_0\mathcal{D}_t^\alpha e_2 = &e_3 + \beta_2 + \psi_2^T \theta_2 + \rho_1 - \beta_1^{(1)}(x_1) x_2 - \beta_1^{(1)}(x_1) \psi_1^T \theta_1 \\ &- \beta_1^{(1)}(\hat{\theta}_1) {}_0\mathcal{D}_t^\alpha \hat{\theta}_1 - \beta_1^{(1)}(r) {}_0\mathcal{D}_t^\alpha r - \beta_1^{(1)}({}_0\mathcal{D}_t^\alpha r) {}_0\mathcal{D}_t^{2\alpha} r. \end{aligned} \tag{3.27} $$

Defining

$$ \begin{aligned} \psi_{v,2} = &\left[\psi_2^T, -\beta_1^{(1)}(x_1) \psi_1^T\right]^T, \ \theta_{v,2} = \left[\theta_2^T, \theta_1^T\right]^T, \\ \zeta_2 = &-\beta_1^{(1)}(x_1) x_2 - \beta_1^{(1)}(\hat{\theta}_1) {}_0\mathcal{D}_t^\alpha \hat{\theta}_1 - \beta_1^{(1)}(r) {}_0\mathcal{D}_t^\alpha r - \beta_1^{(1)}({}_0\mathcal{D}_t^\alpha r) {}_0\mathcal{D}_t^{2\alpha} r, \end{aligned} $$

then (3.27) can be rewritten as

$$ {}_0\mathcal{D}_t^\alpha e_2 = e_3 + \beta_2 + \psi_{v,2}^T \theta_{v,2} + \zeta_2 + \rho_1. \tag{3.28} $$

Let $\overline{W}_2 = W_1 + \frac{1}{2} {e_2}^2$, then its fractional derivative is

$$ {}_0\mathcal{D}_t^\alpha \overline{W}_2 \leq e_1 e_2 - c_1 {e_1}^2 + \iota_1 \tilde{\theta}_1^T \hat{\theta}_1 + e_2 \left[e_3 + \beta_2 + \psi_{v,2}^T \theta_{v,2} + \zeta_2 + \rho_1\right]. \tag{3.29} $$

Design virtual control law as

$$ \beta_2 = -e_1 - c_2 e_2 - \psi_{v,2}^T \hat{\theta}_{v,2} - \zeta_2 - \hat{\rho}_1, \tag{3.30} $$

where $c_2 > 0$ is the design parameter and $\hat{\theta}_{v,2}$ and $\hat{\rho}_1$ respectively represent the estimates of $\theta_{v,2}$ and $\bar{\rho}_1$, then it gives

$$ {}_0\mathcal{D}_t^\alpha \overline{W}_2 \leq -c_1 e_1^2 - c_2 e_2^2 + e_2 e_3 + \iota_1 \tilde{\theta}_1^T \hat{\theta}_1 + e_2 \psi_{v,2}^T \tilde{\theta}_{v,2} + e_2 \tilde{\rho}_1, \tag{3.31} $$

in which $\tilde{\theta}_{v,2} = \theta_{v,2} - \hat{\theta}_{v,2}$ and $\tilde{\rho}_1 = \bar{\rho}_1 - \hat{\rho}_1$.

Let $W_2 = \overline{W}_2 + \frac{1}{2} \tilde{\theta}_{v,2}^T \Gamma_2^{-1} \tilde{\theta}_{v,2} + \frac{1}{2\nu_1} \tilde{\rho}_1^2$, where matrix Γ_2 is positive definite and $\nu_1 > 0$, then

$$ \begin{aligned} {}_0\mathcal{D}_t^\alpha W_2 \leq &-c_1 e_1^2 - c_2 e_2^2 + e_2 e_3 + \iota_1 \tilde{\theta}_1^T \hat{\theta}_1 + e_2 \psi_{v,2}^T \tilde{\theta}_{v,2} \\ &+ e_2 \tilde{\rho}_1 - \tilde{\theta}_{v,2}^T \Gamma_2^{-1} {}_0\mathcal{D}_t^\alpha \hat{\theta}_{v,2} - \frac{1}{\nu_1} \tilde{\rho}_1 {}_0\mathcal{D}_t^\alpha \hat{\rho}_1. \end{aligned} \tag{3.32} $$

By designing the adaptive laws as

$$ \begin{cases} {}_0\mathcal{D}_t^\alpha \hat{\theta}_{v,2} = e_2 \Gamma_2 \psi_{v,2} - \iota_1 \Gamma_2 \hat{\theta}_{v,2} \\ {}_0\mathcal{D}_t^\alpha \hat{\rho}_1 = e_2 \nu_1 - \iota_2 \nu_1 \hat{\rho}_1 \end{cases} \tag{3.33} $$

where $\frac{1}{2} < \iota_2 < 1$, (3.32) can be computed as

$$ {}_0\mathcal{D}_t^\alpha W_2 \leq -c_1e_1^2 - c_2e_2^2 + e_2e_3 + \iota_1\tilde{\theta}_1^T\hat{\theta}_1 + \iota_1\tilde{\theta}_{v,2}^T\hat{\theta}_{v,2} + \iota_2\tilde{\rho}_1\hat{\rho}_1. \tag{3.34} $$

Step i ($i = 3, \cdots, n-1$): The design for the third to $n-1$th steps follows the design process given in Step 2. The virtual controllers β_i and adaptive laws are designed respectively as

$$ \beta_i = -e_{i-1} - c_ie_i - \psi_{v,i}^T\hat{\theta}_{v,i} - \zeta_i - \hat{\rho}_{i-1} \tag{3.35} $$

$$ \begin{cases} {}_0\mathcal{D}_t^\alpha\hat{\theta}_{v,i} = e_i\Gamma_i\psi_{v,i} - \iota_1\Gamma_i\hat{\theta}_{v,i} \\ {}_0\mathcal{D}_t^\alpha\hat{\rho}_{i-1} = e_i\nu_{i-1} - \iota_2\nu_{i-1}\hat{\rho}_{i-1} \end{cases} \tag{3.36} $$

where c_i and ν_{i-1} are positive parameters, matrix Γ_i is positive definite, $\hat{\theta}_{v,i}$ and $\hat{\rho}_{i-1}$ are respectively the estimates of $\theta_{v,i}$ and $\bar{\rho}_{i-1}$, and

$$ \begin{aligned} \psi_{v,i} &= \left[\psi_i^T, -\beta_{i-1}^{(1)}(x_{i-1})\psi_{i-1}^T, \cdots, -\beta_{i-1}^{(1)}(x_1)\psi_1^T\right]^T, \\ \theta_{v,i} &= \left[\theta_i^T, \theta_{i-1}^T, \cdots, \theta_1^T\right]^T, \\ \zeta_i &= -\sum_{j=1}^{i-1}\beta_{i-1}^{(1)}(x_j)x_{j+1} - \sum_{j=1}^{i-1}\beta_{i-1}^{(1)}(\hat{\theta}_{v,j}){}_0\mathcal{D}_t^\alpha\hat{\theta}_{v,j} - \sum_{j=0}^{i-1}\beta_{i-1}^{(1)}({}_0\mathcal{D}_t^{j\alpha}r){}_0\mathcal{D}_t^{(j+1)\alpha}r \end{aligned} $$

with $|\rho_{i-1}(t)| \leq \bar{\rho}_{i-1}$ and ${}_0\mathcal{D}_t^0 r = r$. Similar to Step 2, $\bar{\rho}_{i-1}$ stands for the unknown upper bound of error $\rho_{i-1}(t)$.

Let $W_i = W_{i-1} + \frac{1}{2}e_i^2 + \frac{1}{2}\tilde{\theta}_{v,i}^T\Gamma_i^{-1}\tilde{\theta}_{v,i} + \frac{1}{2\nu_{i-1}}\tilde{\rho}_{i-1}^2$ where $\tilde{\rho}_{i-1} = \bar{\rho}_{i-1} - \hat{\rho}_{i-1}$ and $\tilde{\theta}_{v,i} = \theta_{v,i} - \hat{\theta}_{v,i}$, then its fractional derivative is

$$ {}_0\mathcal{D}_t^\alpha W_i \leq e_ie_{i+1} - \sum_{j=1}^{i}c_je_j^2 + \iota_1\sum_{j=1}^{i}\tilde{\theta}_{v,j}^T\hat{\theta}_{v,j} + \iota_2\sum_{j=1}^{i-1}\tilde{\rho}_j\hat{\rho}_j. \tag{3.37} $$

Step n: The fractional-order derivative of e_n is

$$ {}_0\mathcal{D}_t^\alpha e_n = q(u) + \psi_{v,n}^T\theta_{v,n} + d + \zeta_n + \rho_{n-1}, \tag{3.38} $$

in which

$$ \begin{aligned} \psi_{v,n} &= \left[\psi_n^T, -\beta_{n-1}^{(1)}(x_{n-1})\psi_{n-1}^T, \cdots, -\beta_{n-1}^{(1)}(x_1)\psi_1^T\right]^T, \\ \theta_{v,n} &= \left[\theta_n^T, \theta_{n-1}^T, \cdots, \theta_1^T\right]^T, \\ \zeta_n &= -\sum_{j=1}^{n-1}\beta_{n-1}^{(1)}(x_j)x_{j+1} - \sum_{j=1}^{n-1}\beta_{n-1}^{(1)}(\hat{\theta}_{v,j}){}_0\mathcal{D}_t^\alpha\hat{\theta}_{v,j} - \sum_{j=0}^{n-1}\beta_{n-1}^{(1)}({}_0\mathcal{D}_t^{j\alpha}r){}_0\mathcal{D}_t^{(j+1)\alpha}r. \end{aligned} $$

Consider the Lyapunov function $\overline{W}_n = W_{n-1} + \frac{1}{2}{e_n}^2$ and property that $e_nd \leq |e_n|D \leq 0.2785\lambda D + e_n\tanh(\frac{e_n}{\lambda})D$ with $\lambda > 0$ [191], then its fractional derivative satisfies

$$ \begin{aligned} {}_0\mathcal{D}_t^\alpha\overline{W}_n \leq & e_ne_{n-1} - \sum_{j=1}^{n-1}c_je_j^2 + \iota_1\sum_{j=1}^{n-1}\tilde{\theta}_{v,j}^T\hat{\theta}_{v,j} + \iota_2\sum_{j=1}^{n-2}\tilde{\rho}_j\hat{\rho}_j + e_nq(u) + e_n\zeta_n \\ & + 0.2785\lambda D + e_n\tanh\left(\frac{e_n}{\lambda}\right)D + e_n\psi_{v,n}^T\theta_{v,n} + e_n\rho_{n-1}. \end{aligned} \tag{3.39} $$

From Lemma 3.1 it can be noted that

$$|e_n| \cdot |q(u) - u| \leq \delta |e_n u| + u_{\min}|e_n|, \tag{3.40}$$

which consequently gives

$$e_n q(u) \leq |e_n q(u)| \leq e_n u + \delta |e_n u| + u_{\min}|e_n|. \tag{3.41}$$

Design the virtual controller β_n and the control input u respectively as

$$\beta_n = -e_{n-1} - c_n e_n - \psi_{v,n}^T \hat{\theta}_{v,n} - \zeta_n - \hat{\rho}_{n-1} - \tanh\left(\frac{e_n}{\lambda}\right)\hat{D} \tag{3.42}$$

and

$$u = -\tanh\left(\frac{e_n u_n}{\lambda}\right) u_n, \tag{3.43}$$

where $u_n = \frac{-\beta_n + \mu \tanh\left(\frac{\mu e_n}{\lambda}\right)}{1-\delta}$, $|\rho_{n-1}| \leq \bar{\rho}_{n-1}$, $c_n > 0$, $\mu \geq u_{\min}$, and $\hat{\theta}_{v,n}$, $\hat{\rho}_{n-1}$, and $\hat{D}$ respectively denote the estimations of $\theta_{v,n}$, $\bar{\rho}_{n-1}$, and D, thus (3.41) turns into

$$\begin{aligned} e_n q(u) \leq & (\delta - 1) e_n u_n \tanh\left(\frac{e_n u_n}{\lambda}\right) + u_{\min}|e_n| \\ \leq & (\delta - 1) e_n u_n + 0.2785\lambda(1-\delta) + \mu|e_n|. \end{aligned} \tag{3.44}$$

Substituting (3.42) to (3.44) into (3.39) results in

$$\begin{aligned} {}_0\mathcal{D}_t^\alpha \overline{W}_n \leq & e_n e_{n-1} - \sum_{j=1}^{n-1} c_j e_j^2 + \iota_1 \sum_{j=1}^{n-1} \tilde{\theta}_{v,j}^T \hat{\theta}_{v,j} + \iota_2 \sum_{j=1}^{n-2} \tilde{\rho}_j \hat{\rho}_j + e_n \beta_n \\ & - \mu e_n \tanh\left(\frac{\mu e_n}{\lambda}\right) + 0.2785\lambda(1-\delta) + |\mu e_n| + 0.2785\lambda D \\ & + e_n \tanh\left(\frac{e_n}{\lambda}\right) D + e_n \psi_{v,n}^T \theta_{v,n} + e_n \zeta_n + e_n \rho_{n-1} \\ \leq & -\sum_{j=1}^{n} c_j e_j^2 + \iota_1 \sum_{j=1}^{n-1} \tilde{\theta}_{v,j}^T \hat{\theta}_{v,j} + \iota_2 \sum_{j=1}^{n-2} \tilde{\rho}_j \hat{\rho}_j + e_n \tilde{\rho}_{n-1} \\ & + e_n \psi_{v,n}^T \tilde{\theta}_{v,n} + e_n \tanh\left(\frac{e_n}{\lambda}\right) \tilde{D} + 0.2785\lambda(2 - \delta + D), \end{aligned} \tag{3.45}$$

where $\tilde{\theta}_{v,n} = \theta_{v,n} - \hat{\theta}_{v,n}$, $\tilde{\rho}_{n-1} = \bar{\rho}_{n-1} - \hat{\rho}_{n-1}$, and $\tilde{D} = D - \hat{D}$. Let $W_n = \overline{W}_n + \frac{1}{2}\tilde{\theta}_{v,n}^T \Gamma^{-1}{}_n \tilde{\theta}_{v,n} + \frac{1}{2\nu_{n-1}}\tilde{\rho}_{n-1}^2 + \frac{1}{2\eta}\tilde{D}^2$, in which matrix Γ_n is positive definite, $\eta > 0$, and $\nu_{n-1} > 0$, and design the adaptive laws as

$$\begin{cases} {}_0\mathcal{D}_t^\alpha \hat{\theta}_{v,n} = e_n \Gamma_n \psi_{v,n} - \iota_1 \Gamma_n \hat{\theta}_{v,n} \\ {}_0\mathcal{D}_t^\alpha \hat{\rho}_{n-1} = e_n \nu_{n-1} - \iota_2 \nu_{n-1} \hat{\rho}_{n-1} \\ {}_0\mathcal{D}_t^\alpha \hat{D} = e_n \eta \tanh\left(\frac{e_n}{\lambda}\right) - \iota_3 \hat{D} \eta \end{cases} \tag{3.46}$$

with $\frac{1}{2} < \iota_3 < 1$, then the fractional derivative of W_n satisfies

$$\begin{aligned} {}_0\mathcal{D}_t^\alpha W_n \leq & -\sum_{j=1}^{n} c_j e_j^2 + \iota_1 \sum_{j=1}^{n} \tilde{\theta}_{v,j}^T \hat{\theta}_{v,j} + \iota_2 \sum_{j=1}^{n-1} \tilde{\rho}_j \hat{\rho}_j + \iota_3 \tilde{D}\hat{D} + 0.557\lambda \\ & + 0.2785\lambda D. \end{aligned} \tag{3.47}$$

3.4.2 Stability Analysis

Theorem 3.1 *Consider the closed-loop system contains the fractional-order system (3.1) and adaptive backstepping controller with adaptive laws (3.23), (3.36), and (3.46) and control law (3.43) under the initial conditions with the satisfaction of conditions in Lemma 3.2. The boundedness of all the closed-loop signals is ensured and the tracking error e_1 satisfies*

$$\lim_{t\to\infty}|e_1| \leq \sqrt{\frac{2F_2}{F_1}}, \tag{3.48}$$

where

$$F_1 = \min\Big\{2c_1,\cdots,2c_n,\frac{1}{\lambda_{\max}(\Gamma_1^{-1})},\cdots,\frac{1}{\lambda_{\max}(\Gamma_n^{-1})},\nu_1,\cdots,\nu_{n-1},\eta\Big\},$$

$$F_2 = 0.557\lambda + 0.2785\lambda D + \frac{\iota_1^2}{4\iota_1-2}\sum_{j=1}^{n}\|\theta_{v,j}\|^2 + \frac{\iota_2^2}{4\iota_2-2}\sum_{j=1}^{n-1}\bar{\rho}_j^2 + \frac{\iota_3^2}{4\iota_3-2}D^2.$$

Proof: In accordance with Young's inequality,

$$\begin{aligned}\tilde{\theta}_{v,j}^T\hat{\theta}_{v,j} \leq& \frac{a}{2}\tilde{\theta}_{v,j}^T\tilde{\theta}_{v,j} + \frac{1}{2a}\theta_{v,j}^T\theta_{v,j} - \tilde{\theta}_{v,j}^T\tilde{\theta}_{v,j}\\ =& \frac{1}{2a}\theta_{v,j}^T\theta_{v,j} - \frac{2-a}{2}\tilde{\theta}_{v,j}^T\tilde{\theta}_{v,j}\end{aligned} \tag{3.49}$$

where $j = 1,\cdots,n$ and $a = \frac{2\iota_1-1}{\iota_1}$, then it gives

$$\iota_1\tilde{\theta}_{v,j}^T\hat{\theta}_{v,j} \leq \frac{\iota_1^2}{4\iota_1-2}\theta_{v,j}^T\theta_{v,j} - \frac{\tilde{\theta}_{v,j}^T\Gamma_j^{-1}\tilde{\theta}_{v,j}}{2\lambda_{\max}(\Gamma_j^{-1})}. \tag{3.50}$$

Similarly, we have

$$\iota_2\tilde{\rho}_k\hat{\rho}_k \leq \frac{\iota_2^2}{4\iota_2-2}\bar{\rho}_k^2 - \frac{1}{2}\tilde{\rho}_k^2,\ k = 1,\cdots,n-1, \tag{3.51}$$

$$\iota_3\tilde{D}\hat{D} \leq \frac{\iota_3^2}{4\iota_3-2}D^2 - \frac{1}{2}\tilde{D}^2. \tag{3.52}$$

As a result, (3.47) becomes

$$\begin{aligned}{}_0\mathcal{D}_t^\alpha W_n \leq& -\sum_{j=1}^{n}c_je_j^2 + 0.557\lambda + 0.2785\lambda D - \frac{1}{2}\sum_{j=1}^{n-1}\tilde{\rho}_j^2 - \frac{1}{2}\tilde{D}^2 + \frac{\iota_3^2}{4\iota_3-2}D^2\\ &-\sum_{j=1}^{n}\frac{\tilde{\theta}_{v,j}^T\Gamma_j^{-1}\tilde{\theta}_{v,j}}{2\lambda_{\max}(\Gamma_j^{-1})} + \frac{\iota_1^2}{4\iota_1-2}\sum_{j=1}^{n}\|\theta_{v,j}\|^2 + \frac{\iota_2^2}{4\iota_2-2}\sum_{j=1}^{n-1}\bar{\rho}_j^2\\ \leq& -F_1W_n + F_2.\end{aligned} \tag{3.53}$$

Thus ${}_0\mathcal{D}_t^\alpha W_n \leq 0$ when $W_n \geq \frac{F_2}{F_1}$, indicating the set of initial conditions is an invariant set with proper selection of parameters. Additionally, choosing the integer number $n = 1$ in Lemma 1.1 and taking fractional integral on both sides of inequality (3.53) gives

$$\begin{aligned}W_n(t) \leq& W_n(0)\mathrm{E}_\alpha(-F_1t^\alpha) + F_2t^\alpha\mathrm{E}_{\alpha,\alpha+1}(-F_1t^\alpha)\\ =& W_n(0)\mathrm{E}_\alpha(-F_1t^\alpha) + \frac{F_2}{F_1} + F_2t^\alpha\mathrm{o}\Big(\frac{1}{|-F_1t^\alpha|^2}\Big).\end{aligned} \tag{3.54}$$

From (3.54), it can be known that W_n is bounded and therefore, $e_i\,(i = 1,\cdots,n)$, $\hat{\theta}_{v,i}$, $\hat{\rho}_k\,(k = 1,\cdots,n-1)$, and $\hat{D}$ are bounded. Consequently, the boundedness of x_i can be guaranteed from equations (3.17) and (3.18) and the boundedness of β_i. Together with the real control signal u given in (3.43), it can be proved that u is bounded. Therefore, all the closed-loop signals are bounded. In addition, from (3.54) it gives that $W_n(t) \leq \frac{F_2}{F_1}$ when $t \to \infty$, indicating that the tracking error e_1 asymptotically converges to an adjustable small boundary $\sqrt{\frac{2F_2}{F_1}}$ that can be tuned by choosing the design parameters c_i, Γ_i, ν_k, λ, $\iota_l \, (l = 1, 2, 3)$, and η. ■

Remark 3.3 *It can be observed from the definitions of F_1 and F_2 given below (3.48) that larger c_i, Γ_i, ν_k, and η can result in larger F_1 and smaller F_2 can be gained by reducing the values of ι_l and λ. Therefore the tunable bound $\sqrt{\frac{2F_2}{F_1}}$ can be adjusted smaller by increasing the values of c_i, Γ_i, ν_k, and η and designing ι_l and λ to be smaller. However, larger c_i, Γ_i, ν_k, and η may lead to an increased control force. As a result, the design parameters chosen when constructing the controller for a particular system must take into account the trade-off between the control effort and the control performance.*

3.5 Simulation Studies

Two fractional nonlinear systems will be studied as simulation studies to illustrate the proposed control scheme.

3.5.1 PLL in Fractional-Order

The fractional-order PLL [3] with quantized input is given as follows.

$$\begin{cases} {}_0\mathcal{D}_t^{0.715} x_1 = x_2 + \psi_1(x_1)\theta_1 \\ {}_0\mathcal{D}_t^{0.715} x_2 = Kq(u) + \psi_2(\overline{x}_2)\theta_2 + Kd \end{cases} \tag{3.55}$$

where $y = x_1$, $\psi_1(x_1) = 1$,$\psi_2(\overline{x}_2) = -2.77x_2$, and $K = 7.25$. Here $\theta_1 = 0$, $\theta_2 = 1$, the disturbance $d = 5[\sin(20t) + \cos(20t)] + G(t)$ are unknown where $G(t)$ denotes the Gaussian noise with mean and variance respectively equal to 0.1 and 1, and the reference signal is a sinusoidal waveform $r = \sin(2t)$. The initial values of states are assigned as $x(0) = [-0.1, 0.5]^T$ and Fig. 3.1 shows that without control input, the output of (3.55) cannot track the reference signal.

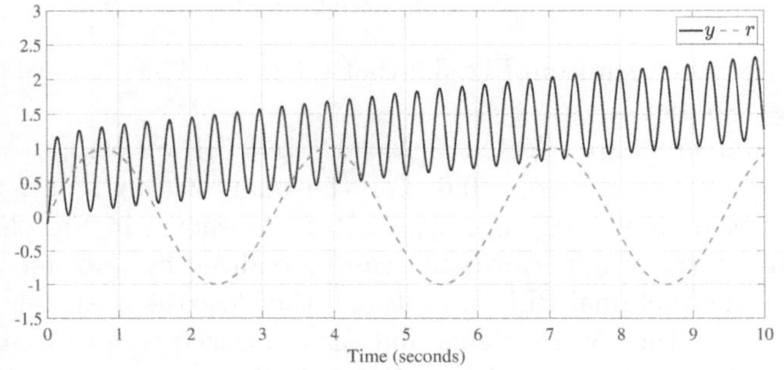

FIGURE 3.1: Reference signal and system output of (3.55) without control input.

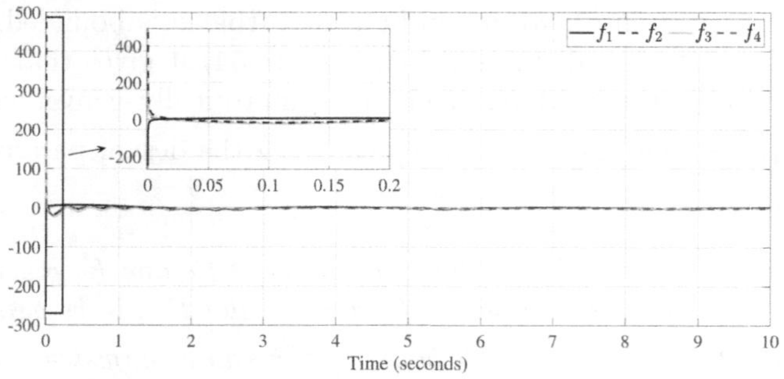

FIGURE 3.2: Boundedness of f_i $(i = 1, \cdots, 4)$.

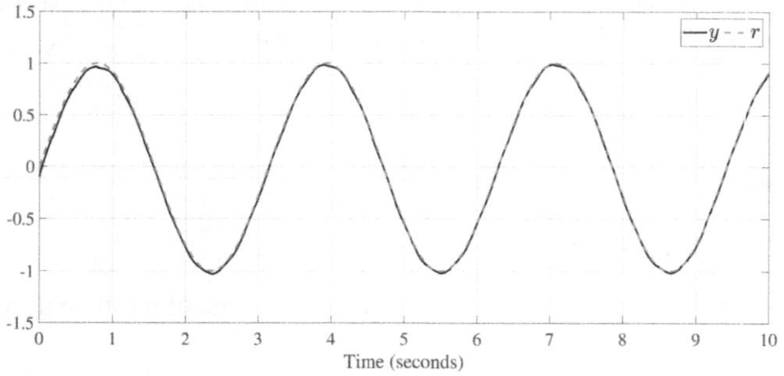

FIGURE 3.3: Reference signal and system output of (3.55) under the proposed control scheme.

According to (3.20) and the definition given in (3.4), it can be known that condition (3.7) is satisfied. Here we define

$$f_1 = \frac{\beta_1 - \beta_1^{(1)}(x_1)x_1}{\Gamma(1-\alpha)(t^\alpha + \epsilon)}, \tag{3.56a}$$

$$f_2 = \frac{\beta_1 - \beta_1^{(1)}(\hat{\theta}_1)\hat{\theta}_1}{\Gamma(1-\alpha)(t^\alpha + \epsilon)}, \tag{3.56b}$$

$$f_3 = \frac{\beta_1 - \beta_1^{(1)}(r)r}{\Gamma(1-\alpha)(t^\alpha + \epsilon)}, \tag{3.56c}$$

$$f_4 = \frac{\beta_1 - \beta_1^{(1)}({}_0\mathcal{D}_t^\alpha r){}_0\mathcal{D}_t^\alpha r}{\Gamma(1-\alpha)(t^\alpha + \epsilon)}, \tag{3.56d}$$

with $\epsilon = 0.001$. It can be seen from Fig. 3.2 that all f_i $(i = 1, \cdots, 4)$ are bounded, implying condition (3.8) is also met.

The control parameters are selected as $\lambda = 0.2$, $\mu = 0.1$, $c_1 = 15$, $c_2 = 10$, $\Gamma_1 = \Gamma_2 = I$, $\eta = \nu_1 = 1$, and $\iota_1 = \iota_2 = \iota_3 = 0.6$. For the quantization input, the corresponding parameters are chosen as $\delta = 0.2$ and $u_{\min} = 0.1$. As shown in Fig. 3.3 to Fig. 3.5 that not only good output tracking performance can be realized by implementing our proposed control scheme to the fractional PLL, but also all the closed-loop signals are guaranteed to be bounded. Moreover, the control signal and the quantized control input are displayed in Fig. 3.6 and Fig. 3.7 respectively, and from these two figures we can note that both control input and quantized control input are bounded, thus further confirming that all signals in the closed-loop system stay bounded.

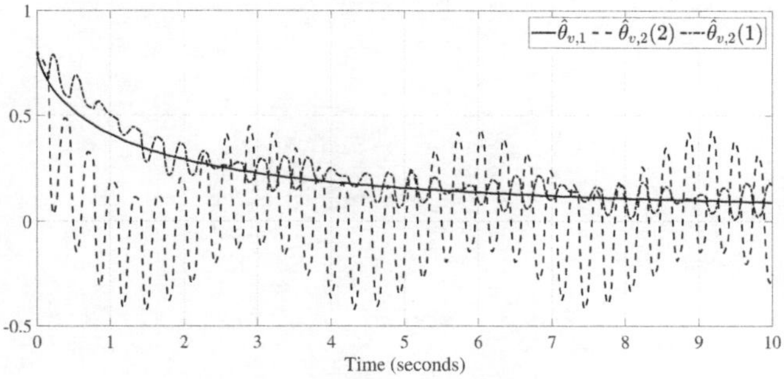

FIGURE 3.4: Estimates $\hat{\theta}_{v,1}$ and $\hat{\theta}_{v,2}$.

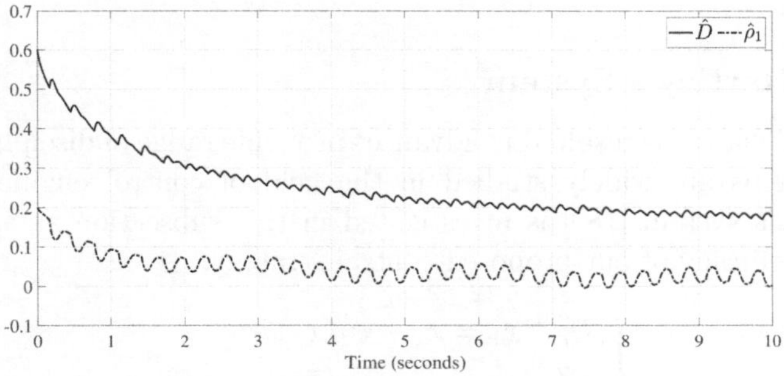

FIGURE 3.5: Estimates $\hat{D}$ and $\hat{\rho}_1$.

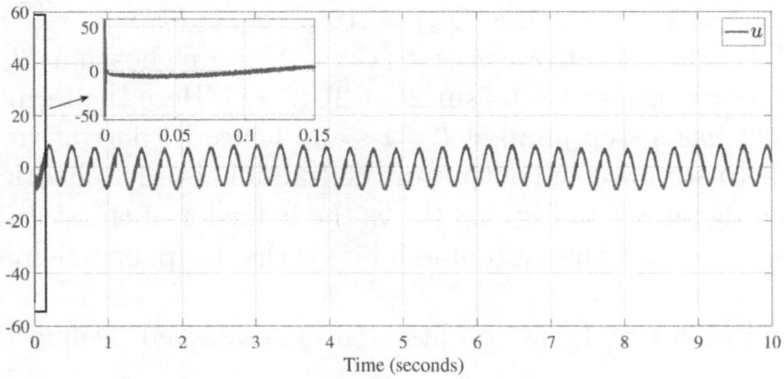

FIGURE 3.6: Control input u.

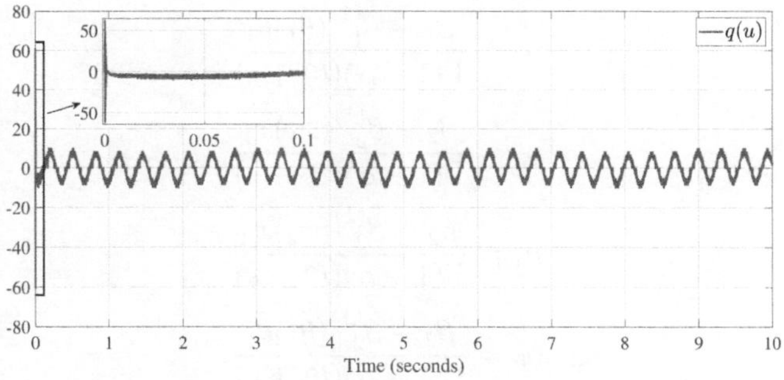

FIGURE 3.7: Quantized input $q(u)$.

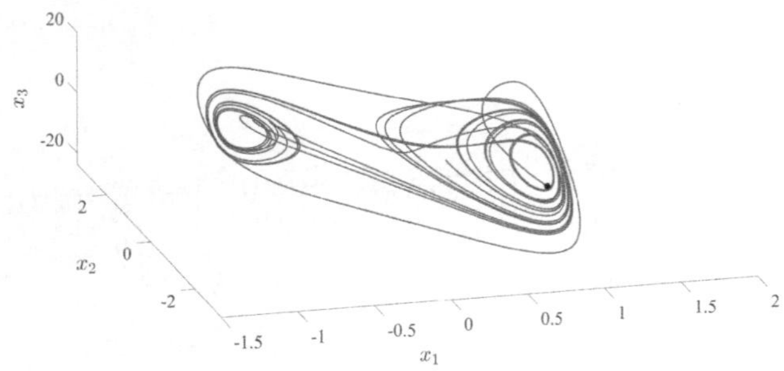

FIGURE 3.8: State variables of the open-loop system (3.57) with black point indicating initial value.

3.5.2 Chua-Hartley's System

The emergence of chaos has resulted in advances in a wide range of disciplines [192, 193, 194], and chaotic systems are widely studied in the field of control engineering. The following Chua-Hartley's system [189] is investigated in this subsection as another example to demonstrate the efficacy of our proposed control strategy.

$$\begin{cases} {}_0\mathcal{D}_t^{0.98} x_1 = x_2 + \psi_1(x_1)\theta_1 \\ {}_0\mathcal{D}_t^{0.98} x_2 = x_3 + \psi_2(\overline{x}_2)\theta_2 \\ {}_0\mathcal{D}_t^{0.98} x_3 = q(u) + \psi_3(\overline{x}_3)\theta_3 + d \end{cases} \tag{3.57}$$

where $y = x_1$, $\psi_1(x_1) = \frac{10}{7}(x_1 - x_1^3)$, $\psi_2(\overline{x}_2) = 10x_1 - x_2$, $\psi_3(\overline{x}_3) = -\frac{100}{7}x_2$, and initial value $x(0) = [0.8,\ -2,\ 1]^T$. The actual values of θ_i $(i = 1, 2, 3)$ are chosen as $\theta_i = 1$ for simulation and the external disturbance is $d = 0.5\sin(2t) + 3\mathrm{U}(t-1)$. Here the term $3\mathrm{U}(t-1)$ indicates that the disturbance has a step jump at $t = 1$ s and hence its derivative is unbounded. The control objective is to stabilize the output of the fractional-order chaotic system (3.57) and hence the reference signal $r = 0$. Fig. 3.8 shows the behavior of the state variables in (3.57), from which can be seen that the system exhibits a chaotic phenomenon without a control signal.

From (3.20) and (3.30) we know condition (3.7) is satisfied. Define

$$p_1 = \frac{\beta_1 - \beta_1^{(1)}(x_1)x_1}{\Gamma(1-\alpha)(t^\alpha + \epsilon)}, \tag{3.58a}$$

$$p_2 = \frac{\beta_1 - \beta_1^{(1)}(\hat{\theta}_1)\hat{\theta}_1}{\Gamma(1-\alpha)(t^\alpha + \epsilon)}, \tag{3.58b}$$

$$p_3 = \frac{\beta_2 - \beta_2^{(1)}(x_1)x_1}{\Gamma(1-\alpha)(t^\alpha + \epsilon)}, \tag{3.58c}$$

$$p_4 = \frac{\beta_2 - \beta_2^{(1)}(x_2)x_2}{\Gamma(1-\alpha)(t^\alpha + \epsilon)}, \tag{3.58d}$$

$$p_5 = \frac{\beta_2 - \beta_2^{(1)}(\hat{\theta}_1)\hat{\theta}_1}{\Gamma(1-\alpha)(t^\alpha + \epsilon)}, \tag{3.58e}$$

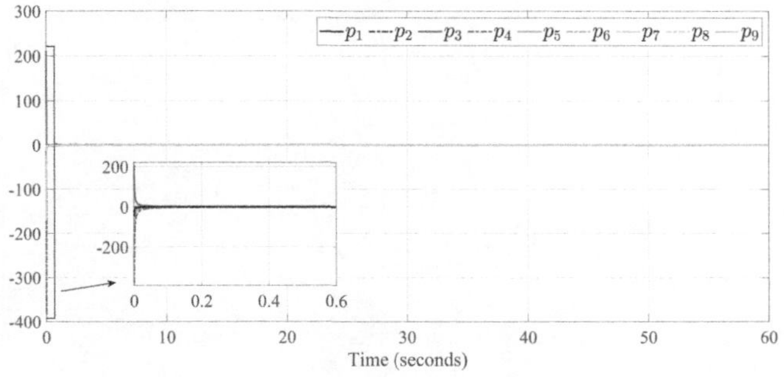

FIGURE 3.9: Boundedness of p_i $(i = 1, \cdots, 9)$.

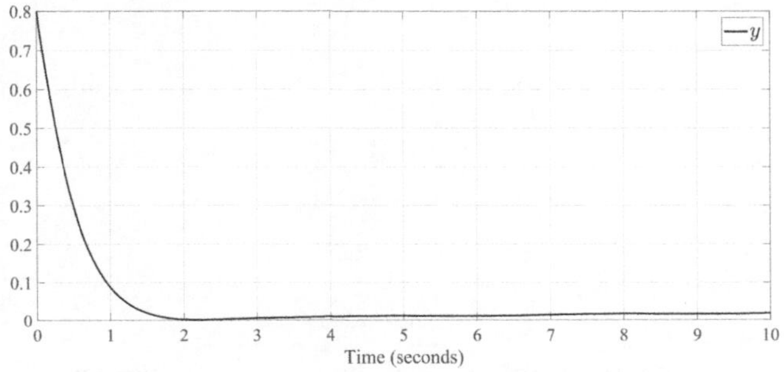

FIGURE 3.10: System output of (3.57) under proposed control scheme.

$$p_6 = \frac{\beta_2 - \beta_2^{(1)}(\hat{\theta}_{v,2}(2))\hat{\theta}_{v,2}(2)}{\Gamma(1-\alpha)(t^\alpha + \epsilon)}, \tag{3.58f}$$

$$p_7 = \frac{\beta_2 - \beta_2^{(1)}(\hat{\theta}_{v,2}(1))\hat{\theta}_{v,2}(1)}{\Gamma(1-\alpha)(t^\alpha + \epsilon)}, \tag{3.58g}$$

$$p_8 = \frac{\beta_2 - \beta_2^{(1)}(e_2)e_2}{\Gamma(1-\alpha)(t^\alpha + \epsilon)}, \tag{3.58h}$$

$$p_9 = \frac{\beta_2 - \beta_2^{(1)}(\hat{\rho}_1)\hat{\rho}_1}{\Gamma(1-\alpha)(t^\alpha + \epsilon)}, \tag{3.58i}$$

where $\epsilon = 0.001$, and the boundedness of (3.58a) to (3.58i) is shown in Fig. 3.9.

The quantization parameters are set to $\delta = 0.2$ and $u_{\min} = 0.1$ and the control parameters are established as $\lambda = 0.2$, $\mu = 0.1$, $c_1 = 3$, $c_2 = 4$, $c_3 = 10$, $\Gamma_1 = \Gamma_2 = \Gamma_3 = 0.1I$, $\eta = 0.1$, $\nu_1 = \nu_2 = 1$, and $\iota_1 = \iota_2 = \iota_3 = 0.51$. From Fig. 3.10, it is noticed that the proposed adaptive controller removes the chaotic behaviors and can drive the system output to a small vicinity around the origin. Besides, Fig. 3.11 and Fig. 3.12 show the control input and the quantized one accordingly, from where it can be seen that both signals remain bounded. Furthermore, Fig. 3.13 and Fig. 3.14 showcase that under the proposed controller, the estimates of unknown parameters $\hat{\theta}_{v,i}$ $(i = 1, 2, 3)$, $\hat{D}$, $\hat{\rho}_1$, and $\hat{\rho}_2$ are also bounded. All the simulation results illustrate the efficacy of the investigated control strategy and verify the obtained theoretical results.

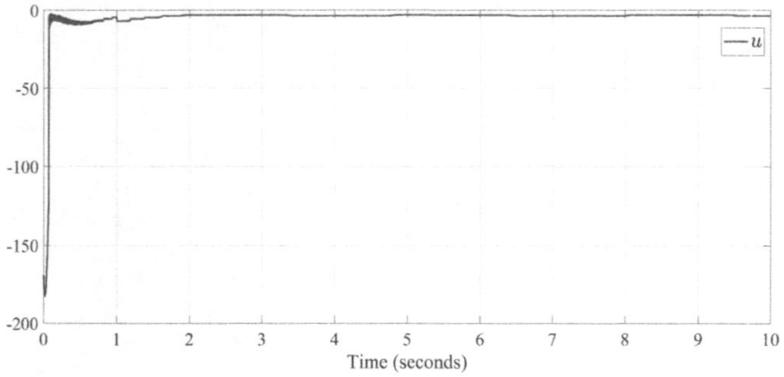

FIGURE 3.11: Control input u of (3.57).

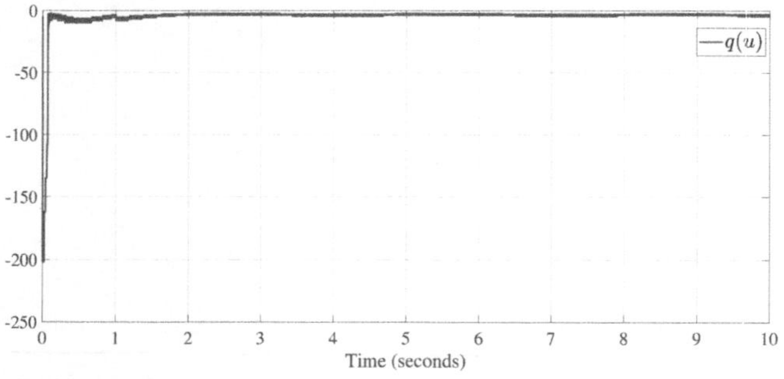

FIGURE 3.12: Quantized input $q(u)$ of (3.57).

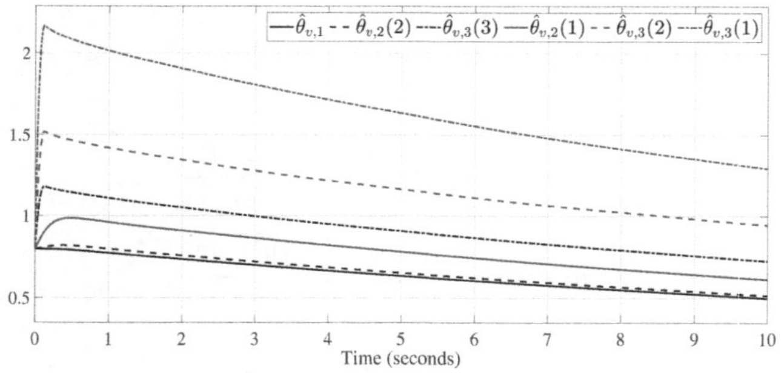

FIGURE 3.13: Estimates $\hat{\theta}_{v,i}$ $(i = 1, 2, 3)$ of (3.57).

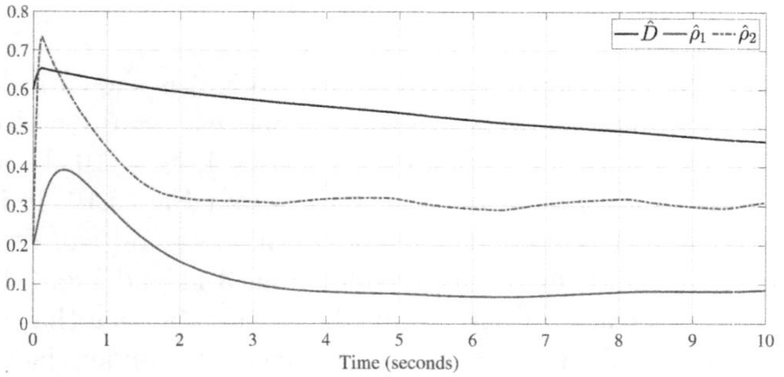

FIGURE 3.14: Estimates $\hat{D}$, $\hat{\rho}_1$, and $\hat{\rho}_2$ of (3.57).

3.6 Conclusions

In this chapter, we propose an analog chain rule computation method to calculate the fractional derivatives of virtual control signals that exist in the adaptive backstepping controller design procedure for fractional nonlinear systems. It is rigorously demonstrated that, when using the proposed quantized adaptive control input, the output of the associated closed-loop uncertain nonlinear fractional system in the presence of unknown external disturbances is able to accurately track the reference signal while also keeping all closed-loop signals remain bounded, thus ensuring system stability. The efficacy of the proposed control method is finally substantiated by simulated examples.

4

Adaptive Backstepping Fractional-Order Control for Incommensurate Uncertain Fractional-Order Nonlinear Systems with External Disturbance

In this chapter, we will focus on constructing a smooth adaptive backstepping controller for a class of incommensurate fractional-order nonlinear systems with uncertainties and time-varying disturbance to achieve that the system output asymptotically tracks a predefined signal, while ensuring the global stability of the resulting closed-loop system. Moreover, same as in the previous chapters, rigorous theoretical analysis based on actual fractional-order systems is provided in this chapter, which shows the difference from the existing results that are established on the transformed integer-order systems rather than the original fractional-order ones.

4.1 Introduction

In the previous two chapters, the backstepping-based adaptive control problem for commensurate nonlinear fractional-order systems is investigated. However, it is of great importance yet challenge to investigate the incommensurate fractional-order system, which is a universal form and can be observed in many practical and industrial scenarios such as fractional-order PLL [56, 3, 57] and chaotic systems [195, 196]. Nevertheless, few results about the backstepping control of such systems have been reported, as mentioned in the last paragraph of Subsection 1.3.1.

To the best of our knowledge, it is still open to construct a smooth adaptive backstepping controller for incommensurate fractional-order nonlinear systems with uncertainties and time-varying disturbance to achieve that the system output asymptotically tracks a predefined signal, while ensuring global stability of the resulting closed-loop system. By virtue of a smooth controller, continuous control effort can be well implemented on the actual actuators, thereby avoiding chattering and also potentially extending the lifespan of devices. Such issues are successfully addressed in this chapter. Both simulation and experimental tests are provided to illustrate the effectiveness of our proposed control scheme.

In comparison with existing results, the contributions of this chapter are mainly threefold.

1. Compared with the backstepping-based adaptive control schemes for fractional-order nonlinear systems in [97, 98, 99, 101, 104, 106, 1, 112, 197] and Chapter 2

DOI: 10.1201/9781003620372-4

and Chapter 3 where commensurate fractional-order systems are studied, incommensurate fractional-order nonlinear systems are considered in this chapter.

2. Note that the results presented in [125] and [126] are established for systems without uncertainties and disturbances. In contrast, our output asymptotic tracking property and global system stability are achieved via the proposed control scheme in the presence of parametric uncertainties and unknown time-varying external disturbances. Moreover, unlike the discontinuous controller constructed in [127], a smooth control signal is obtained in this chapter.

3. Different from the results in [93, 94, 122, 123, 124, 2] that are established on the transformed integer-order systems rather than the original fractional-order ones, rigorous theoretical analysis based on actual fractional-order systems is provided in this chapter.

Besides, fractional-order plants with integer-order controllers are investigated in Chapter 2 and Chapter 3, while in this chapter, the fractional-order plant is controlled under a fractional-order controller, as will be illustrated later.

The outline of this chapter is structured as follows. Problem formulation is provided in Section 4.2. An adaptive controller is designed in Section 4.3 with stability analysis. In Section 4.4, the control strategy is demonstrated by simulation and experimental studies and finally, the chapter is concluded in Section 4.5.

4.2 Problem Formulation

A class of nth-order uncertain incommensurate fractional-order nonlinear systems with time-varying disturbance in strict feedback form is considered in this chapter:

$$\begin{cases} \dot{x}_1(t) = x_2(t) + \phi_1^T(\overline{x}_1(t))\theta_1 \\ \dot{x}_2(t) = x_3(t) + \phi_2^T(\overline{x}_2(t))\theta_2 \\ \quad\vdots \\ \dot{x}_{n-1}(t) = x_n(t) + \phi_{n-1}^T(\overline{x}_{n-1}(t))\theta_{n-1} \\ {}_0\mathcal{D}_t^\alpha x_n(t) = g(x(t))u(t) + \phi_n^T(x(t))\theta_n + d(t) \\ y(t) = x_1(t) \end{cases} \tag{4.1}$$

where $u(t) \in \mathbb{R}$ and $y(t) \in \mathbb{R}$ represent the control signal and system output respectively, $\alpha \in (0,1]$, $\overline{x}_i(t) = [x_1(t), x_2(t), \cdots, x_i(t)]^T \in \mathbb{R}^i$ and $x(t) = [x_1(t), x_2(t), \cdots, x_n(t)]^T \in \mathbb{R}^n$ denote the measurable state vectors, $\phi_i(\overline{x}_i(t)) \in \mathbb{R}^p$ $(i = 1, 2, \cdots, n)$ are known nonlinear smooth functions, $\theta_i \in \mathbb{R}^p$ are unknown constant vectors, $g(x(t))$ is a known non-zero smooth nonlinear function and $d(t)$ is an unknown time-varying disturbance.

Remark 4.1 *The system (4.1) describes a class of incommensurate fractional-order systems such as fractional-order PLL [56, 3, 57] and chaotic systems [195, 196], which will be investigated as numerical and experimental studies in Section 4.4 to verify the established control scheme.*

The control objective in this chapter is to design an adaptive controller for the class of incommensurate fractional-order systems given in (4.1) so that the corresponding closed-loop system is globally asymptotically stable in the sense that all the closed-loop signals are bounded and the system output $y(t)$ asymptotically tracks a given reference signal $y_r(t)$, i.e., $\lim_{t\to\infty}[y(t) - y_r(t)] = 0$.

Assumption 4.1 *The* $(1-\alpha)$*th-order Caputo-type fractional derivative of the external disturbance* ${}_0\mathcal{D}_t^{1-\alpha}d(t)$ *is bounded by an unknown bound* D.

Assumption 4.2 *The reference signal* $y_r(t)$ *and its first* n *derivatives* $y_r^{(i)}$ $i = 1, \cdots, n$ *are known, bounded, and smooth.*

Remark 4.2 *As analyzed in the next section and stated in Remark 4.3, Assumption 4.1 is needed for compensating for the external disturbance when designing control input. Note that a broad class of disturbances can be covered under this assumption. For instance, Assumption 4.1 is satisfied if disturbance* $d(t)$ *is bounded when* $\alpha = 1$*, and the disturbances in the simulation and experimental studies also satisfy such an assumption. Assumption 4.2 is required for all standard backstepping-based adaptive tracking control [86], even for integer-order systems.*

4.3 Design of Adaptive Backstepping Controller and Stability Analysis

4.3.1 Adaptive Controller Design

An adaptive controller is designed based on the backstepping design procedure to achieve the above objective.

Defining virtual errors as: $z_1 = x_1 - y_r$, $z_i = x_i - y_r^{(i-1)} - \alpha_{i-1}\,(i = 2, \ldots, n)$, where $\alpha_j\,(j = 1, 2, \ldots, n-1)$ is the virtual control signal that will be designed later.

Step 1: Firstly, consider the Lyapunov function $\overline{V}_1 = \frac{1}{2}{z_1}^2$, then its time derivative is

$$\dot{\overline{V}}_1 = z_1\dot{z}_1 = z_1z_2 + z_1\alpha_1 + z_1\phi_1^T\theta_1. \tag{4.2}$$

Design the virtual control signal α_1 as

$$\alpha_1 = -c_1z_1 - \phi_1^T\hat{\theta}_1, \tag{4.3}$$

where $\hat{\theta}_1$ stands for the estimate of θ_1 and c_1 is a positive design parameter. Defining $\tilde{\theta}_1 = \theta_1 - \hat{\theta}_1$, then (4.2) becomes

$$\dot{\overline{V}}_1 = z_1z_2 - c_1z_1^2 - z_1\phi_1^T\hat{\theta}_1 + z_1\phi_1^T\theta_1 = z_1z_2 - c_1z_1^2 + z_1\phi_1^T\tilde{\theta}_1. \tag{4.4}$$

Let $V_1 = \overline{V}_1 + \frac{1}{2}\tilde{\theta}_1^T\Gamma_1^{-1}\tilde{\theta}_1$, where Γ_1 is a positive definite matrix, then the time derivative of V_1 is

$$\dot{V}_1 = \dot{\overline{V}}_1 + \tilde{\theta}_1^T\Gamma_1^{-1}\dot{\tilde{\theta}}_1 = z_1z_2 - c_1z_1^2 + z_1\phi_1^T\tilde{\theta}_1 - \tilde{\theta}_1^T\Gamma_1^{-1}\dot{\hat{\theta}}_1. \tag{4.5}$$

By designing adaptive law as

$$\dot{\hat{\theta}}_1 = z_1\Gamma_1\phi_1, \tag{4.6}$$

then (4.5) becomes

$$\dot{V}_1 = z_1z_2 - c_1z_1^2 + z_1\phi_1^T\tilde{\theta}_1 - \tilde{\theta}_1^T\Gamma_1^{-1}z_1\Gamma_1\phi_1 = z_1z_2 - c_1z_1^2. \tag{4.7}$$

Step i $(i = 2, \cdots, n-1)$: The time derivative of z_i is

$$\dot{z}_i = \dot{x}_i - y_r^{(i)} - \dot{\alpha}_{i-1} = x_{i+1} + \phi_i^T\theta_i - y_r^{(i)} - \dot{\alpha}_{i-1} = z_{i+1} + \alpha_i + \phi_{v,i}^T\theta_{v,i} + \xi_i, \tag{4.8}$$

where

$$\begin{aligned}
\theta_{v,i} =& \big[\theta_i^T, \theta_{i-1}^T, \cdots, \theta_1^T\big]^T, \\
\phi_{v,i} =& \Big[\phi_i^T, \Big(-\frac{\partial\alpha_{i-1}}{\partial x_{i-1}}\Big)\phi_{i-1}^T, \cdots, \Big(-\frac{\partial\alpha_{i-1}}{\partial x_2}\Big)\phi_2^T, \Big(-\frac{\partial\alpha_{i-1}}{\partial x_1}\Big)\phi_1^T\Big]^T, \\
\xi_i =& \sum_{j=2}^{i}\Big(-\frac{\partial\alpha_{i-1}}{\partial x_{j-1}}\Big)x_j + \sum_{j=1}^{i-1}\Big(-\frac{\partial\alpha_{i-1}}{\partial\hat{\theta}_{v,j}}\Big)\dot{\hat{\theta}}_{v,j} + \sum_{j=1}^{i-1}\Big(-\frac{\partial\alpha_{i-1}}{\partial y_r^{(j-1)}}\Big)y_r^{(j)},
\end{aligned}$$

with $\hat{\theta}_{v,j}$ standing for the estimate of $\theta_{v,j}$, $j = 1, 2, \cdots, i-1$.

Design virtual control signal α_i as

$$\alpha_i = -z_{i-1} - c_i z_i - \phi_{v,i}^T\hat{\theta}_{v,i} - \xi_i, \tag{4.9}$$

where $c_i > 0$, $\hat{\theta}_{v,i}$ represents the estimate of $\theta_{v,i}$ with its adaptive law designed as

$$\dot{\hat{\theta}}_{v,i} = z_i\Gamma_i\phi_{v,i}, \tag{4.10}$$

in which Γ_i is positive definite.

Define $\tilde{\theta}_{v,i} = \theta_{v,i} - \hat{\theta}_{v,i}$ and let $V_i = V_{i-1} + \frac{1}{2}{z_i}^2 + \frac{1}{2}\tilde{\theta}_{v,i}^T\Gamma_i^{-1}\tilde{\theta}_{v,i}$, then

$$\dot{V}_i = \dot{V}_{i-1} + z_i\dot{z}_i - \tilde{\theta}_{v,i}^T\Gamma_i^{-1}\dot{\hat{\theta}}_{v,i}. \tag{4.11}$$

Since $\dot{V}_{i-1} = z_{i-1}z_i - \sum_{j=1}^{i-1} c_j {z_j}^2$, thus we get

$$\dot{V}_i = z_i z_{i+1} - \sum_{j=1}^{i} c_j {z_j}^2. \tag{4.12}$$

Step n: According to Property 1.1, from (4.1) we know that

$$\dot{x}_n = {}_0\mathcal{D}_t^{1-\alpha}\,{}_0\mathcal{D}_t^{\alpha}x_n = {}_0\mathcal{D}_t^{1-\alpha}\big[g(x)u\big] + {}_0\mathcal{D}_t^{1-\alpha}\phi_n^T\theta_n + {}_0\mathcal{D}_t^{1-\alpha}d. \tag{4.13}$$

Then the time derivative of z_n is

$$\begin{aligned}
\dot{z}_n =& \dot{x}_n - y_r^{(n)} - \dot{\alpha}_{n-1} \\
=& {}_0\mathcal{D}_t^{1-\alpha}\big[g(x)u\big] + {}_0\mathcal{D}_t^{1-\alpha}\phi_n^T\theta_n + {}_0\mathcal{D}_t^{1-\alpha}d - y_r^{(n)} - \dot{\alpha}_{n-1} \\
=& {}_0\mathcal{D}_t^{1-\alpha}\big[g(x)u\big] + \phi_{v,n}^T\theta_{v,n} + \xi_n + {}_0\mathcal{D}_t^{1-\alpha}d,
\end{aligned} \tag{4.14}$$

where

$$\begin{aligned}
\theta_{v,n} =& \big[\theta_n^T, \theta_{n-1}^T, \cdots, \theta_1^T\big]^T, \\
\phi_{v,n} =& \Big[{}_0\mathcal{D}_t^{1-\alpha}\phi_n^T, \Big(-\frac{\partial\alpha_{n-1}}{\partial x_{n-1}}\Big)\phi_{n-1}^T, \cdots, \Big(-\frac{\partial\alpha_{n-1}}{\partial x_1}\Big)\phi_1^T\Big]^T, \\
\xi_n =& \sum_{j=2}^{n}\Big(-\frac{\partial\alpha_{n-1}}{\partial x_{j-1}}\Big)x_j + \sum_{j=1}^{n-1}\Big(-\frac{\partial\alpha_{n-1}}{\partial\hat{\theta}_{v,j}}\Big)\dot{\hat{\theta}}_{v,j} + \sum_{j=1}^{n-1}\Big(-\frac{\partial\alpha_{n-1}}{\partial y_r^{(j-1)}}\Big)y_r^{(j)} - y_r^{(n)}.
\end{aligned}$$

The control input is designed as

$$\begin{cases} {}_0\mathcal{D}_t^{1-\alpha} h = -z_{n-1} - c_n z_n - \phi_{v,n}^T \hat{\theta}_{v,n} - \xi_n - sg(z_n, \epsilon)\hat{D} \\ u = \dfrac{h}{g(x)} \end{cases} \tag{4.15}$$

in which $h(t)$ is a controller state, function $\epsilon(t) = e^{-at}$ with $a > 0$, c_n is a positive design parameter, and $\hat{\theta}_{v,n}$ and $\hat{D}$ indicate the estimates of $\theta_{v,n}$ and D, respectively.

Considering the Lyapunov function $\overline{V}_n = V_{n-1} + \frac{1}{2}{z_n}^2$, then its time derivative is

$$\begin{aligned} \dot{\overline{V}}_n =& z_{n-1}z_n - \sum_{j=1}^{n-1} c_j {z_j}^2 + z_n\Big[- z_{n-1} - c_n z_n - \phi_{v,n}^T \hat{\theta}_{v,n} - \xi_n - sg(z_n, \epsilon)\hat{D} \\ & + \phi_{v,n}^T \theta_{v,n} + \xi_n + {}_0\mathcal{D}_t^{1-\alpha} d\Big] \\ \leq & z_{n-1}z_n - \sum_{j=1}^{n-1} c_j {z_j}^2 + z_n\Big[- z_{n-1} - c_n z_n - \phi_{v,n}^T \hat{\theta}_{v,n} - \xi_n - sg(z_n, \epsilon)\hat{D} \\ & + \phi_{v,n}^T \theta_{v,n} + \xi_n\Big] + |z_n| D \\ =& - \sum_{j=1}^{n} c_j {z_j}^2 + z_n \phi_{v,n}^T \tilde{\theta}_{v,n} - z_n sg(z_n, \epsilon)\hat{D} + |z_n| D \end{aligned} \tag{4.16}$$

where $\tilde{\theta}_{v,n} = \theta_{v,n} - \hat{\theta}_{v,n}$. According to Lemma 2.1, since $|z_n| < \epsilon + z_n sg(z_n, \epsilon)$, hence $|z_n| D \leq \epsilon D + z_n sg(z_n, \epsilon)D$. Therefore, (4.16) can be expressed as

$$\begin{aligned} \dot{\overline{V}}_n \leq & - \sum_{j=1}^{n} c_j {z_j}^2 + z_n \phi_{v,n}^T \tilde{\theta}_{v,n} - z_n sg(z_n, \epsilon)\hat{D} + \epsilon D + z_n sg(z_n, \epsilon) D \\ =& - \sum_{j=1}^{n} c_j {z_j}^2 + z_n \phi_{v,n}^T \tilde{\theta}_{v,n} + z_n sg(z_n, \epsilon)\tilde{D} + \epsilon D \end{aligned} \tag{4.17}$$

in which $\tilde{D} = D - \hat{D}$. Let $V_n = V_{n-1} + \frac{1}{2}{z_n}^2 + \frac{1}{2}\tilde{\theta}_{v,n}^T {\Gamma_n}^{-1} \tilde{\theta}_{v,n} + \frac{1}{2\lambda}\tilde{D}^2$, where Γ_n is positive definite and $\lambda > 0$, then

$$\begin{aligned} \dot{V}_n =& \dot{V}_{n-1} + z_n \dot{z}_n - \tilde{\theta}_{v,n}^T {\Gamma_n}^{-1} \dot{\hat{\theta}}_{v,n} - \frac{1}{\lambda}\tilde{D}\dot{\hat{D}} \\ \leq & - \sum_{j=1}^{n} c_j {z_j}^2 + z_n \phi_{v,n}^T \tilde{\theta}_{v,n} + z_n sg(z_n, \epsilon)\tilde{D} + \epsilon D - \tilde{\theta}_{v,n}^T {\Gamma_n}^{-1} \dot{\hat{\theta}}_{v,n} - \frac{1}{\lambda}\tilde{D}\dot{\hat{D}}. \end{aligned} \tag{4.18}$$

Design adaptive laws as

$$\begin{cases} \dot{\hat{\theta}}_{v,n} = z_n \Gamma_n \phi_{v,n}, \\ \dot{\hat{D}} = \lambda z_n sg(z_n, \epsilon), \end{cases} \tag{4.19}$$

finally we have

$$\begin{aligned} \dot{V}_n \leq & - \sum_{j=1}^{n} c_j {z_j}^2 + z_n \phi_{v,n}^T \tilde{\theta}_{v,n} + z_n sg(z_n, \epsilon)\tilde{D} + \epsilon D - \tilde{\theta}_{v,n}^T {\Gamma_n}^{-1} z_n \Gamma_n \phi_{v,n} - \frac{\tilde{D}\lambda z_n sg(z_n, \epsilon)}{\lambda} \\ =& - \sum_{j=1}^{n} c_j {z_j}^2 + \epsilon D. \end{aligned} \tag{4.20}$$

Remark 4.3 *From Lemma 2.1, it can be noticed that $sg(z_n, \epsilon)$ in (4.15) is differentiable and, thus, it is a smooth function for $t \geq 0$, which further implies that the term ${}_0\mathcal{D}_t^{1-\alpha}h$ as well as the control signal u are smooth. Due to the fact that $z_n {}_0\mathcal{D}_t^{1-\alpha}d \leq |z_n|D$ shown in (4.16), the compensation for the unknown disturbance can be realized by dealing with the bound of ${}_0\mathcal{D}_t^{1-\alpha}d$, i.e., D.*

4.3.2 Stability Analysis

After designing the adaptive controller, the main result of this chapter is to be established and presented in Theorem 4.1 below.

Theorem 4.1 *The closed-loop system consisting of incommensurate fractional system (4.1) and adaptive controller contains control law (4.15) and adaptive laws (4.6), (4.10), and (4.19) is globally asymptotically stable in the sense that all the signals in the closed-loop system are uniformly bounded and the asymptotic output tracking is realized, i.e., $\lim_{t\to\infty}[y(t) - y_r(t)] = 0$.*

Proof: Taking integration on both sides of (4.20) results in

$$\begin{aligned} V_n(t) \leq & V_n(0) + D\int_0^t e^{-a\tau}\mathrm{d}\tau - \sum_{j=1}^{n}\int_0^t c_j z_j^2 \mathrm{d}\tau \leq V_n(0) + D\int_0^t e^{-a\tau}\mathrm{d}\tau \\ \leq & V_n(0) + \frac{D}{a}. \end{aligned} \tag{4.21}$$

In light of (4.21), $V_n(t)$ is bounded, hence every signal in $V_n(t)$ is bounded. From (4.3) and the smoothness of ϕ_1, we know that α_1 is bounded. By the definition of z_2, it follows that x_2 is bounded. Following similar procedure, the boundedness of $\alpha_i\,(i = 2, \cdots, n-1)$, ${}_0\mathcal{D}_t^{1-\alpha}h$, and $x_i\,(i = 3, \cdots, n)$ is ensured.

According to Definition 1.13, for a function $f(t)$, if ${}_{t_0}\mathcal{D}_t^{\alpha}f(t)$ is continuous on $[t_0,\, t]$ with $t > t_0$ and $\alpha \in [0,\, 1]$, then ${}_{t_0}\mathcal{D}_t^{\alpha}f(t)$ is β-order integrable on $[t_0,\, t]$ $(\beta \geq 0)$. Since ${}_0\mathcal{D}_t^{1-\alpha}h$ is bounded and smooth, thus it is $(1-\alpha)$-order integrable on $[0,\, t]$, which implies $h(t)$ is bounded on this time interval. Besides, in accordance with Lemma 1.3 and Definition 1.13, we have

$$\begin{aligned} {}_0I_t^{1-\alpha}\,{}_0\mathcal{D}_t^{1-\alpha}h = & h(t) - h(0) = \frac{1}{\Gamma(1-\alpha)}\int_0^t \frac{{}_0\mathcal{D}_t^{1-\alpha}h(\tau)}{(t-\tau)^{1-1+\alpha}}\mathrm{d}\tau \\ = & \frac{1}{\Gamma(1-\alpha)}\int_0^t {}_0\mathcal{D}_t^{1-\alpha}h(\tau)(t-\tau)^{-\alpha}\mathrm{d}\tau \\ = & \frac{1}{\Gamma(1-\alpha)}\left[{}_0\mathcal{D}_t^{1-\alpha}h(t) * t^{-\alpha}\right] \end{aligned} \tag{4.22}$$

where $*$ represents the convolution operator. Due to the fact that ${}_0\mathcal{D}_t^{1-\alpha}h$ is bounded and $t^{-\alpha}$ will monotonically decrease towards 0 as time goes to infinity, hence $h(t)$ is continuously bounded. As a result, from (4.15), it is notable that the control input u is bounded. Therefore, from system (4.1) and (4.13), we get that the first-order derivatives of $x_i\,(i = 1, \cdots, n)$ are bounded, which indicates that all these states are uniformly continuous. According to Assumption 4.2, since y_r and $y_r^{(i)}$, $i = 1, \cdots, n$ are uniformly continuous, hence $z_1 = x_1 - y_r$ is uniformly continuous and consequently, α_1 is uniformly continuous. Subsequently, $z_i = x_i - y_r^{(i-1)} - \alpha_{i-1}\,(i = 2, \cdots, n)$ can be recursively proved to be uniformly continuous. From (4.21) we know that $\lim_{t\to\infty}\sum_{j=1}^{n}\int_0^t c_j z_j^2 \mathrm{d}\tau$ is bounded, therefore by Barbalat's lemma, we have $\lim_{t\to\infty} z_j = 0\,(j = 1, \cdots, n)$, and it further implies that the system output y tracks the reference signal y_r asymptotically. ■

Remark 4.4 *Note that if $\alpha = 1$ in (4.1), then the αth-order Caputo derivative of x_n will become its conventional first-order derivative [8]. Therefore, different from the control schemes investigated in such as [198] and [24] where fractional-order controllers are designed for integer-order systems, the proposed fractional-order adaptive controller will become an integer-order one when $\alpha = 1$ and the corresponding results in this chapter will turn into those for integer-order cases.*

Remark 4.5 *Different from the control schemes in the previous chapter where commensurate uncertain fractional-order nonlinear systems are investigated, control law (4.15) involves fractional integral to address the incommensurate property of the considered fractional-order systems. With the help of (4.13) and the designed control scheme, the overall closed-loop system stability analysis can be conducted based on the traditional integer Lyapunov method. However, this brings difficulty in proving the boundedness of the control signal (4.15), whereas the works in Chapter 2 and Chapter 3 are free from such a challenge. As shown in (4.22) and its context, theoretical analysis is provided in this chapter to overcome such a difficulty.*

4.4 Simulation and Experimental Studies

4.4.1 Simulation Illustration

A third-order incommensurate fractional-order nonlinear chaotic system in (4.23) below is presented in this subsection as a simulation study for illustrating the proposed control scheme. The reference signal y_r is set to be 0 for this third-order chaotic system, that is, the control objective is to asymptotically stabilize the fractional-order chaotic system

$$\begin{cases} \dot{x}_1(t) = -x_2(t) + \phi_1(x_1(t))\theta_1 \\ \dot{x}_2(t) = x_3(t) + \phi_2(x_1(t), x_2(t))\theta_2 \\ {}_0\mathcal{D}_t^\alpha x_3(t) = u(t) + \phi_3(x_1(t), x_2(t), x_3(t))\theta_3 + d(t) \\ y(t) = x_1(t). \end{cases} \tag{4.23}$$

For simulation studies, $\alpha = 0.94$, $\phi_1(x_1) = 1$, $\phi_2(x_1, x_2) = 1$, $\phi_3(x_1, x_2, x_3) = x_1x_2 - x_3$, $x(0) = [-0.444,\ 1.111,\ -0.4]^T$, $\theta_1 = 0.9$, $\theta_2 = 0.4$, $\theta_3 = 1$, and $d(t) = 0.5\sin(2t)$. The control parameters are designed as $c_1 = c_2 = c_3 = 2$, $\Gamma_1 = \Gamma_2 = \Gamma_3 = I$, and $\lambda = 5$.

As can be observed from Fig. 4.1, the system without any control effort will exhibit chaotic phenomenon, as elaborated in [195] and [196]. From Fig. 4.2 and Fig. 4.3, it is seen that with the proposed control method, not only the chaotic behaviors of the original uncontrolled system can be eliminated but also the output signal and the virtual errors will all be asymptotically stabilized to 0 by the smooth control input u shown in Fig. 4.4. The results displayed in Fig. 4.5 verify that all the signals, including parameter estimates in the closed-loop system, are bounded. As a comparison, the dynamic surface control scheme proposed in [2] is also applied to control the system in (4.23). The results are shown in Fig. 4.6, from which it can be noticed that the system output signal can merely be driven to a small region around 0.

4.4.2 Experimental Verification

In this subsection, experimental tests are conducted to verify the proposed control algorithm on a fractional-order PLL. PLL generates periodic output to lock the incoming signal and it

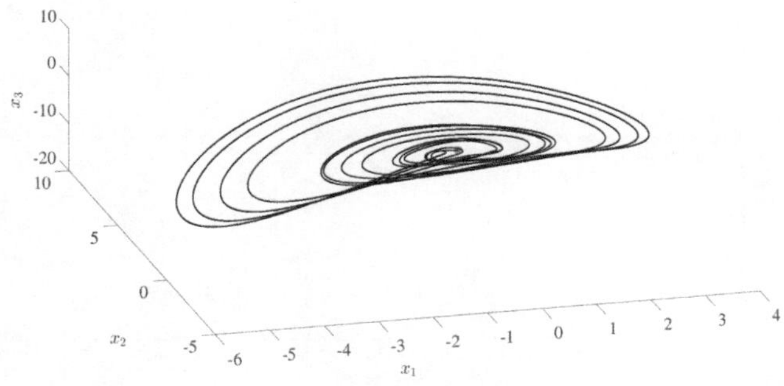

FIGURE 4.1: State variables of the uncontrolled system (4.23) with *red dot* representing initial condition.

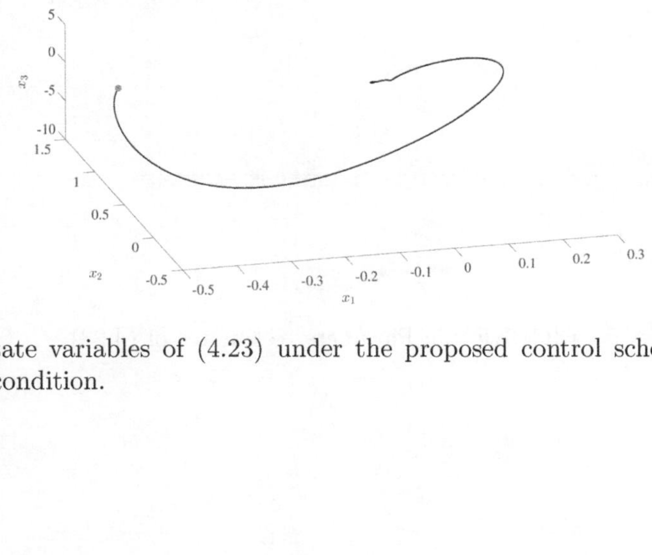

FIGURE 4.2: State variables of (4.23) under the proposed control scheme with *red dot* denoting initial condition.

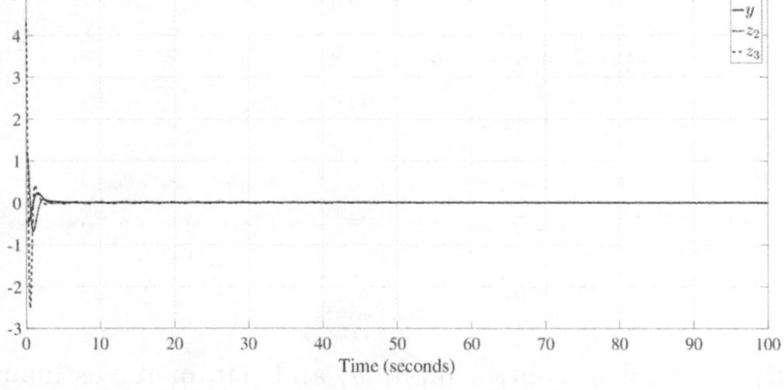

FIGURE 4.3: Output signal y and virtual errors z_2 and z_3 of (4.23).

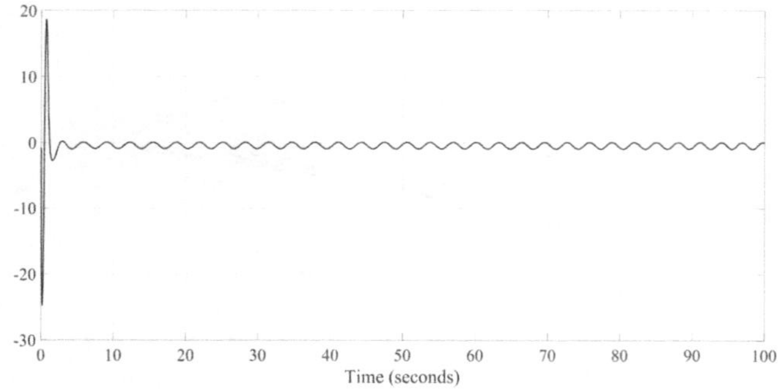

FIGURE 4.4: Control input u of (4.23).

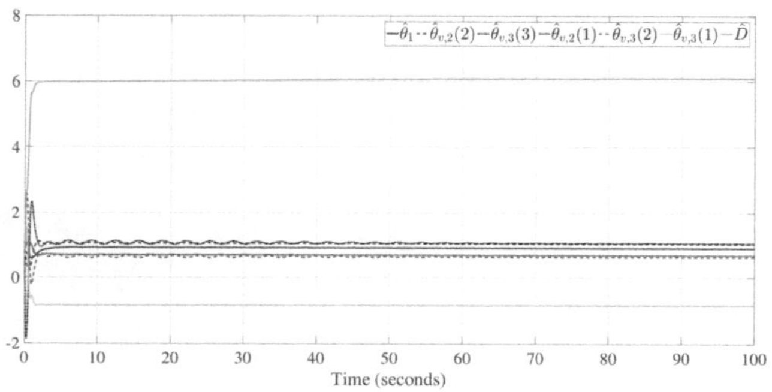

FIGURE 4.5: Parameter estimates of (4.23).

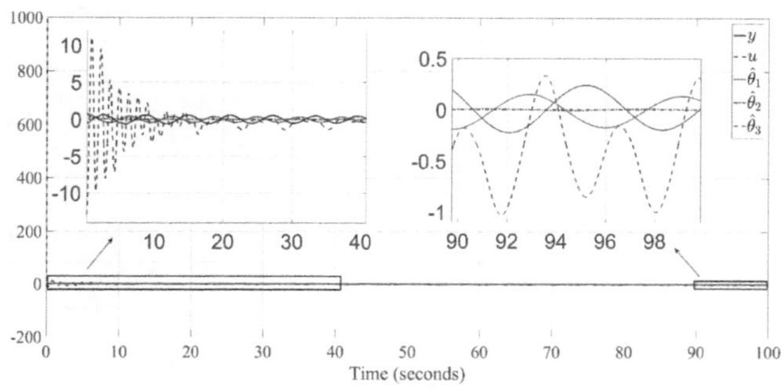

FIGURE 4.6: Output signal y, control input u, and parameter estimates of (4.23) under control scheme in [2].

can be found in a variety of applications, for example, clock clean-up circuit, demodulator circuit, tracking generator, etc. [3]. As presented in [56, 3, 57] that a fractional-order PLL, which contains a phase detector, a fractional-order loop filter, and a voltage controlled oscillator, has wider bandwidth and higher capture range compared with an integer-order PLL. We build a fractional-order PLL in RT-lab OP5600, of which the block diagram is displayed in Fig. 4.7. The experimental platform involving hardware setup is presented in Fig. 4.8. By converting the model of fractional-order PLL from transfer functions to a state space form, the dynamics of such fractional-order PLL can be described as

$$\begin{cases} \dot{x}_1(t) = x_2(t) + \phi_1(x_1(t))\theta_1 \\ {}_0\mathcal{D}_t^{\alpha} x_2(t) = Ku(t) + \phi_2(x_1(t), x_2(t))\theta_2 + Kd(t) \\ y(t) = x_1(t) \end{cases} \tag{4.24}$$

where the system parameters are $\alpha = 0.43$, $K = K_d K_f K_0 = 7250$, $b = 2.77$, $\phi_1(x_1) = 1$, $\phi_2(x_1, x_2) = -bx_2$, $\theta_1 = 0$, and $\theta_2 = 1$. In (4.24), K_d, K_f, and K_0 stand for the sensitivity of phase detector, fractional-order loop filter, and voltage controlled oscillator, respectively. The external disturbance to the system is a sinusoidal signal expressed as $d(t) = 5[\sin(200\pi t) + \cos(200\pi t)]$.

The incoming reference signal y_r of the fractional-order PLL is a sinusoidal voltage waveform with an amplitude of 1 V and a frequency of 50 Hz. To apply our proposed control scheme, the relevant control parameters are chosen as $c_1 = 4000$, $c_2 = 5000$, $\Gamma_1 = 100I$, $\Gamma_2 = [100, 0; 0, 0.01]$, and $\lambda = 50$. The experimental test results are shown in Fig. 4.9, from which we can observe that the PLL output can track the reference voltage perfectly, even with a smooth control signal generated from our control law. On the contrary, as shown in Fig. 4.10, the PLL output can hardly track the reference voltage by replacing our proposed control scheme in Fig. 4.7 with the dynamic surface control method raised in [2].

To make another comparison based on the same experimental platform with identical system parameters and disturbance, experimental test results of the fractional-order PLL in [3], whose block diagram is given in Fig. 4.11, are provided in Fig. 4.12. We can notice that the PLL output in [3] cannot track the given reference voltage, while asymptotic tracking can be ensured with the control scheme investigated in this chapter.

All the results presented in this section demonstrate the effectiveness of the proposed scheme and verify the theoretical results established.

Remark 4.6 *Due to the incommensurate structure, the controller development and system stability analysis for incommensurate fractional nonlinear systems are much more complicated compared to the commensurate counterparts, and the current studies on the control of incommensurate fractional-order nonlinear systems remain less, based on which, highlights the contribution of this chapter. Nevertheless, it is worth extending the proposed method to a more general class of incommensurate fractional systems [105, 199, 200], i.e., the order of fractional derivatives of different system states are different from each other instead of merely the last state, in order to expand the applicability of our developed control scheme.*

4.5 Conclusions

In this chapter, a class of incommensurate fractional-order nonlinear systems is investigated and an adaptive backstepping control scheme with smooth control input is designed. The

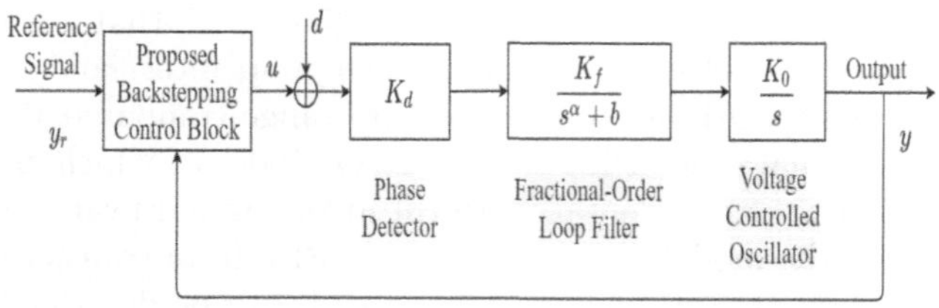

FIGURE 4.7: Block diagram of fractional-order PLL with the proposed control scheme.

FIGURE 4.8: Experimental platform.

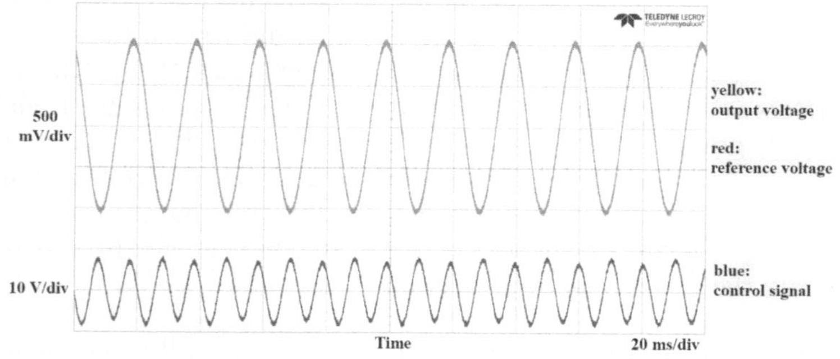

FIGURE 4.9: Experimental waveforms of steady-state response of fractional-order PLL.

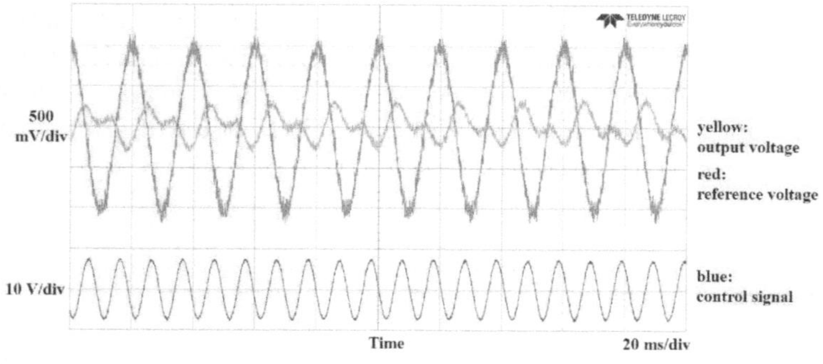

FIGURE 4.10: Experimental waveforms of steady-state response of fractional-order PLL under control scheme in [2].

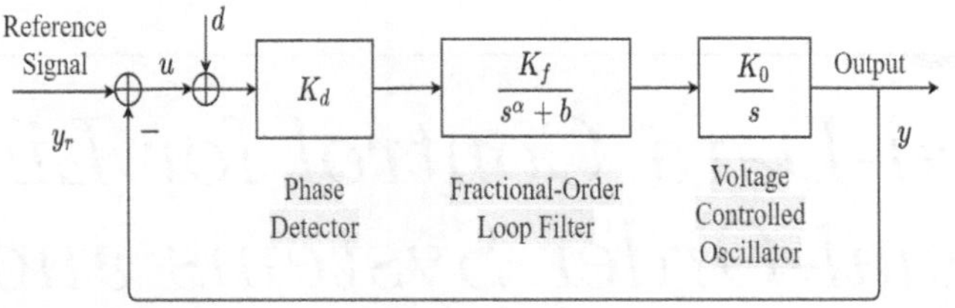

FIGURE 4.11: Block diagram of fractional-order PLL in [3].

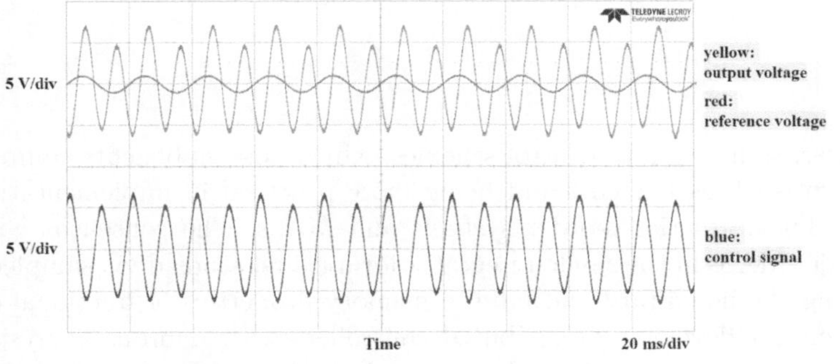

FIGURE 4.12: Experimental waveforms of steady-state response of fractional-order PLL in [3].

considered systems are perturbed by system uncertainties and time-varying external disturbance with unknown bound. Adaptive laws are designed to handle uncertainties and external disturbance. With the proposed smooth control law, it is proved that the resulting closed-loop fractional-order system is globally stable in the sense that all the closed-loop signals are bounded and perfect asymptotic output tracking is also guaranteed. The efficiency of the proposed technique is demonstrated by a simulation study on a third-order fractional chaotic system and an experimental study on a fractional-order PLL with comparison to that in [3].

5

Sampled-Data Control for Linear Fractional-Order Systems and MASs in the Single-Integrator Form

In this chapter, sampled-data control schemes, which possess benefits compared with continuous control such as low cost and being more practical in implementation, will be introduced for linear fractional-order systems and MASs. The consensus control of linear fractional-order MASs in the single-integrator form is first studied via sampled-data control. By considering the hereditary and infinite memory properties of fractional-order calculus, a novel scheme for developing distributed controllers with rigorous analysis is presented. Different from existing research works where the unique characteristics of fractional calculus are not taken into account, our resulting fractional-order system is strictly analyzed by considering the time evolution of the system states that relies on not only their initial values but also all the historical control signals. The corresponding challenge of establishing the global boundedness of the studied controller consisting of a weighted sum of previous control inputs is overcome. It can be ensured that the closed-loop system is globally stable and asymptotic consensus for fractional-order MASs is realized with our proposed control scheme. Secondly, a finite-dimensional sampled-data-based control scheme for linear single-integrator form fractional-order systems is proposed in this chapter to guarantee the corresponding closed-loop system is globally ISS. Instruction for the finite-dimensional controller on choosing the control signal storage space to meet predefined control accuracy is also provided.

5.1 Introduction

In the previous three chapters, backstepping-based adaptive control methods, which are continuous control, have been introduced for nonlinear uncertain fractional-order systems with respect to the commensurate case and the incommensurate case.

Motivated by the discussions in Subsection 1.3.2, we address the consensus issue with the consideration of unique characteristics of fractional-order calculus. With our proposed control design scheme, it is shown that the closed-loop system is globally stable and asymptotic consensus for fractional-order multi-agent systems is achieved. Different from existing research works, the resulting fractional-order system is rigorously analyzed by considering time evolution of the system states which depends on their initial values at t_0 and also all the previous control signals. A new challenge that is to establish the global boundedness of the proposed distributed controllers which contain all the previous control inputs is overcome in this chapter in Section 5.2. Simulation studies illustrate the effectiveness of the proposed control scheme and also reveal its accuracy in achieving asymptotic consensus compared to an existing approach.

DOI: 10.1201/9781003620372-5

Furthermore, since the proposed controller in Section 5.2 contains the weighted sum of all the previous control signals while the current control input has weaker relation with older control signal, thus in practice, finite recent steps of control signals might be applied for computing the current control signal to reduce the required signal storage space and ensure certain performance accuracy in the meantime, which further indicates an interesting topic to investigate the finite-dimensional sampled-data control of fractional-order systems and its trade-off between control performance and storage size. In Section 5.3, we consider such a problem. A finite-dimensional sampled-data-based control scheme for continuous-time linear single-integrator form fractional-order systems is proposed to guarantee the global stability of the corresponding closed-loop systems in the sense that with any initial condition, the closed-loop states converge to an ultimate bound around the origin under control signals with bounded magnitudes. The controller design procedure is developed upon the argumentation investigated in Section 5.2. Nevertheless, as the main difference, finite-dimensional controllers, which are computationally tractable and also, far more applicable for practical applications, are proposed in this section. Also, the stability condition in Section 5.2 that should be satisfied at each sampling instant for achieving the global stability of the closed-loop systems is relaxed in Section 5.3, as presented in Proposition 5.1.

In summary, the main contributions of this chapter are as follows.

1. Considering the unique hereditary and infinite memory properties of fractional-order calculus, a sampled-data-based control is designed in Section 5.2 that can realize asymptotic consensus for fractional-order MASs in single-integrator form and ensure global stability in the sense that all signals in the closed-loop systems remain globally bounded.
2. A finite-dimensional sampled-data-based control strategy is proposed in Section 5.3 for linear single-integrator form fractional-order systems to guarantee the overall stability of the resultant closed-loop system and a guide on choosing the control signal storage space to fulfill a predefined control accuracy will be provided.

The rest of the chapter is organized as follows. The design for sampled-data-based consensus control scheme for linear fractional-order MASs in single-integrator form is given in Section 5.2. Section 5.3 establishes the finite-dimensional sampled-data-based control strategy for linear single-integrator form fractional-order systems.

5.2 Sampled-Data Control-Based Consensus of Linear Fractional-Order MASs

5.2.1 Problem Formulation

A group of N fractional-order agents are governed by

$$ {}_{t_0}\mathcal{D}_t^{\alpha} x_i(t) = u_i(t), \text{ for } i = 1, 2, \cdots, N \tag{5.1} $$

where the fractional-orders of all the states are equal to $\alpha \in (0, 1]$, $x_i \in \mathbb{R}$ and $u_i \in \mathbb{R}$ represent the measurable state and control input of the ith agent, respectively.

Remark 5.1 *All the agents in this section are in one-dimensional space for convenience. The results established can be easily extended to n-dimensional space by applying the Kronecker product.*

Remark 5.2 *The usefulness of single integrator fractional-order MASs is evidenced by their ability to more accurately simulate natural phenomena, for instance, the synchronized movement of agents in fractional surroundings like macromolecular liquids and porous materials [201] where the stress-strain relationship is governed by single integrator fractional-order dynamics, and the coordination among fractional supercapacitors [202, 203] for energy distribution and storage.*

The control problem in this section is to design a distributed controller u_i for each agent described in (5.1) to achieve the following objectives: 1) All the signals in the closed-loop systems are globally bounded; 2) Asymptotic consensus for fractional-order systems (5.1) is ensured, i.e., $\lim_{t\to\infty}\|x_i(t) - x_j(t)\| = 0, \forall i, j = 1, 2, \cdots, N$, and additionally, $\lim_{t\to\infty} x_i(t) = \frac{\sum_{i=1}^{N} x_i(t_0)}{N}$.

Suppose that the communications among the N agents can be represented by a directed graph $\mathcal{G} \triangleq (\mathcal{V}, \mathcal{E})$ where $\mathcal{V} = \{1, 2, \cdots, N\}$ means the set of indexes (or vertices) corresponding to each agent, $\mathcal{E} \subseteq \mathcal{V} \times \mathcal{V}$ is the set of edges between two distinct agents. An edge $(i, j) \in \mathcal{E}$ denotes that agent j can obtain information from agent i, but not necessarily vice versa. In this case, agent i is called a neighbor of agent j and we indicate the set of neighbors for agent i as $\mathcal{N}_i$. In this section, $(i, i) \notin \mathcal{E}$ and $i \notin \mathcal{N}_i$ since self edges (i, i) are not allowed. $A = [a_{ij}] \in \mathbb{R}^{N\times N}$ is the connectivity matrix with $a_{ij} = 1$ if $(j, i) \in \mathcal{E}$ and $a_{ij} = 0$ if $(j, i) \notin \mathcal{E}$. Throughout this section, the diagonal elements $a_{ii} = 0$. An in-degree matrix $\triangle$ is introduced as $\triangle = \mathrm{diag}(\triangle_i) \in \mathbb{R}^{N\times N}$ with $\triangle_i = \sum_{j\in\mathcal{N}_i} a_{ij}$ being the ith row sum of A. Therefore, the Laplacian matrix of $\mathcal{G}$ is defined as $\mathcal{L} = \triangle - A$. A digraph is strongly connected if there is a directed path that connects any two arbitrary nodes of the graph and is balanced if for all $i \in \mathcal{V}$, $\sum_{j\neq i} a_{ij} = \sum_{j\neq i} a_{ji}$.

Assumption 5.1 *The digraph $\mathcal{G}$ is strongly connected and balanced.*

For the development and analysis of the proposed control scheme, the following lemma is first established.

Lemma 5.1 *For $\forall j \in \mathbb{N}$ and $0 < \alpha < 1$, the following results hold*
1) $0 < |(j + 1)^\alpha - 2j^\alpha + (j - 1)^\alpha| < 1$,
2) $\lim_{j\to\infty}|(j + 1)^\alpha - 2j^\alpha + (j - 1)^\alpha| = 0$.

Proof: Define

$$f(j) = (j + 1)^\alpha - 2j^\alpha + (j - 1)^\alpha \tag{5.2}$$

and

$$g(s) = s^\alpha - (s - 1)^\alpha. \tag{5.3}$$

It can be shown from (5.3) that, for $\forall s \in \mathbb{R}_{>1}$,

$$\begin{aligned} & g(s) > 0, \\ & \frac{\mathrm{d}g(s)}{\mathrm{d}s} = \alpha s^{\alpha-1} - \alpha(s - 1)^{\alpha-1} < 0. \end{aligned} \tag{5.4}$$

From (5.4), we can obtain that $g(s)$ is monotonically decreasing and $\lim_{s\to\infty} g(s) = 0$, which further implies that $f(j) = [g(j + 1) - g(j)] < 0$ and it will monotonically tend to 0 as j tends to ∞. Additionally, it can be easily checked that $|f(1)| < 1$ for $0 < \alpha < 1$ and $|f(j + 1)| < |f(j)|$. ■

5.2.2 Distributed Controller Design and Stability Analysis

5.2.2.1 Distributed Controller Design

To achieve the above objectives, a distributed controller is designed for each local agent based on periodic sampled-data control technology. The sampling instants are described by a discrete-time sequence $\{t_k\}$ with $t_0 < t_1 < \cdots < t_k < \cdots$ and $t_{k+1} - t_k = h$, where $h > 0$ is the sampling period.

For $t \in [t_0, t_1)$, let $x(t) = [x_1(t), x_2(t), \cdots, x_N(t)]^T$ and $\underline{u}_0 = [u_{1,0}, u_{2,0}, \cdots, u_{N,0}]^T$ where $u_{i,0}$ denotes the control signal for the ith agent within this time interval. According to Lemma 1.4, the value of $x(t_1)$ can be computed as follows

$$x(t_1) = x(t_0) + \frac{1}{\Gamma(\alpha)} \int_{t_0}^{t_1} (t_1 - \tau)^{\alpha-1} \underline{u}_0 \mathrm{d}\tau = x(t_0) + \underline{u}_0 \frac{h^\alpha}{\Gamma(\alpha+1)}. \tag{5.5}$$

By designing the controller for the ith agent as

$$u_{i,0} = \gamma \sum_{j \in \mathcal{N}_i} a_{ij} \big(x_j(t_0) - x_i(t_0)\big), \tag{5.6}$$

where $\gamma > 0$, the closed-loop systems in this sampling period can be expressed as

$$x(t_1) = x(t_0) - \gamma \mathcal{L} x(t_0) \frac{h^\alpha}{\Gamma(\alpha+1)}. \tag{5.7}$$

Now for $t \in [t_k, t_{k+1})\,(k \in \mathbb{N})$, define $\underline{u}_k = [u_{1,k}, u_{2,k}, \cdots, u_{N,k}]^T$. Since the fractional-order derivative of x depends on all the historical values of $\underline{u}$, thus not only the value of $\underline{u}$ at present instant but also all of its previous values are needed to determine the future behavior of fractional-order system. Therefore, $x(t_{k+1})$ should be expressed in terms of $x(t_0)$ and $\underline{u}_j (j = 0, 1, \cdots, k)$ as follows

$$\begin{aligned} x(t_{k+1}) =& x(t_0) + \frac{1}{\Gamma(\alpha)} \int_{t_0}^{t_1} (t_{k+1} - \tau)^{\alpha-1} \underline{u}_0 \mathrm{d}\tau + \cdots + \frac{1}{\Gamma(\alpha)} \int_{t_k}^{t_{k+1}} (t_{k+1} - \tau)^{\alpha-1} \underline{u}_k \mathrm{d}\tau \\ =& x(t_k) + \sum_{l=1}^{k} \frac{[(l+1)h]^\alpha - 2(lh)^\alpha + [(l-1)h]^\alpha}{\Gamma(\alpha+1)} \underline{u}_{k-l} + \frac{\underline{u}_k}{\Gamma(\alpha+1)} h^\alpha. \end{aligned} \tag{5.8}$$

Then based on (5.8), we design the distributed controller for the ith agent as

$$u_{i,k} = \gamma \sum_{j \in \mathcal{N}_i} a_{ij} \big(x_j(t_k) - x_i(t_k)\big) - \sum_{l=1}^{k} [(l+1)^\alpha - 2l^\alpha + (l-1)^\alpha] u_{i,(k-l)}, \tag{5.9}$$

which results in the following closed-loop systems

$$x(t_{k+1}) = x(t_k) - \gamma \mathcal{L} x(t_k) \frac{h^\alpha}{\Gamma(\alpha+1)}. \tag{5.10}$$

Remark 5.3 *Existing studies on sampled-data control of fractional-order MASs divide the whole time interval into sampling intervals $[t_k, t_{k+1})$, $k \in \mathbb{N}_0$ and solve the fractional-order equations on each sampling interval by treating $x(t_k)$ as the initial value for this corresponding time period. Through this way, the closed-loop fractional-order systems are simply modeled by simplified discrete-time systems in such a way that $x(t_{k+1})$ is expressed as a function only of $x(t_k)$ and $\underline{u}_k$, as is done in the integer-order derivative systems discretization. However, due to the hereditary and infinite memory properties of fractional-order derivative,* *when converting the actual initial value $x(t_0)$ into $x(t_k)$ for solving fractional-order equations, the second term $\sum_{l=1}^{k}\frac{[(l+1)h]^{\alpha}-2(lh)^{\alpha}+[(l-1)h]^{\alpha}}{\Gamma(\alpha+1)}\underline{u}_{k-l}$ on the right-hand side of (5.8) cannot be neglected. On the contrary, the control scheme design and system analysis in this chapter are carried out strictly by bearing the unique properties of fractional-order calculus in mind, which specifically can be seen from (5.8), (5.9) and the boundedness analysis of control signals given later.*

Remark 5.4 *Note that the second term on the right-hand side of (5.8) depends on all the previous control signals $\underline{u}_j$, $j = 0, 1, \cdots, k-1$, and thus it cannot be assumed to be bounded before establishing system stability, giving rise to a challenge in controller design and analysis. Such a challenge is overcome in our proposed controller in (5.9) which consists of two parts. The first part $u_{i1,k} = \gamma\sum_{j\in N_i} a_{ij}(x_j(t_k) - x_i(t_k))(k \in \mathbb{N}_0)$ is designed for achieving consensus of fractional-order MASs and the second part $u_{i2,k} = -\sum_{l=1}^{k}[(l+1)^{\alpha} - 2l^{\alpha} + (l-1)^{\alpha}]u_{i,(k-l)}$ which exists in $t \geq t_1$ aims at compensating for the effect caused by the hereditary and infinite memory properties of fractional-order calculus. As can be observed from Lemma 5.1, although $f(l) = (l+1)^{\alpha} - 2l^{\alpha} + (l-1)^{\alpha}$ is an increasing function with respect to l, its value always remains in the interval $(-1, 0)$ with $0 < \alpha < 1$. Also, $\lim_{l\to\infty} f(l) = 0$. Therefore, $u_{i2,k} = |f(1)|u_{i,(k-1)} + |f(2)|u_{i,(k-2)} + \cdots + |f(k)|u_{i,0}$. Clearly, $u_{i2,k}$ is designed in such a way that the control signals at earlier time instants are less weighted. For example, the weight $|f(k)|$ of $u_{i,0}$ will tend to 0 as k increases. This is reasonable for practical situations of fractional-order systems. It can be observed from (5.7) and (5.10) that our designed distributed controllers allow us to analyze the stability of the closed-loop systems in the same way as that of the integer-order discrete-time closed-loop systems, which will be demonstrated in the following.*

5.2.2.2 Stability Analysis

The main result of this section is presented in the following theorem, where a stability criterion is given.

Theorem 5.1 *Consider the closed-loop systems consisting of fractional systems (5.1) and sampled-data-based distributed controllers (5.6) and (5.9). All the signals in the closed-loop systems are globally bounded and asymptotic consensus is achieved, i.e., $\lim_{t\to\infty}\|x_i(t) - x_j(t)\| = 0$, $\forall i, j = 1, 2, \cdots, N$, if the design parameters h and γ satisfy*

$$0 < \gamma\frac{h^{\alpha}}{\Gamma(\alpha+1)} < \frac{1}{\triangle_{max}} \tag{5.11}$$

and

$$\left[\beta^{k+1} + \sum_{l=1}^{k+1}|f(l)|\right] \leq 1, \tag{5.12}$$

in which $\triangle_{max}$ is the maximum degree of the graph $\mathcal{G}$, $f(l)$ is defined in (5.2) and

$$\beta = 1 - \gamma\lambda_2(\mathcal{L}_s)\frac{h^{\alpha}}{\Gamma(\alpha+1)}, \tag{5.13}$$

where $\mathcal{L}_s = \frac{\mathcal{L}+\mathcal{L}^T}{2}$ and $\lambda_2(\mathcal{L}_s)$ represents the second smallest eigenvalue of $\mathcal{L}_s$. Besides, since the digraph is balanced, asymptotic average-consensus can be achieved, i.e., $\lim_{t\to\infty} x_i(t) = \frac{\sum_{i=1}^{N} x_i(t_0)}{N}$.

Proof: As mentioned above, the main challenge is how to achieve global stability of the resulting systems in the presence of the second term on the right-hand side of (5.8). For this purpose, an additional control action is proposed in (5.9), in order to compensate for the effects of this term. However, with this new control term which is the weighted sum of previous control signals, it is difficult to show the boundedness of control inputs. To overcome this difficulty, we first establish the following relationship

$$\|\underline{u}_k\| \leq \|\underline{u}_0\|, \text{ for all } k \geq 0. \tag{5.14}$$

Now we define error vectors as

$$e(t_k) = \mathcal{L}x(t_k),\ k \in \mathbb{N}_0. \tag{5.15}$$

Then the proof of (5.14) is completed through mathematical induction as detailed below.

Step 1: According to (5.6), (5.9) and (5.15), $\underline{u}_0$ and $\underline{u}_1$ can be respectively expressed as follows

$$\underline{u}_0 = -\gamma e(t_0), \tag{5.16}$$

$$\underline{u}_1 = -\gamma e(t_1) + |f(1)|\underline{u}_0. \tag{5.17}$$

Therefore, we have

$$\begin{aligned}\|\underline{u}_1\| =& \left\| -\gamma e(t_1) + |f(1)|\underline{u}_0 \right\| \leq \gamma\|e(t_1)\| + |f(1)| \cdot \|\underline{u}_0\| \\ \leq& \gamma\beta\|e(t_0)\| + |f(1)| \cdot \|\underline{u}_0\| = \left[\beta + |f(1)|\right] \cdot \|\underline{u}_0\|.\end{aligned} \tag{5.18}$$

By designing γ and h in such a way that $\left[\beta + |f(1)|\right] \leq 1$, then we have $\|\underline{u}_1\| \leq \|\underline{u}_0\|$.

Step 2: Assuming that $\|\underline{u}_k\| \leq \|\underline{u}_0\|$ holds for $k > 1$. According to (5.9) and (5.15), $\|\underline{u}_{k+1}\|$ can be expressed as

$$\begin{aligned}\|\underline{u}_{k+1}\| =& \left\| -\gamma e(t_{k+1}) + \sum_{l=1}^{k+1} |f(l)|\underline{u}_{k+1-l} \right\| \\ \leq& \gamma\|e(t_{k+1})\| + \sum_{l=1}^{k+1} |f(l)| \cdot \|\underline{u}_{k+1-l}\|.\end{aligned} \tag{5.19}$$

Since $\|e(t_{k+1})\| \leq \beta\|e(t_k)\| \leq \cdots \leq \beta^{k+1}\|e(t_0)\|$, therefore (5.19) becomes

$$\|\underline{u}_{k+1}\| \leq \gamma\beta^{k+1}\|e(t_0)\| + \sum_{l=1}^{k+1} |f(l)| \cdot \|\underline{u}_{k+1-l}\|. \tag{5.20}$$

Since for $1 \leq l \leq k+1$, $\|\underline{u}_{k+1-l}\| \leq \|\underline{u}_0\|$ holds, hence

$$\|\underline{u}_{k+1}\| \leq \beta^{k+1}\|\underline{u}_0\| + \sum_{l=1}^{k+1} |f(l)| \cdot \|\underline{u}_0\| = \left[\beta^{k+1} + \sum_{l=1}^{k+1} |f(l)|\right] \cdot \|\underline{u}_0\|. \tag{5.21}$$

If γ and h are chosen to satisfy (5.12), then we have (5.14). According to Lemma 5.1, $|f(l)| < 1$ for all $l \in \mathbb{N}$ and it will monotonically tend to 0 as $l \to \infty$. Hence it is worth noting that Lemma 5.1 is the premise that (5.12) holds for all $k \geq 0$.

Since $\|\underline{u}_0\| = \gamma\|e(t_0)\|$ is bounded, therefore the global boundedness of all control signals is guaranteed.

Furthermore, (5.10) can be rewritten as

$$x(t_{k+1}) = Px(t_k) \tag{5.22}$$

where $P = I_N - \gamma\frac{h^\alpha}{\Gamma(\alpha+1)}\mathcal{L}$ is Perron matrix of digraph $\mathcal{G}$ with $\gamma\frac{h^\alpha}{\Gamma(\alpha+1)}$ treated as an integrated gain. Under condition (5.11), all the eigenvalues of P are within the unit circle, which further implies the global boundedness of $x(t)$.

Also, from (5.10) it can be noticed that the ith agent in the closed-loop systems is described as

$$x_i(t_{k+1}) = x_i(t_k) + \gamma\frac{h^\alpha}{\Gamma(\alpha+1)}\sum_{j\in\mathcal{N}_i} a_{ij}\big(x_j(t_k) - x_i(t_k)\big). \tag{5.23}$$

Hence, under Assumption 5.1, the proof of the condition (5.11) for reaching asymptotic consensus follows from that of Theorem 2 in [204] and thus the theorem is proved. Besides from (5.18) and (5.20), it can be seen that β is a lower bound of the decaying rate of $\|e(t_k)\|$, which denotes the lower bound of the consensus convergence rate of the fractional MASs under the proposed control scheme. ■

5.2.3 Simulation Example

In this subsection, an example is presented to demonstrate the proposed design control scheme and verify the established theoretical results. By comparing to an existing scheme in [4], it is revealed that the proposed controller can achieve asymptotic consensus in a more precise way.

Consider a group of five fractional-order agents with the following dynamics

$${}_{t_0}\mathcal{D}_t^\alpha x_i(t) = u_i(t),\ i = 1, 2, \cdots, 5, \tag{5.24}$$

where $\alpha = 0.9$ and initial values of states are $x(t_0) = [4.5, 5, 6, 1.5, -1]^T$. The connection weights of the graph are $a_{14} = a_{15} = a_{21} = a_{23} = a_{31} = a_{32} = a_{42} = a_{43} = a_{54} = 1$ and other entries of A are equal to zero.

Based on (5.11) to (5.13), the designed control parameter and sampling period are respectively selected as $\gamma = 0.15$ and $h = 0.85$ s. The simulation results are shown in Fig. 5.1 to Fig. 5.3. From Fig. 5.1, asymptotic consensus and average-consensus are realized with $\lim_{t\to\infty} x_i(t) = x_{\text{final}} = \frac{\sum_{i=1}^5 x_i(t_0)}{5} = 3.2$. The value of $\|\underline{u}\|^2$ can be observed from Fig. 5.2, which shows the boundedness of all control signals. Furthermore, for the purpose of verifying the condition for ensuring all the control signals are globally bounded, the value of $\left[\beta^{k+1} + \sum_{l=1}^{k+1}|f(l)|\right]$ is given in Fig. 5.3, from which it can be noticed that inequality (5.12) holds for $\forall k \in \mathbb{N}_0$ with the selected design parameters.

To better illustrate the accuracy and effectiveness of our proposed control algorithm, a comparative simulation study between the control scheme in [4] and in this section is implemented under the same control parameters. The mean absolute error $r(t) = \frac{1}{5}\sum_{i=1}^5|x_i(t) - x_{\text{final}}|$ with the proposed scheme and control method in [4] are displayed in Fig. 5.4. It can be seen from Fig. 5.1 and Fig. 5.4 that the mean absolute error with control scheme in [4] is larger compared to that with the control scheme in this section. Furthermore, as can be observed in Fig. 5.2, the control signals under the control scheme in [4] share similar magnitude with our proposed control inputs. Besides, by comparing the areas under the solid line and dash line in Fig. 5.2, it can be noticed that the accumulated control efforts in the proposed scheme are comparable to those in [4], which exemplifies the effectiveness of the proposed schemes. All these results verify our theoretical results established. More importantly, we have provided a solid analysis of the fractional-order system by taking its hereditary and infinite memory properties into account.

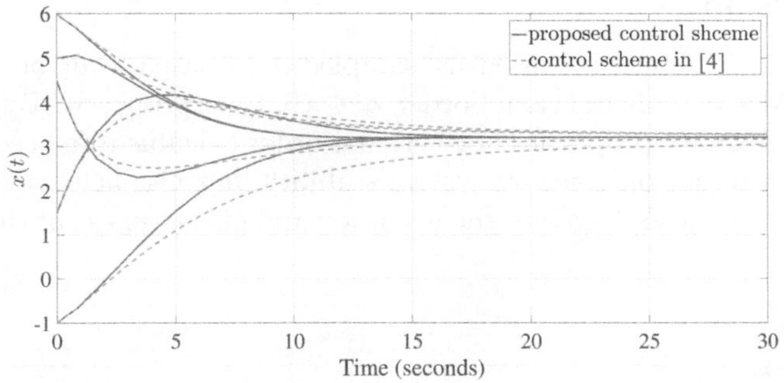

FIGURE 5.1: States of (5.24).

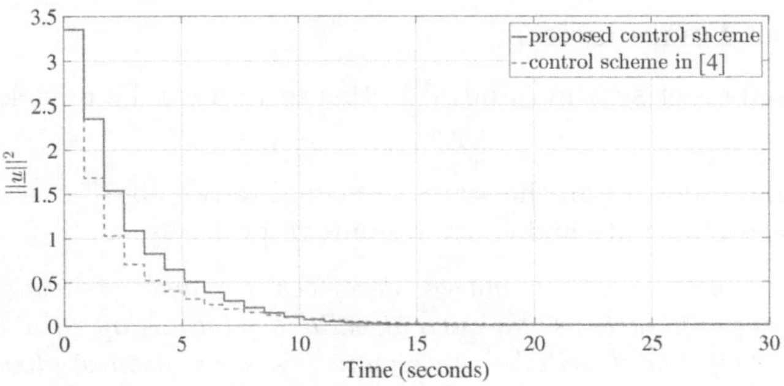

FIGURE 5.2: $\|\underline{u}\|^2$ of controllers in this section and in [4].

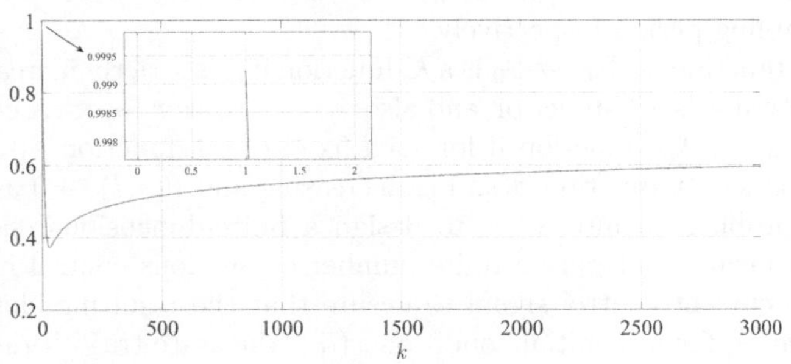

FIGURE 5.3: Value of $\left[\beta^{k+1} + \sum_{l=1}^{k+1} |f(l)|\right]$ under the proposed control scheme.

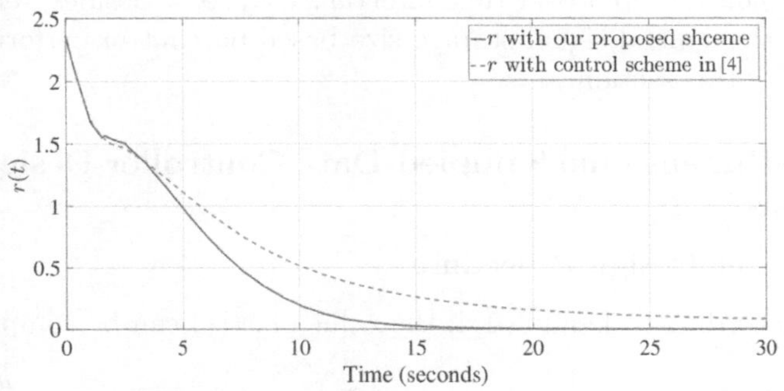

FIGURE 5.4: Mean absolute error r with proposed scheme and scheme in [4].

5.2.4 Conclusions

In this section, a distributed consensus sampled-data-based control scheme for single-integrator form MASs with fractional-order $\alpha \in (0,1)$ is proposed. By taking the hereditary and infinite memory properties of fractional-order calculus into account, a new control scheme is designed to not only ensure system stability, but also achieve asymptotic consensus. Simulation results also illustrate the accuracy and effectiveness of the proposed control algorithm.

5.3 Finite-Dimensional Sampled-Data Control of Linear Fractional-Order Systems

5.3.1 Problem Formulation

A class of fractional-order systems studied in this section can be modeled as

$$ {}_{t_0}\mathcal{D}_t^{\alpha} x(t) = u(t) \tag{5.25}$$

where the fractional-orders of all the states are equal to $\alpha \in (0,1]$, and $x \in \mathbb{R}^n$ and $u \in \mathbb{R}^n$ represent the measurable state and control input, respectively.

Remark 5.5 *The system (5.25) indeed describes a class of realistic processes with fractional-order property such as the current-voltage relationship of a fractional-order capacitor [39, 203, 205], which will be investigated as simulation studies in this section to verify the established control scheme.*

A periodic sampling approach will be implemented in this section for controller design with a discrete-time sequence $\{t_k\}$ $(k \in \mathbb{N}_0)$ and $t_{k+1} - t_k = h > 0$ standing for the sampling instants and sampling period respectively.

A continuous function $\gamma : \mathbb{R}_0 \to \mathbb{R}_0$ is a $\mathcal{K}$-function if it is strictly increasing and $\gamma(0) = 0$; it is a $\mathcal{K}_\infty$-function if it is a $\mathcal{K}$-function and also $\gamma(r) \to \infty$ as $r \to \infty$. A continuous function $\beta : \mathbb{R}_0 \times \mathbb{R}_0 \to \mathbb{R}_0$ is a $\mathcal{KL}$-function if for each fixed $t \geq 0$, function $\beta(\cdot, t)$ is a $\mathcal{K}$-function and, for each fixed $s \geq 0$, function $\beta(s, \cdot)$ is decreasing and $\beta(s,t) \to 0$ as $t \to \infty$.

The control problem of interest is to design a finite-dimensional sampled-data-based controller, which means that only a finite number of previous control inputs are required to determine the current control signal to ensure that the resulting closed-loop system is globally ISS. That is, for any initial condition $x(t_0)$, the state trajectories satisfy

$$\|x(t_k)\| \leq \beta(\|x(t_0)\|, t_k) + \gamma(\|u\|_\infty), \; k \in \mathbb{N}_0 \tag{5.26}$$

where function β is a $\mathcal{KL}$-function and γ is a $\mathcal{K}$-function. Here $\|u\|_\infty$ represents the supremum norm of the control input over time interval $t \in [t_0, \infty)$. Besides, rendering a guideline on how to select the control signal storage size based on control performance is the other control objective of this section.

5.3.2 Finite-Dimensional Sampled-Data Controller Design and Stability Analysis

5.3.2.1 Controller Design Procedure

For $t \in [t_0, t_1)$, according to Lemma 1.4, the value of $x(t_1)$ can be computed as

$$x(t_1) = x(t_0) + \frac{1}{\Gamma(\alpha)} \int_{t_0}^{t_1} (t_1 - \tau)^{\alpha - 1} u_0 \mathrm{d}\tau = x(t_0) + u_0 \frac{h^\alpha}{\Gamma(\alpha + 1)} \tag{5.27}$$

where u_0 denotes the control signal within this time interval.

By designing u_0 as

$$u_0 = \bar{K}x(t_0) \tag{5.28}$$

with design matrix $\bar{K} \in \mathbb{R}^{n\times n}$, the closed-loop systems in this sampling period can be expressed as

$$x(t_1) = \left[I_n + \frac{h^\alpha}{\Gamma(\alpha+1)}\bar{K}\right]x(t_0). \tag{5.29}$$

For $t \in [t_k, t_{k+1})$ $(k \in \mathbb{N})$, let u_k be the control signal for the corresponding time interval. Since the fractional-order derivative of x depends on all the historical values of u, hence $x(t_{k+1})$ should be expressed as follows

$$\begin{aligned} x(t_{k+1}) =& x(t_0) + \frac{1}{\Gamma(\alpha)}\int_{t_0}^{t_1}(t_{k+1}-\tau)^{\alpha-1}u_0 \mathrm{d}\tau + \cdots + \frac{1}{\Gamma(\alpha)}\int_{t_k}^{t_{k+1}}(t_{k+1}-\tau)^{\alpha-1}u_k \mathrm{d}\tau \\ =& x(t_k) + \frac{h^\alpha}{\Gamma(\alpha+1)}u_k + \sum_{l=1}^{k}\frac{[(l+1)h]^\alpha - 2(lh)^\alpha + [(l-1)h]^\alpha}{\Gamma(\alpha+1)}u_{k-l}. \end{aligned} \tag{5.30}$$

Then based on (5.30), we design u_k as

$$u_k = \begin{cases} \bar{K}x(t_k) - \sum_{l=1}^{k}[(l+1)^\alpha - 2l^\alpha + (l-1)^\alpha]u_{k-l}, \text{ when } k \le L \\ \bar{K}x(t_k) - \sum_{l=1}^{L}[(l+1)^\alpha - 2l^\alpha + (l-1)^\alpha]u_{k-l}, \text{ when } k > L \end{cases} \tag{5.31}$$

where $L \in \mathbb{N}$ is a design parameter that determines the number of previous control data needed to be stored for computing the current control input. Unlike Section 5.2, u_k does not depend on all the previous control inputs and thus can be more easily and practically implemented.

As a result, the closed-loop systems for the case that $k \le L$ can be expressed as

$$x(t_{k+1}) = \left[I_n + \frac{h^\alpha}{\Gamma(\alpha+1)}\bar{K}\right]x(t_k). \tag{5.32}$$

For the case when $k > L$, with the controller designed in (5.31), the closed-loop systems can be deducted as

$$\begin{aligned} x(t_{k+1}) =& x(t_k) + \frac{h^\alpha}{\Gamma(\alpha+1)}\bar{K}x(t_k) + \sum_{l=L+1}^{k}[(l+1)^\alpha - 2l^\alpha + (l-1)^\alpha]\frac{h^\alpha}{\Gamma(\alpha+1)}u_{k-l} \\ =& \left[I_n + \frac{h^\alpha}{\Gamma(\alpha+1)}\bar{K}\right]x(t_k) + \sum_{l=L+1}^{k}\frac{h^\alpha f(l)}{\Gamma(\alpha+1)}u_{k-l}, \end{aligned} \tag{5.33}$$

in which $f(l)$ is defined in (5.2).

5.3.2.2 Stability Analysis

Before demonstrating the stability analysis for the closed-loop fractional-order systems, a proposition with respect to function $f(l)$ in (5.2) is presented, which later will be applied to establish our first result of this section.

Proposition 5.1 *For any scalar* $\sigma \in (0,\ 1 - |f(1)|]$, *we have*

$$\sigma^k + \sum_{l=1}^{k}|f(l)| < 1,\ \forall k \in \mathbb{N}_{>1}. \tag{5.34}$$

Proof: Firstly, the establishment of $\sigma + |f(1)| \leq 1$ indicates that $\sigma \leq 2^\alpha - 1$. Then according to induction, it is assumed that for $k = m$,

$$\sigma^m + \sum_{l=1}^{m} |f(l)| < 1 \tag{5.35}$$

holds for $m \in \mathbb{N}_{>1}$. Hence for $k = m + 1$, we have

$$\begin{aligned} \sigma^{m+1} + \sum_{l=1}^{m+1} |f(l)| =& \sigma^{m+1} + 1 - (m+2)^\alpha + (m+1)^\alpha \\ \leq& (2^\alpha - 1)\sigma^m + 1 - (m+2)^\alpha + (m+1)^\alpha. \end{aligned} \tag{5.36}$$

Since inequality (5.35) can be re-expressed as $\sigma^m + 1 - (m+1)^\alpha + m^\alpha < 1$, thus (5.36) turns into

$$\begin{aligned} \sigma^{m+1} + \sum_{l=1}^{m+1} |f(l)| <& (2^\alpha - 1)\big[(m+1)^\alpha - m^\alpha\big] + 1 - (m+2)^\alpha + (m+1)^\alpha \\ =& (2m+2)^\alpha - (2m)^\alpha - \big[(m+2)^\alpha - m^\alpha\big] + 1. \end{aligned} \tag{5.37}$$

Define a function as

$$g(x) = (x+2)^\alpha - x^\alpha, \ \alpha \in (0, 1). \tag{5.38}$$

Note that for $x \in \mathbb{R}_{>1}$, $g(x) > 0$ and its derivative with respect to x is less than 0, as shown in (5.39).

$$\frac{\mathrm{d}g(x)}{\mathrm{d}x} = \frac{\alpha}{(x+2)^{1-\alpha}} - \frac{\alpha}{x^{1-\alpha}} < 0. \tag{5.39}$$

Thus $g(x)$ will decrease monotonically for $x \in \mathbb{R}_{>1}$.

According to (5.38) and (5.39), the right-hand side of (5.37) will be less than 1 and hence it further results in

$$\sigma^{m+1} + \sum_{l=1}^{m+1} |f(l)| < 1. \tag{5.40}$$

Therefore, as long as $\sigma + |f(1)| \leq 1$ is satisfied, then inequality (5.34) can be ensured for $\forall k \in \mathbb{N}_{>1}$. ■

The first result of this section is now ready to be presented in the following theorem, where a stability criterion is given.

Theorem 5.2 *Consider the closed-loop system consisting of fractional-order systems (5.25) and sampled-data-based finite-dimensional controllers (5.28) and (5.31). The closed-loop system is ISS in the sense of (5.26) if the design parameter h and matrix $\bar{K}$ satisfy*

$$c + |f(1)| \leq 1, \tag{5.41}$$

where

$$c = \left\| I_n + \frac{h^\alpha}{\Gamma(\alpha+1)}\bar{K} \right\|. \tag{5.42}$$

Proof: We first establish the following relationship to ensure the global boundedness of the control signal

$$\|u_k\| \leq \|u_0\|, \text{ for all } k \geq 0. \tag{5.43}$$

From (5.31), the designed controller contains a weighted sum of finite number of previous control inputs. To establish inequality (5.43), firstly we consider the case when L is the integer variable k, or in other words, the current control input involves the weighted sum of all the previous control data. Then under an approach similar to that proposed in Section 5.2, the proof of (5.43) is completed through mathematical induction as detailed below.

Step 1: According to (5.28) and (5.31), we have

$$\begin{aligned}\|u_1\| \leq& \|\bar{K}x(t_1)\| + |f(1)|\|u_0\| \leq \left\| I_n + \frac{h^\alpha}{\Gamma(\alpha+1)}\bar{K} \right\| \|\bar{K}x(t_0)\| + |f(1)|\|u_0\| \\ =& \left[c + |f(1)|\right]\|u_0\|. \end{aligned} \tag{5.44}$$

By designing h and $\bar{K}$ such that (5.41) holds gives $\|u_1\| \leq \|u_0\|$.

Step 2: Assuming that $\|u_k\| \leq \|u_0\|$ holds for $k > 1$. Therefore, according to (5.31), $\|u_{k+1}\|$ can be expressed as

$$\begin{aligned}\|u_{k+1}\| =& \left\| \bar{K}x(t_{k+1}) + \sum_{l=1}^{k+1} |f(l)| u_{k+1-l} \right\| \leq c^{k+1}\|\bar{K}x(t_0)\| + \sum_{l=1}^{k+1} |f(l)|\|u_{k+1-l}\| \\ \leq& \left[c^{k+1} + \sum_{l=1}^{k+1} |f(l)|\right]\|u_0\|. \end{aligned} \tag{5.45}$$

By (5.41) and Proposition 5.1, we have $c^{k+1} + \sum_{l=1}^{k+1}|f(l)| < 1$, thus (5.43) holds.

Now consider the case when L is a constant integer. For $k \leq L$, the proof follows the above analysis corresponding to the situation that L constantly equals k. Then for $k > L$, we can get $\|u_k\| \leq \left\|\bar{K}x(t_k) + \sum_{l=1}^{L}|f(l)|u_{k-l}\right\| \leq \left\|\bar{K}x(t_k) + \sum_{l=1}^{k}|f(l)|u_{k-l}\right\|$. In light of the proof for the case that the current control signal contains the weighted sum of all the historical values of control inputs, (5.43) can be easily obtained with finite-dimensional controller.

Apparently, $c = \|I_n + \frac{h^\alpha}{\Gamma(\alpha+1)}\bar{K}\| < 1$ is the premise to ensure the establishment of (5.41), which conforms the stability condition for the discrete-time closed-loop systems with the form shown in (5.32). For the case when $k \leq L$, the Euclidean norm of $x(t_{k+1})$ can be calculated as

$$\begin{aligned}\|x(t_{k+1})\| =& \left\| \left[I_n + \frac{h^\alpha}{\Gamma(\alpha+1)}\bar{K}\right]x(t_k) \right\| \\ \leq& c\|x(t_k)\| \leq c^2\|x(t_{k-1})\| \leq \cdots \leq c^{k+1}\|x(t_0)\|. \end{aligned} \tag{5.46}$$

Then for case when $k > L$, computing the Euclidean norm on both sides of (5.33) gives

$$\begin{aligned}\|x(t_{k+1})\| \leq& c^{k+1}\|x(t_0)\| + \sum_{j=L+1}^{k} c^{k-j} \sum_{l=L+1}^{j} \frac{h^\alpha|f(l)|}{\Gamma(\alpha+1)}\|u_{j-l}\| \\ \leq& c^{k+1}\|x(t_0)\| + \sum_{j=L+1}^{k} c^{k-j} \sum_{l=L+1}^{j} \frac{h^\alpha|f(l)|}{\Gamma(\alpha+1)}\|u_0\|. \end{aligned} \tag{5.47}$$

Defining $r(k)_L = \sum_{j=L+1}^{k} c^{k-j} \sum_{l=L+1}^{j} \frac{h^\alpha|f(l)|}{\Gamma(\alpha+1)}$, then (5.47) can be rewritten as

$$\|x(t_{k+1})\| \leq c^{k+1}\|x(t_0)\| + r(k)_L\|u_0\| = \beta(\|x(t_0)\|, t_{k+1}) + \gamma(\|u_0\|) \tag{5.48}$$

where function $\beta : \mathbb{R}_0 \times \mathbb{R}_0 \to \mathbb{R}_0$ is a $\mathcal{KL}$-function and $\gamma : \mathbb{R}_0 \to \mathbb{R}_0$ is a $\mathcal{K}$-function. Since (5.43) holds for $\forall k \geq 0$, hence the closed-loop fractional-order systems can be ensured to be ISS for any initial condition $x(t_0) \in \mathbb{R}^n$. ■

Remark 5.6 *Subjecting to the case of an infinite-dimensional control scheme where L is always equal to k, then the asymptotic stability of closed-loop fractional-order systems (5.32) can be achieved with controllers (5.28) and (5.31). Moreover, as demonstrated in the next subsection, the bound for the closed-loop state trajectories will decrease with the increment in the requirement of control accuracy.*

5.3.3 Selection of Control Signal Storage Size

Define an error signal as $e(t_k) = x(t_k) - x^*(t_k)$, $\forall k \in \mathbb{N}_0$. Here $x^*(t_k)$ denotes the closed-loop state under the control of an infinite-dimensional controller that is given in (5.51) below. In this section, L should be designed to meet that the upper bound of the error signal should be less than a given index ϵ, which is described as follows:

$$\sup\{\|e(t_{k+1})\| : k \in \mathbb{N}_0\} < \epsilon, \text{ where } \epsilon > 0. \tag{5.49}$$

The goal for this subsection is to provide a quantitative sufficient condition for choosing L to ensure (5.49) is satisfied, as stated in the following theorem.

Theorem 5.3 *Given any scalar $\epsilon > 0$, then L can be chosen to ensure (5.49) as long as the inequality below is satisfied*

$$1 - \frac{\epsilon(1-c)\Gamma(\alpha+1)}{h^\alpha \|\bar{K}x(t_0)\|} < \sum_{l=1}^{L} |f(l)|, \tag{5.50}$$

in which c is defined in (5.42).

Proof: Consider employing an infinite-dimensional control scheme that L continuously equals k, then the corresponding closed-loop systems can be described as

$$x^*(t_{k+1}) = \left[I_n + \frac{h^\alpha}{\Gamma(\alpha+1)}\bar{K}\right] x^*(t_k),\ x^*(t_0) = x(t_0). \tag{5.51}$$

Then the evolution of $e(t_k)$ is given as follows

$$e(t_{k+1}) = \begin{cases} 0, & k \le L \\ \sum_{j=L+1}^{k} \left[I_n + \frac{h^\alpha \bar{K}}{\Gamma(\alpha+1)}\right]^{k-j} \sum_{l=L+1}^{j} \frac{h^\alpha f(l) u_{j-l}}{\Gamma(\alpha+1)}, & k > L. \end{cases} \tag{5.52}$$

According to (5.52), the Euclidean norm of $e(t_{k+1})$ can be computed as

$$\begin{aligned} \|e(t_{k+1})\| &\le \sum_{j=L+1}^{k} \left\| I_n + \frac{h^\alpha}{\Gamma(\alpha+1)}\bar{K} \right\|^{k-j} \sum_{l=L+1}^{j} \frac{h^\alpha |f(l)|}{\Gamma(\alpha+1)} \|u_{j-l}\| \\ &\le \sum_{j=L+1}^{k} c^{k-j} \sum_{l=L+1}^{j} \frac{h^\alpha |f(l)|}{\Gamma(\alpha+1)} \|u_0\|. \end{aligned} \tag{5.53}$$

As h and $\bar{K}$ meet condition (5.41), which insinuates the satisfaction of inequality (5.34) with $\sigma = c$, thus it gives

$$\sum_{l=L+1}^{k} |f(l)| \le 1 - c^k - \sum_{l=1}^{L} |f(l)|, \text{ for } k > L. \tag{5.54}$$

Therefore the term $r(k)_L$ on the right-hand side of (5.53) can be computed as

$$\begin{aligned} r(k)_L =& \frac{h^\alpha}{\Gamma(\alpha+1)} \sum_{j=L+1}^{k} c^{k-j} \sum_{l=L+1}^{j} |f(l)| \\ \leq& \frac{h^\alpha}{\Gamma(\alpha+1)} \sum_{j=L+1}^{k} c^{k-j} \left[1 - \sum_{l=1}^{L} |f(l)| - c^j\right] \\ \leq& \frac{h^\alpha}{\Gamma(\alpha+1)} \sum_{j=L+1}^{k} c^{k-j} \left[1 - \sum_{l=1}^{L} |f(l)|\right] \\ \leq& \frac{h^\alpha}{\Gamma(\alpha+1)} \left[1 - \sum_{l=1}^{L} |f(l)|\right] \frac{1}{1-c}. \end{aligned} \tag{5.55}$$

By combining (5.53) and (5.55), condition (5.49) can be written as

$$\frac{h^\alpha}{\Gamma(\alpha+1)} \left[1 - \sum_{l=1}^{L} |f(l)|\right] \frac{1}{1-c} \|\bar{K}x(t_0)\| < \epsilon. \tag{5.56}$$

Hence, given $\epsilon > 0$ with design parameters h and $\bar{K}$ satisfying inequality (5.41), the smallest value of L can be obtained by solving (5.56), which consequently results in (5.50). ■

Remark 5.7 *For smaller ϵ, which implies a requirement on more precise control accuracy, the storage space L for preserving the historical values of control inputs should be increased correspondingly. This is also in line with the trade-off between control performance and storage space in practical applications.*

5.3.4 Illustrative Example

An example is now used to demonstrate the proposed finite-dimensional control scheme and verify the established theoretical results. In accordance with [39, 203, 205], the current-voltage relationship of a fractional-order capacitor is defined as

$${}_{t_0}\mathcal{D}_t^\alpha v(t) = \frac{i(t)}{C_\alpha} \tag{5.57}$$

where $v(t)$ and $i(t)$ are the voltage and current of the capacitor and C_α represents the capacitance with unit $\mathrm{F/(second)}^{1-\alpha}$.

By defining system state $x(t) = v(t)$ and control signal $u(t) = i(t)$, the proposed control scheme will be utilized to discharge the fractional-order capacitor described in (5.57) with $\alpha = 0.5$, $C_\alpha = 1\ \mathrm{F/(second)}^{1-\alpha}$, and initial voltage $x(0) = 2.5$ V for simulation studies. A controllable current source is considered as the controller based on the proposed control law to realize fast discharge of the fractional-order capacitor, as shown in Fig. 5.5. The sampling period and control parameter are respectively designed as $h = 0.1$ s and $\bar{K} = -2.1$. As can be seen from Fig. 5.6, compared to the circuit that discharges the fractional-order capacitor by replacing the controllable current source with a resistance $R = 1\ \Omega$, faster discharge speed can be achieved under the proposed controller.

According to (5.50), the smallest value of L corresponding to a certain given ϵ can be computed, as reflected in Table 5.1.

Simulation results of capacitor responses are given in Fig. 5.6 and Fig. 5.7, from which we can notice that ISS is ensured for the closed-loop circuit under the proposed finite-dimensional control strategy. The values of $\|u\|^2$ under different L can also be observed

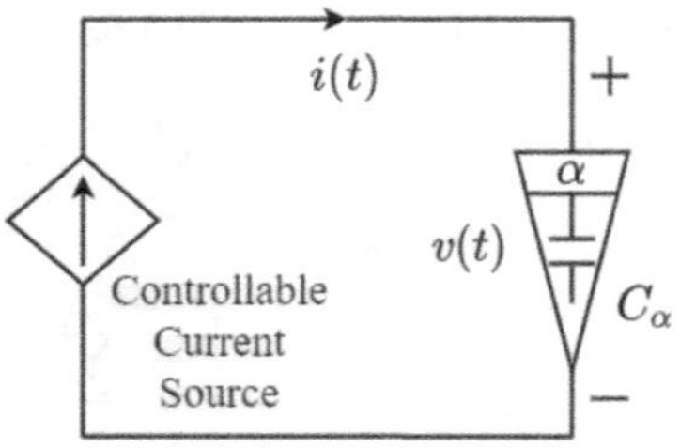

FIGURE 5.5: The discharge circuit of a fractional-order capacitor.

TABLE 5.1: Correspondence between ϵ and the smallest L.

ϵ	0.5	0.3	0.1
Smallest value of L	7	20	181

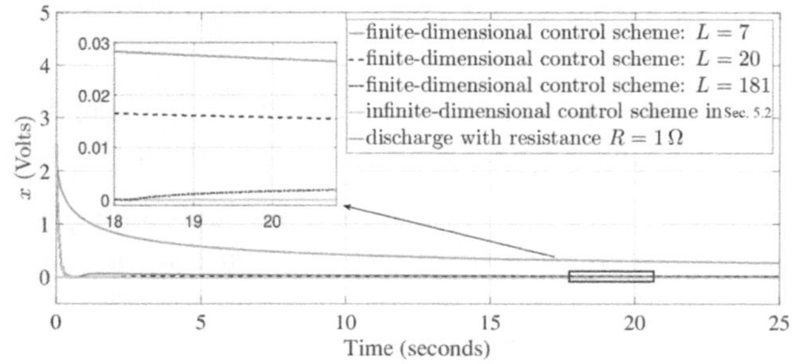

FIGURE 5.6: The trajectories of x under different L.

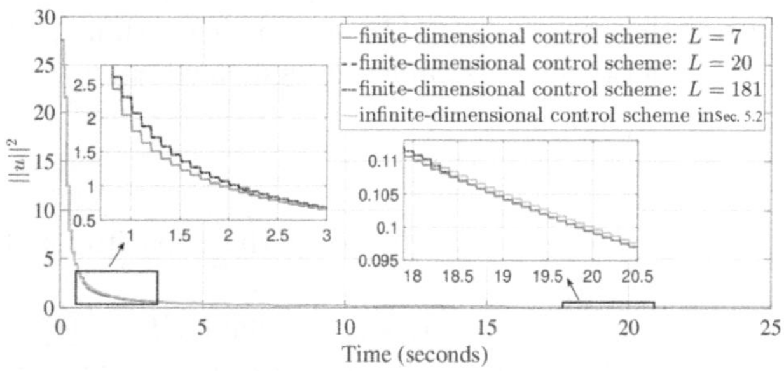

FIGURE 5.7: Values of $||u||^2$ under different L.

from Fig. 5.7. These results verify the boundedness of all the system signals. Moreover, since only a limited number of recent values of the capacitor current are stored to compute the present current in our designed control scheme, the corresponding cumulative control effort will decrease compared with the one using a control method presented in Section 5.2 that contains all the previous control inputs.

With the minimum values of L displayed in Table 5.1, the condition (5.49) corresponding to distinct ϵ will always be satisfied. From this condition, it is notable that the bound of the error signal should be satisfied for all sampling instants except for the initial time

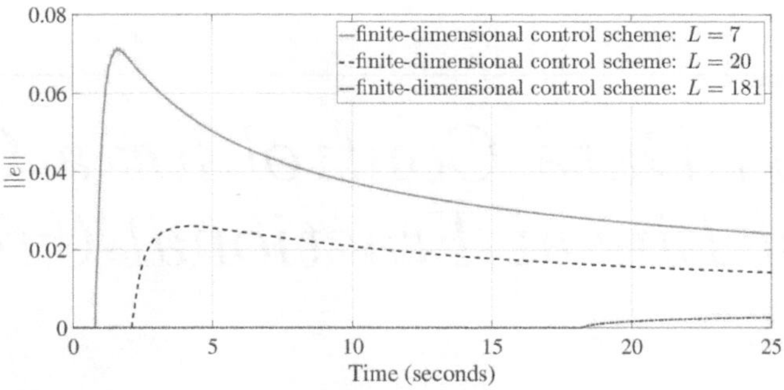

FIGURE 5.8: Values of $\|e\|$ under different L.

instant. Therefore, not only at steady-state but also for the whole transition process, the boundedness of the error signal can be guaranteed. Results illustrated in Fig. 5.8 verify that the upper bound of $\|e\|$ in Theorem 5.3 is generally satisfied. All these results presented in this subsection verify the effectiveness of the proposed control scheme and the theoretical results established.

5.3.5 Conclusions

In this section, a procedure to design a finite-dimensional sampled-data-based controller for a class of linear single-integrator form continuous-time fractional-order systems that guarantees global ISS of the associated closed-loop system is introduced. A quantitative condition for deciding the storage size of previous control inputs in order to compute the current control signal is also presented to meet certain control performance. All the theoretical results in this section are verified by simulation studies.

6

Sampled-Data Control for a General Class of Linear Fractional-Order Systems

The control scheme presented in the previous chapter is only valid for fractional-order systems with dynamics in single-integrator form. Hence it will be interesting and more practically significant to extend the results to a more general class of fractional-order linear systems. In this chapter, based on the sampled-data control method, the stabilization problem for a general class of linear continuous-time fractional systems whose solution contains the Mittag-Leffler function that does not obey the basic exponentiation identity will be addressed. By considering the infinite memory and hereditary characteristics of fractional-order calculus, a sampled-data controller that guarantees the resulting closed-loop system to be asymptotically stable is proposed. Simulation examples are presented to demonstrate the effectiveness of the proposed controller and verify the established results.

6.1 Introduction

Different from integer-order systems that have been well-studied for decades, the sampled-data control for fractional-order systems has only received attention in recent years. The consensus problem for linear fractional-order MASs, whose agents are in single-integrator form, is investigated in [138] and [4] via sampled-data control. For linear fractional-order double-integrator MASs, conditions for achieving consensus and containment under sampled-data control schemes are presented respectively in [133, 134, 135, 136] and [73, 140, 141]. A sampled-data control protocol for linear fractional-order MASs where the coupling matrix can be non-Laplacian is studied in [137]. Nevertheless, none of the above-mentioned work considers the infinite memory and hereditary characteristics of fractional-order calculus when designing control laws. This motivated the works in Chapter 5, where the characteristics of fractional-order calculus are rigorously taken into account by containing the weighted sum of the historical input values into the current control signal.

However, in Chapter 5, only a class of linear fractional-order systems in the form of single-integrator is considered. In this chapter, we handle the sampled-data control issue for a more general class of linear fractional-order systems. It is worth noting that the sampled-data controllers designed in Chapter 5 are based on discretizing the solutions of single-integrator fractional-order differentiation equations. Nevertheless, it is challenging to apply this design procedure to a general class of linear fractional-order systems whose solution expression contains the Mittag-Leffler function that does not obey the basic exponentiation identity. To overcome such a difficulty, a novel sampled-data-based control scheme will be proposed in this chapter. With the designed scheme, the asymptotic stability of the resulting

DOI: 10.1201/9781003620372-6

closed-loop system is ensured. In addition, it can be proved that our proposed controller can also be applied to the class of systems investigated in Chapter 5, which implies that its established result is a special case of the one in this chapter.

The chapter is organized as follows. Problem formulation is presented in Section 6.2. The design of the proposed sampled-data-based controller is stated in detail in Section 6.3 with analysis provided. Numerical examples are given in Section 6.4 to demonstrate the effectiveness of the established results. Finally, the chapter is summarized in Section 6.5.

6.2 Problem Formulation

A class of linear continuous-time fractional-order systems studied in this section can be described as

$${}_0\mathcal{D}_t^\alpha x(t) = Ax(t) + Bu(t), \tag{6.1}$$

where $\alpha \in (0, 1]$, initial time t_0 is 0, $x \in \mathbb{R}^n$ and $u \in \mathbb{R}^m$ respectively represent the system state and control input, and $A \in \mathbb{R}^{n\times n}$ and $B \in \mathbb{R}^{n\times m}$ are known constant matrices.

Assumption 6.1 *The matrix $B \in \mathbb{R}^{n\times m}$ is full row rank.*

Remark 6.1 *Assumption 6.1 is the requirement for the existence of the right inverse matrix $P = B^T(BB^T)^{-1}$ of matrix B, which will be used in the controller design procedure as shown in the next section. Such an assumption is met for any system with its number of inputs m not less than its number of outputs n. Existing studies on such a class of systems can be found in for instance [206, 207, 208, 209, 210].*

The control objective is to design a sampled-data-based control law for the general class of fractional-order linear systems depicted in (6.1) to ensure that all signals are bounded and the resulting closed-loop system is asymptotically stable, i.e., $\lim_{t\to\infty}\|x(t)\| = 0$.

Remark 6.2 *In accordance with Lemma 1.5, the solution of the class of linear fractional-order systems (6.1) is described as*

$$x(t) = \int_0^t (t-\tau)^{\alpha-1}\mathrm{E}_{\alpha,\alpha}\big(A(t-\tau)^\alpha\big)Bu(\tau)\mathrm{d}\tau + x(0)\mathrm{E}_\alpha\big(At^\alpha\big), \tag{6.2}$$

where $x(0)$ is the initial condition. Unlike the exponential function which owns the property that $\mathrm{e}^{at+bt} = \mathrm{e}^{at}\mathrm{e}^{bt}$ $(a, b \in \mathbb{R})$, such a property does not hold for the Mittag-Leffler function. Therefore, it is difficult to directly design a sampled-data control scheme for such class of systems based on the expression of the solution given in (6.2). Inspired by the numerical methods for solving fractional-order differential equations [211], a novel sampled-data-based controller design method will be investigated in this chapter, which is different from the ones proposed in the previous chapter that are based on discretizing the solution expressions of the systems under consideration. On the other hand, the same as Chapter 5, the infinite memory and hereditary characteristics of fractional-order calculus will also be taken into account in this chapter when designing the sampled-data controller.

Remark 6.3 *Compared with the class of systems studied in Section 5.3 that the trajectories of system states will not diverge without control inputs, or in other words, the designed control laws in Section 5.3 cannot handle unstable fractional-order systems, a more general class of linear fractional-order systems are considered in this chapter. Corresponding comparison studies will be provided in Section 6.4 for illustration.*

Remark 6.4 *It should be mentioned that the fractional-order lithium-ion battery model established in [37] shares the same form as the one described in (6.1). Also, as to be shown in Section 6.4, the voltage-current relationship of the fractional-order capacitor [39, 203, 205] is a special case of (6.1). Therefore, it is of practical value to study the class of linear fractional-order systems as shown in equation (6.1).*

6.3 Sampled-Data Controller Design and Stability Analysis

6.3.1 Controller Design Procedure

A periodic sampling scheme is to be implemented for designing the controller in this chapter. According to the definition of Caputo fractional-order derivative, details of which can be found in Definition 1.11, by denoting the sampling period as $T_s > 0$, then at sampling moment $t = mT_s$ $(m \in \mathbb{N})$, the left-hand side of (6.1) is

$$\begin{aligned}
{}_0^C\mathcal{D}_t^{\alpha}x(t) =& \frac{1}{\Gamma(1-\alpha)}\int_0^t \frac{x^{(1)}(\tau)}{(t-\tau)^{\alpha}}\mathrm{d}\tau = \frac{1}{\Gamma(1-\alpha)}\sum_{h=0}^{m-1}\int_{hT_s}^{(h+1)T_s}\frac{x^{(1)}(\tau)}{(mT_s-\tau)^{\alpha}}\mathrm{d}\tau \\
=& \sum_{h=0}^{m-1}\int_{hT_s}^{(h+1)T_s}\left[\frac{x\big((h+1)T_s\big)-x(hT_s)}{T_s}-\frac{x^{(3)}(s_h)T_s^2}{4!}\right]\frac{(mT_s-\tau)^{-\alpha}}{\Gamma(1-\alpha)}\mathrm{d}\tau \\
=& \sum_{h=0}^{m-1}\left[\frac{x\big((h+1)T_s\big)-x(hT_s)}{T_s}-\frac{x^{(3)}(s_h)T_s^2}{4!}\right]\int_{hT_s}^{(h+1)T_s}\frac{(mT_s-\tau)^{-\alpha}}{\Gamma(1-\alpha)}\mathrm{d}\tau \\
=& \frac{1}{\Gamma(1-\alpha)}\sum_{h=0}^{m-1}\left[\frac{x\big((h+1)T_s\big)-x(hT_s)}{T_s}-\frac{x^{(3)}(s_h)T_s^2}{4!}\right] \\
& \left\{\frac{-\big[mT_s-(h+1)T_s\big]^{1-\alpha}}{1-\alpha}+\frac{(mT_s-hT_s)^{1-\alpha}}{1-\alpha}\right\} \\
=& \frac{1}{\Gamma(2-\alpha)}\sum_{h=0}^{m-1}\left[\frac{x\big((h+1)T_s\big)-x(hT_s)}{T_s^{\alpha}}-\frac{x^{(3)}(s_h)T_s^{3-\alpha}}{4!}\right]\Big[(m-h)^{1-\alpha} \\
& -(m-h-1)^{1-\alpha}\Big]
\end{aligned} \tag{6.3}$$

where $s_h \in \big[hT_s, (h+1)T_s\big]$ $(h \in \mathbb{N}_0)$.

In accordance with Taylor's theorem, it gives

$$x\big((h-\iota)T_s\big) = x(hT_s) - \iota T_s x^{(1)}(hTs) + \frac{(\iota T_s)^2}{2!}x^{(2)}(hT_s) - \frac{(\iota T_s)^3}{3!}x^{(3)}(hT_s), \tag{6.4}$$

with $\iota = 1,\,2,\,3$ and x being three times differentiable. Note that in this chapter, $x^{(3)}(s_h)$ is numerically calculated based on (6.4) and backward difference method as follows to ensure the causality of the designed controller

$$x^{(3)}(s_h) = \frac{1}{T_s^3}\Big[-x\big((h-3)T_s\big)+3x\big((h-2)T_s\big)-3x\big((h-1)T_s\big)+x(hT_s)\Big], \tag{6.5}$$

and in the computation of $x^{(3)}(s_i)$ $(i = 0, 1, 2)$, the states before the initial time $t = 0$ are considered to be all equal to $x(0)$ according to Remark 1.2, i.e., $x(l) = x(0)$, $\forall l < 0$.

According to (6.1) and (6.3), we have

$$\sum_{h=0}^{m-1}\left[\frac{x((h+1)T_s)-x(hT_s)}{\Gamma(2-\alpha)T_s^{\alpha}}-\frac{x^{(3)}(s_h)T_s^{3-\alpha}}{\Gamma(2-\alpha)4!}\right]\left[(m-h)^{1-\alpha}-(m-h-1)^{1-\alpha}\right]$$
$$=Ax(mT_s)+Bu(mT_s^-), \tag{6.6}$$

where the control signal u is designed to be left continuous, that is, at the sampling moment mT_s, $u(mT_s^-)=u\big((m-1)T_s\big)$.

The sampling period T_s is designed to ensure that $\frac{I_n}{T_s^{\alpha}\Gamma(2-\alpha)}-A$ is invertible to guarantee the existence of the expressions for the values of $x(mT_s)$. For the case when $m=1$, the following equation can be obtained according to (6.6):

$$\frac{\big[x(T_s)-x(0)\big]}{T_s^{\alpha}\Gamma(2-\alpha)}-\frac{T_s^{3-\alpha}x^{(3)}(s_0)}{4!\Gamma(2-\alpha)}=Ax(T_s)+Bu(T_s^-). \tag{6.7}$$

Define matrix $G=\left[\frac{I_n}{T_s^{\alpha}\Gamma(2-\alpha)}-A\right]^{-1}$. Then (6.7) can be written as

$$x(T_s)=GBu(T_s^-)+G\frac{x(0)}{T_s^{\alpha}\Gamma(2-\alpha)}+G\frac{T_s^{3-\alpha}x^{(3)}(s_0)}{4!\Gamma(2-\alpha)}. \tag{6.8}$$

By designing $u(T_s^-)=u(0)$ as

$$u(0)=-\bar{K}x(0)-\frac{T_s^{3-\alpha}}{4!\Gamma(2-\alpha)}Px^{(3)}(s_0), \tag{6.9}$$

in which $\bar{K}\in\mathbb{R}^{m\times n}$ is a matrix with positive elements to be designed and $P=B^T(BB^T)^{-1}$ which exists from Assumption 6.1, then it yields

$$\begin{aligned}x(T_s)=&-GB\bar{K}x(0)-G\frac{T_s^{3-\alpha}}{4!\Gamma(2-\alpha)}BPx^{(3)}(s_0)+G\frac{x(0)}{T_s^{\alpha}\Gamma(2-\alpha)}+G\frac{T_s^{3-\alpha}x^{(3)}(s_0)}{4!\Gamma(2-\alpha)}\\=&\left[\frac{G}{T_s^{\alpha}\Gamma(2-\alpha)}-GB\bar{K}\right]x(0).\end{aligned} \tag{6.10}$$

Now for $m=2$, according to (6.6), we have

$$\begin{aligned}\left[\frac{I_n}{T_s^{\alpha}\Gamma(2-\alpha)}-A\right]x(2T_s)=&\frac{x(T_s)}{T_s^{\alpha}\Gamma(2-\alpha)}-\frac{2^{1-\alpha}-1}{T_s^{\alpha}\Gamma(2-\alpha)}x(T_s)+\frac{2^{1-\alpha}-1}{T_s^{\alpha}\Gamma(2-\alpha)}x(0)\\&+Bu(2T_s^-)+\frac{T_s^{3-\alpha}}{4!\Gamma(2-\alpha)}\left[x^{(3)}(s_0)(2^{1-\alpha}-1)+x^{(3)}(s_1)\right],\end{aligned} \tag{6.11}$$

which can be further expressed as

$$\begin{aligned}x(2T_s)=&GBu(2T_s^-)+G\frac{T_s^{3-\alpha}}{4!\Gamma(2-\alpha)}\left[x^{(3)}(s_0)(2^{1-\alpha}-1)+x^{(3)}(s_1)\right]\\&+G\frac{2-2^{1-\alpha}}{T_s^{\alpha}\Gamma(2-\alpha)}x(T_s)+G\frac{2^{1-\alpha}-1}{T_s^{\alpha}\Gamma(2-\alpha)}x(0).\end{aligned} \tag{6.12}$$

Since the fractional derivative of the state x is related to its historical values, hence the control signal $u(2T_s^-)=u(T_s)$ is designed as

$$\begin{aligned}u(T_s)=&-\bar{K}x(T_s)-P\frac{T_s^{3-\alpha}}{4!\Gamma(2-\alpha)}\left[x^{(3)}(s_0)(2^{1-\alpha}-1)+x^{(3)}(s_1)\right]\\&-\frac{2^{1-\alpha}-1}{T_s^{\alpha}\Gamma(2-\alpha)}Px(0)\end{aligned} \tag{6.13}$$

to counteract the effect caused by the system state memory values and consequently, the closed-loop system for $m = 2$ is expressed as

$$\begin{aligned} x(2T_s) =& - GB\bar{K}x(T_s) - GBP\frac{T_s^{3-\alpha}}{4!\Gamma(2-\alpha)}\Big[x^{(3)}(s_0)(2^{1-\alpha}-1) + x^{(3)}(s_1)\Big] \\ & - G\frac{2^{1-\alpha}-1}{T_s^{\alpha}\Gamma(2-\alpha)}BPx(0) + G\frac{T_s^{3-\alpha}}{4!\Gamma(2-\alpha)}\Big[x^{(3)}(s_0)(2^{1-\alpha}-1) \\ & + x^{(3)}(s_1)\Big] + G\frac{2-2^{1-\alpha}}{T_s^{\alpha}\Gamma(2-\alpha)}x(T_s) + G\frac{2^{1-\alpha}-1}{T_s^{\alpha}\Gamma(2-\alpha)}x(0) \\ =& \left[\frac{2-2^{1-\alpha}}{T_s^{\alpha}\Gamma(2-\alpha)}G - GB\bar{K}\right]x(T_s). \end{aligned} \tag{6.14}$$

For $m \geq 3$, in accordance with (6.6), the value of $x(mT_s)$ can be calculated as

$$\begin{aligned} x(mT_s) =& GBu(mT_s^-) + G\frac{x\big((m-1)T_s\big)}{T_s^{\alpha}\Gamma(2-\alpha)} + G\frac{T_s^{3-\alpha}}{4!\Gamma(2-\alpha)}\sum_{h=0}^{m-1}\big[(m-h)^{1-\alpha} \\ & - (m-1-h)^{1-\alpha}\big]x^{(3)}(s_h) + G\sum_{h=0}^{m-2}\frac{\big[(m-h)^{1-\alpha} - (m-1-h)^{1-\alpha}\big]}{T_s^{\alpha}\Gamma(2-\alpha)} \\ & \Big[x(hT_s) - x\big((h+1)T_s\big)\Big] \\ =& GBu(mT_s^-) + G\frac{2-2^{1-\alpha}}{T_s^{\alpha}\Gamma(2-\alpha)}x\big((m-1)T_s\big) + G\frac{T_s^{3-\alpha}}{4!\Gamma(2-\alpha)} \\ & \sum_{h=0}^{m-1}\big[(m-h)^{1-\alpha} - (m-1-h)^{1-\alpha}\big]x^{(3)}(s_h) \\ & - G\sum_{h=1}^{m-2}\frac{(h+2)^{1-\alpha} - 2(h+1)^{1-\alpha} + h^{1-\alpha}}{T_s^{\alpha}\Gamma(2-\alpha)}x\big((m-1-h)T_s\big) \\ & + G\frac{m^{1-\alpha} - (m-1)^{1-\alpha}}{T_s^{\alpha}\Gamma(2-\alpha)}x(0). \end{aligned} \tag{6.15}$$

The control input $u(mT_s^-) = u\big((m-1)T_s\big)$ is designed as

$$\begin{aligned} u\big((m-1)T_s\big) =& - \bar{K}x\big((m-1)T_s\big) + \sum_{h=1}^{m-2}\frac{(h+2)^{1-\alpha} - 2(h+1)^{1-\alpha} + h^{1-\alpha}}{T_s^{\alpha}\Gamma(2-\alpha)}Px\big((m-1 \\ & - h)T_s\big) - \frac{m^{1-\alpha} - (m-1)^{1-\alpha}}{T_s^{\alpha}\Gamma(2-\alpha)}Px(0) - \frac{T_s^{3-\alpha}}{4!\Gamma(2-\alpha)}\sum_{h=0}^{m-1}\big[(m-h)^{1-\alpha} \\ & - (m-1-h)^{1-\alpha}\big]Px^{(3)}(s_h). \end{aligned} \tag{6.16}$$

As a result, by substituting (6.16) into (6.15), the closed-loop system can be deducted as

$$\begin{aligned}
x(mT_s) = & -GB\bar{K}x\big((m-1)T_s\big) - \frac{m^{1-\alpha}-(m-1)^{1-\alpha}}{T_s^\alpha\Gamma(2-\alpha)}GBPx(0) \\
& + G\sum_{h=1}^{m-2}\frac{(h+2)^{1-\alpha}-2(h+1)^{1-\alpha}+h^{1-\alpha}}{T_s^\alpha\Gamma(2-\alpha)}BPx\big((m-1-h)T_s\big) \\
& - G\frac{T_s^{3-\alpha}}{4!\Gamma(2-\alpha)}\sum_{h=0}^{m-1}\big[(m-h)^{1-\alpha}-(m-1-h)^{1-\alpha}\big]BPx^{(3)}(s_h) \\
& + G\frac{T_s^{3-\alpha}}{4!\Gamma(2-\alpha)}\sum_{h=0}^{m-1}\big[(m-h)^{1-\alpha}-(m-1-h)^{1-\alpha}\big]x^{(3)}(s_h) \\
& - G\sum_{h=1}^{m-2}\frac{(h+2)^{1-\alpha}-2(h+1)^{1-\alpha}+h^{1-\alpha}}{T_s^\alpha\Gamma(2-\alpha)}x\big((m-1-h)T_s\big) \\
& + G\frac{2-2^{1-\alpha}}{T_s^\alpha\Gamma(2-\alpha)}x\big((m-1)T_s\big) + G\frac{m^{1-\alpha}-(m-1)^{1-\alpha}}{T_s^\alpha\Gamma(2-\alpha)}x(0) \\
= & \left[\frac{2-2^{1-\alpha}}{T_s^\alpha\Gamma(2-\alpha)}G - GB\bar{K}\right]x\big((m-1)T_s\big).
\end{aligned} \tag{6.17}$$

6.3.2 Stability Analysis

The main result of this chapter is given in the following theorem with the stability analysis of the resulting closed-loop system presented in the proof.

Theorem 6.1 *Consider the closed-loop system consisting of linear fractional-order system (6.1) and sampled-data control laws (6.9), (6.13) and (6.16). All the signals in the closed-loop system can be ensured to be bounded and asymptotic stability can be realized, i.e.,* $\lim_{t\to\infty}\|x(t)\| = 0$, *if the design control matrix* $\bar{K}$ *and sampling period* T_s *are chosen according to the following condition*

$$\beta \triangleq \max\left\{\left\|\frac{G}{T_s^\alpha\Gamma(2-\alpha)} - GB\bar{K}\right\|, \left\|\frac{(2-2^{1-\alpha})G}{T_s^\alpha\Gamma(2-\alpha)} - GB\bar{K}\right\|\right\} < 1. \tag{6.18}$$

Proof: With the design parameters $\bar{K}$ and T_s that satisfy the condition given in (6.18), the following inequality can be acquired when $m = 1$:

$$\begin{aligned}
\|x(T_s)\| = & \left\|\left[\frac{G}{T_s^\alpha\Gamma(2-\alpha)} - GB\bar{K}\right]x(0)\right\| \le \left\|\frac{G}{T_s^\alpha\Gamma(2-\alpha)} - GB\bar{K}\right\|\|x(0)\| \\
\le & \beta\|x(0)\|.
\end{aligned} \tag{6.19}$$

For the case when $m \ge 2$, according to (6.14) and (6.17), we can get

$$\begin{aligned}
\|x(mT_s)\| = & \left\|\left[\frac{(2-2^{1-\alpha})G}{T_s^\alpha\Gamma(2-\alpha)} - GB\bar{K}\right]x\big((m-1)T_s\big)\right\| \\
\le & \left\|\frac{(2-2^{1-\alpha})G}{T_s^\alpha\Gamma(2-\alpha)} - GB\bar{K}\right\|\left\|x\big((m-1)T_s\big)\right\| \\
\le & \beta\left\|x\big((m-1)T_s\big)\right\| \le \cdots \le \beta^m\|x(0)\|.
\end{aligned} \tag{6.20}$$

We now consider $u(mT_s^-)$. From (6.20), it can be noticed that

$$\begin{aligned}&\left\|\frac{1}{T_s^3}\Big[-x\big((h-3)T_s\big)+3x\big((h-2)T_s\big)-3x\big((h-1)T_s\big)+x(hT_s)\Big]\right\|\\ \leq&\frac{1}{T_s^3}\Big[\big\|x\big((h-3)T_s\big)\big\|+3\big\|x\big((h-2)T_s\big)\big\|+3\big\|x\big((h-1)T_s\big)\big\|+\|x(hT_s)\|\Big]\leq\frac{8\|x(0)\|}{T_s^3}\end{aligned} \tag{6.21}$$

holds for any $h\in\mathbb{N}_0$. By defining $M=\frac{8\|x(0)\|}{T_s^3}$, then in light of (6.9) and (6.5), it gives

$$\begin{aligned}\|u(0)\|=&\left\|-\bar{K}x(0)-\frac{T_s^{3-\alpha}}{4!\Gamma(2-\alpha)}Px^{(3)}(s_0)\right\|=\left\|-\bar{K}x(0)-\frac{T_s^{3-\alpha}}{4!\Gamma(2-\alpha)}\right.\\ &\left.P\frac{1}{T_s^3}\big[-x(-3T_s)+3x(-2T_s)-3x(-T_s)+x(0)\big]\right\|\\ \leq&\bar{K}\|x(0)\|+\frac{T_s^{3-\alpha}}{4!\Gamma(2-\alpha)}\|P\|M.\end{aligned} \tag{6.22}$$

Similarly, from (6.13), the following inequality for $m=2$ can be deducted

$$\begin{aligned}\|u(T_s)\|\leq&\bar{K}\|x(T_s)\|+\frac{2^{1-\alpha}-1}{T_s^\alpha\Gamma(2-\alpha)}\|P\|\|x(0)\|+\frac{T_s^{3-\alpha}(2^{1-\alpha}-1)}{4!\Gamma(2-\alpha)}\|P\|M+\frac{T_s^{3-\alpha}\|P\|M}{4!\Gamma(2-\alpha)}\\ <&\bar{K}\|x(0)\|+\frac{2^{1-\alpha}-1}{T_s^\alpha\Gamma(2-\alpha)}\|P\|\|x(0)\|+\frac{T_s^{3-\alpha}2^{1-\alpha}}{4!\Gamma(2-\alpha)}\|P\|M.\end{aligned} \tag{6.23}$$

For $m\geq 3$, we have

$$\begin{aligned}u\big((m-1)T_s\big)=&-\bar{K}x\big((m-1)T_s\big)-\frac{m^{1-\alpha}-(m-1)^{1-\alpha}}{T_s^\alpha\Gamma(2-\alpha)}Px(0)+\\ &\sum_{h=1}^{m-2}\frac{(h+2)^{1-\alpha}-2(h+1)^{1-\alpha}+h^{1-\alpha}}{T_s^\alpha\Gamma(2-\alpha)}Px\big((m-1-h)T_s\big)\\ &-\frac{T_s^{3-\alpha}}{4!\Gamma(2-\alpha)}\sum_{h=0}^{m-1}\big[(m-h)^{1-\alpha}-(m-1-h)^{1-\alpha}\big]P\\ &\frac{-x\big((h-3)T_s\big)+3x\big((h-2)T_s\big)-3x\big((h-1)T_s\big)+x(hT_s)}{T_s^3}\\ \leq&-\bar{K}x\big((m-1)T_s\big)-\frac{m^{1-\alpha}-(m-1)^{1-\alpha}}{T_s^\alpha\Gamma(2-\alpha)}Px(0)+\\ &\sum_{h=1}^{m-2}\frac{(h+2)^{1-\alpha}-2(h+1)^{1-\alpha}+h^{1-\alpha}}{T_s^\alpha\Gamma(2-\alpha)}Px\big((m-1-h)T_s\big)\\ &+\frac{T_s^{3-\alpha}}{4!\Gamma(2-\alpha)}\sum_{h=0}^{m-1}\big[-(m-h)^{1-\alpha}+(m-1-h)^{1-\alpha}\big]P\\ &\frac{-x\big((h-3)T_s\big)+3x\big((h-2)T_s\big)-3x\big((h-1)T_s\big)+x(hT_s)}{T_s^3}\\ &+\frac{T_s^{3-\alpha}}{4!\Gamma(2-\alpha)}\sum_{h=0}^{m-2}\big[(m-1-h)^{1-\alpha}-(m-2-h)^{1-\alpha}\big]P+F\\ &\frac{-x\big((h-3)T_s\big)+3x\big((h-2)T_s\big)-3x\big((h-1)T_s\big)+x(hT_s)}{T_s^3}\end{aligned} \tag{6.24}$$

where F is a non-negative constant that ensures $F + \frac{T_s^{3-\alpha}}{4!\Gamma(2-\alpha)} \sum_{h=0}^{m-2} \big[(m-1-h)^{1-\alpha} - (m-2-h)^{1-\alpha}\big] P \frac{1}{T_s^3} \Big[- x\big((h-3)T_s\big) + 3x\big((h-2)T_s\big) - 3x\big((h-1)T_s\big) + x(hT_s)\Big] \geq 0$ for $\forall m \geq 3$. Thus the following relationship about $\big\|u\big((m-1)T_s\big)\big\|$ can be obtained

$$\begin{aligned}
\big\|u\big((m-1)T_s\big)\big\| <& \bar{K}\|x(0)\| + \frac{m^{1-\alpha} - (m-1)^{1-\alpha}}{T_s^{\alpha}\Gamma(2-\alpha)} \|P\| \|x(0)\| \\
&+ \sum_{h=1}^{m-2} \frac{\big[- (h+2)^{1-\alpha} + 2(h+1)^{1-\alpha} - h^{1-\alpha}\big]}{T_s^{\alpha}\Gamma(2-\alpha)} \|P\| \|x(0)\| \\
&+ \frac{T_s^{3-\alpha}}{4!\Gamma(2-\alpha)} \sum_{h=0}^{m-2} \big[- (m-h)^{1-\alpha} + 2(m-1-h)^{1-\alpha} \\
&- (m-2-h)^{1-\alpha}\big] \|P\| M + \frac{T_s^{3-\alpha}}{4!\Gamma(2-\alpha)} \|P\| M + F.
\end{aligned} \tag{6.25}$$

It has been proved in Proposition 5.1 in Section 5.3 that for $\nu \in (0,\, 2^{\alpha} - 1]$, the following condition holds:

$$\nu^{m-1} + \sum_{h=1}^{m-1} \big[- (h+1)^{\alpha} + 2h^{\alpha} - (h-1)^{\alpha}\big] < 1,\ m \in \mathbb{N}_{>1}, \tag{6.26}$$

which implies that $\sum_{h=0}^{m-2} \big[- (m-h)^{1-\alpha} + 2(m-1-h)^{1-\alpha} - (m-2-h)^{1-\alpha}\big] < 1$. Hence, inequality (6.25) turns into

$$\begin{aligned}
\big\|u\big((m-1)T_s\big)\big\| <& \bar{K}\|x(0)\| + \frac{m^{1-\alpha} - (m-1)^{1-\alpha}}{T_s^{\alpha}\Gamma(2-\alpha)} \|P\| \|x(0)\| \\
&+ \sum_{h=1}^{m-2} \frac{\big[- (h+2)^{1-\alpha} + 2(h+1)^{1-\alpha} - h^{1-\alpha}\big]}{T_s^{\alpha}\Gamma(2-\alpha)} \|P\| \|x(0)\| \\
&+ 2\frac{T_s^{3-\alpha}}{4!\Gamma(2-\alpha)} \|P\| M + F \\
=& \bar{K}\|x(0)\| + \frac{2^{1-\alpha} - 1}{T_s^{\alpha}\Gamma(2-\alpha)} \|P\| \|x(0)\| + 2\frac{T_s^{3-\alpha}}{4!\Gamma(2-\alpha)} \|P\| M + F.
\end{aligned} \tag{6.27}$$

As a result, under the condition (6.18), the boundedness of all closed-loop signals is demonstrated in (6.19), (6.20), (6.22), (6.23), and (6.27) at the sampling instants $t = mT_s$ for all $m \in \mathbb{N}$. It can be noted from (6.9), (6.13), and (6.16) that control inputs remain constants between sampling instants, hence clearly the boundedness of the control signal at times other than sampling instants is ensured. From then on, the boundedness of system states at time $t \in \big((m-1)T_s, mT_s\big)$ for all $m \in \mathbb{N}$ will be proven by contradiction. Firstly, suppose there exists a T_p satisfying $0 < T_p < T_s$ such that $\|x(0)\|$ is bounded while $\|x(T_p)\| = \infty$. At $t = T_s$, we have

$$\begin{aligned}
{}_0\mathcal{D}_t^{\alpha} x(t) =& \frac{1}{\Gamma(1-\alpha)} \left[\int_0^{T_p} \frac{x^{(1)}(\tau)}{(T_s - \tau)^{\alpha}} \mathrm{d}\tau + \int_{T_p}^{T_s} \frac{x^{(1)}(\tau)}{(T_s - \tau)^{\alpha}} \mathrm{d}\tau \right] \\
=& \frac{1}{\Gamma(2-\alpha)} \left[\frac{x(T_p) - x(0)}{T_p} - \frac{x^{(3)}(s_0) T_p^2}{4!} \right] \big[T_s^{1-\alpha} - (T_s - T_p)^{1-\alpha}\big] \\
&+ \frac{1}{\Gamma(2-\alpha)} \left[\frac{x(T_s) - x(T_p)}{T_s - T_p} - \frac{x^{(3)}(s_0)(T_s - T_p)^2}{4!} \right] (T_s - T_p)^{1-\alpha}.
\end{aligned} \tag{6.28}$$

Combining (6.28) with (6.1) results in

$$\begin{aligned}&\left[\frac{I_n}{\Gamma(2-\alpha)(T_s-T_p)^\alpha}-A\right]x(T_s)\\&=\frac{x(T_p)\left[(T_s-T_p)^{1-\alpha}-T_s^{1-\alpha}\right]}{\Gamma(2-\alpha)T_p}+\frac{x(T_p)}{\Gamma(2-\alpha)(T_s-T_p)^\alpha}+\frac{x(0)\left[T_s^{1-\alpha}-(T_s-T_p)^{1-\alpha}\right]}{\Gamma(2-\alpha)T_p}\\&\quad+Bu(0)+\frac{x^{(3)}(s_0)}{\Gamma(2-\alpha)4!}\left\{(T_s-T_p)^{3-\alpha}+T_p^2\left[T_s^{1-\alpha}-(T_s-T_p)^{1-\alpha}\right]\right\}.\end{aligned}\tag{6.29}$$

From (6.29) and the boundedness of $u(0)$, it can be noticed that $\|x(T_s)\| = \infty$ if $\|x(T_p)\| = \infty$ while $\|x(0)\|$ is finite, which shows a contradiction to (6.19). Thus the assumption $\|x(T_p)\| = \infty$ is false. Similarly for $m \in \mathbb{N}_{\geq 2}$, assume $\|x((m-1)T_s+T_p)\| = \infty$. At $t = mT_s$ it gives

$$\begin{aligned}{}_0\mathcal{D}_t^\alpha x(t)=&\frac{1}{\Gamma(1-\alpha)}\left[\sum_{h=0}^{m-2}\int_{hT_s}^{(h+1)T_s}\frac{x^{(1)}(\tau)}{(mT_s-\tau)^\alpha}\mathrm{d}\tau+\int_{(m-1)T_s}^{(m-1)T_s+T_p}\frac{x^{(1)}(\tau)}{(mT_s-\tau)^\alpha}\mathrm{d}\tau\right.\\&\left.+\int_{(m-1)T_s+T_p}^{mT_s}\frac{x^{(1)}(\tau)}{(mT_s-\tau)^\alpha}\mathrm{d}\tau\right]\\=&\frac{1}{\Gamma(2-\alpha)}\left\{\sum_{h=0}^{m-2}\left[\frac{x((h+1)T_s)-x(hT_s)}{T_s^\alpha}-\frac{x^{(3)}(s_h)T_s^{3-\alpha}}{4!}\right]\right.\\&\left[(m-h)^{1-\alpha}-(m-h-1)^{1-\alpha}\right]+\left[T_s^{1-\alpha}-(T_s-T_p)^{1-\alpha}\right]\\&\left[\frac{x((m-1)T_s+T_p)-x((m-1)T_s)}{T_p}-\frac{x^{(3)}(s_{m-1})T_p^2}{4!}\right]\\&\left.+\left[\frac{x(mT_s)-x((m-1)T_s+T_p)}{T_s-T_p}-\frac{x^{(3)}(s_{m-1})(T_s-T_p)^2}{4!}\right](T_s-T_p)^{1-\alpha}\right\},\end{aligned}\tag{6.30}$$

then together with (6.1) we can get

$$\begin{aligned}&\left[\frac{I_n}{\Gamma(2-\alpha)(T_s-T_p)^\alpha}-A\right]x(mT_s)\\&=\frac{x((m-1)T_s+T_p)}{\Gamma(2-\alpha)}\left[\frac{1}{(T_s-T_p)^\alpha}-\frac{T_s^{1-\alpha}-(T_s-T_p)^{1-\alpha}}{T_p}\right]+\frac{x((m-1)T_s)}{\Gamma(2-\alpha)T_p}\\&\left[T_s^{1-\alpha}-(T_s-T_p)^{1-\alpha}\right]-\sum_{h=0}^{m-2}\frac{x((h+1)T_s)-x(hT_s)}{\Gamma(2-\alpha)T_s^\alpha}\left[(m-h)^{1-\alpha}-(m\right.\\&\left.-h-1)^{1-\alpha}\right]+\frac{x^{(3)}(s_{m-1})}{\Gamma(2-\alpha)4!}\left\{(T_s-T_p)^{3-\alpha}+T_p^2\left[T_s^{1-\alpha}-(T_s-T_p)^{1-\alpha}\right]\right\}\\&+\sum_{h=0}^{m-2}\frac{x^{(3)}(s_h)T_s^{3-\alpha}}{\Gamma(2-\alpha)4!}\left[(m-h)^{1-\alpha}-(m-h-1)^{1-\alpha}\right]+Bu((m-1)T_s).\end{aligned}\tag{6.31}$$

It can be observed from the boundedness of control inputs and (6.31) that $\|x(mT_s)\|$ will be infinite with infinite $\|x((m-1)T_s+T_p)\|$, which contradicts (6.20). Consequently, such a $T_p \in (0, T_s)$ does not exist and thus the boundedness of system states at other times than sampling instants is also guaranteed. As conclusion, the boundedness of all closed-loop signals is proved. In addition, based on condition (6.18) and from (6.19) and (6.20),

$\lim_{m\to\infty}\|x(mT_s)\| = 0$, i.e., $\lim_{t\to\infty}\|x(t)\| = 0$. Thus the asymptotic stability of the closed-loop system is ensured. ■

6.4 Simulation Studies

Under the proposed control scheme based on periodic sampling, the implementation of the simulations in this section is realized on the basis of (6.6).

6.4.1 Demonstrative Example: SISO System

In this subsection, a example of a SISO fractional-order system is first to be presented to illustrate the proposed sampled-data control scheme, verify the theoretical results, and also examine the sensitivity of the designed control system to parametric perturbations.

Consider the linear SISO fractional-order system given below

$$ {}_0\mathcal{D}_t^{0.7}x(t) = 2x(t) + 2u(t) \tag{6.32}$$

with initial value $x(0) = 3.5$. In accordance with Lemma 1.5 we can notice that when no control signal is applied to (6.32), the solution of this fractional-order system is:

$$x(t) = x(0)\mathrm{E}_{0.7}(2t^{0.7}), \tag{6.33}$$

which indicates that the open-loop system is unstable.

Based on condition (6.18), we can choose control parameter $\bar{K} = 2.1$ and sampling period $T_s = 0.05$ s. In fact,

$$\beta = 0.6889 < 1. \tag{6.34}$$

The corresponding control input is displayed in Fig. 6.1, which shows that the control signal is bounded. Besides, as can be noticed from Fig. 6.2, the closed-loop system is asymptotically stabilized under our proposed controller with $\lim_{t\to\infty}\|x(t)\| = 0$. Note that the scheme in Section 5.3 cannot be applied to system (6.32) as it is unstable.

As confirmed in this simulation study and others to be presented in the following subsections, the resulting closed-loop system under our proposed control scheme is ensured to be asymptotically stable as the designed parameters and system parameters satisfy condition (6.18). Nevertheless, the system parameters normally cannot be accurately acquired in practice, which means that perturbations $\triangle A$ and $\triangle B$ may exist in A and B respectively, yet such uncertainties are unknown when designing control parameters $\bar{K}$ and T_s on the basis of (6.18).

To verify that the proposed control scheme is robust to a certain extent of the uncertainty of system parameters, sensitivity analysis is conducted by accordingly adding $\triangle A$ and $\triangle B$ to system parameters $A = 2$ and $B = 2$ in (6.32). As can be seen from Fig. 6.3 that for $\triangle A \leq 2.19$ (where B is fixed as 2), the designed sampled-data controller with $\bar{K} = 2.1$ and $T_s = 0.05$ s can still achieve the asymptotic stability result stated in Theorem 6.1 without significantly increasing the control cost. In particular, for the worst case when $\triangle A = 2.19$, condition (6.18) is still satisfied with

$$\beta = 0.9980 < 1. \tag{6.35}$$

For system parameter B, it can be noticed from Fig. 6.4 and Fig. 6.5 that under fixed $A = 2$, the designed controller can handle perturbation $\triangle B$ within $[-1.04, 4.65]$. Moreover,

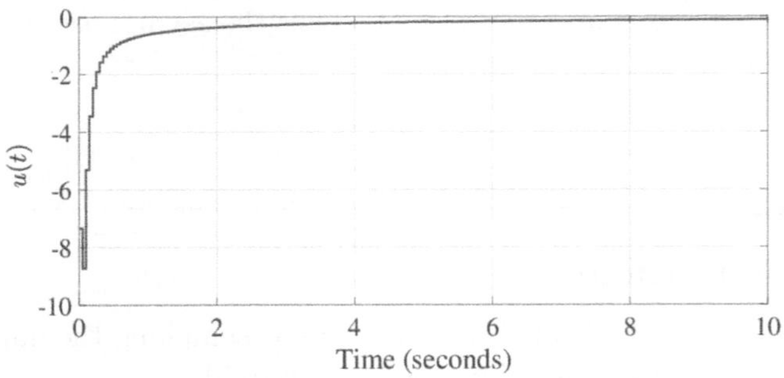

FIGURE 6.1: The control input of (6.32).

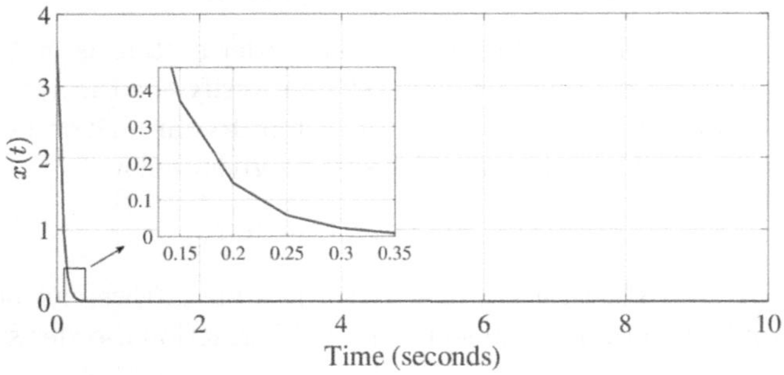

FIGURE 6.2: The system state x of (6.32) under the proposed control scheme.

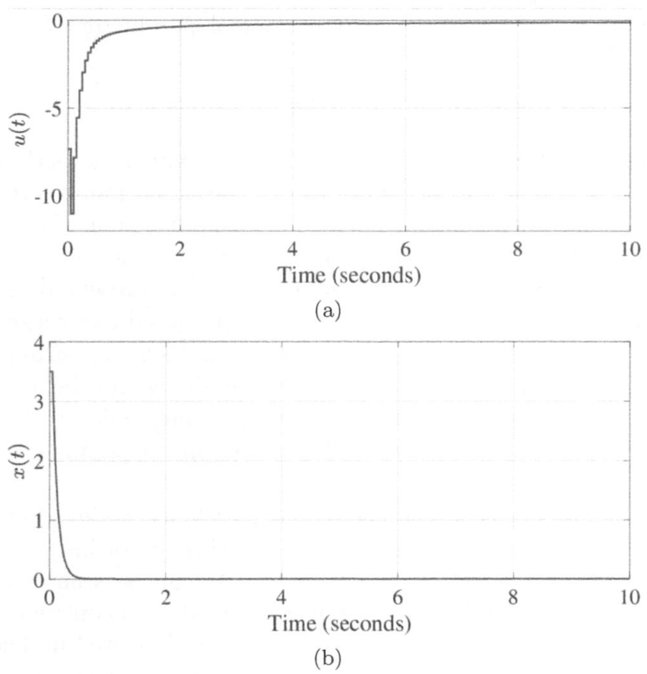

FIGURE 6.3: System response of (6.32) under the proposed control scheme with $\triangle A = 2.19$. (a) Control input. (b) System state.

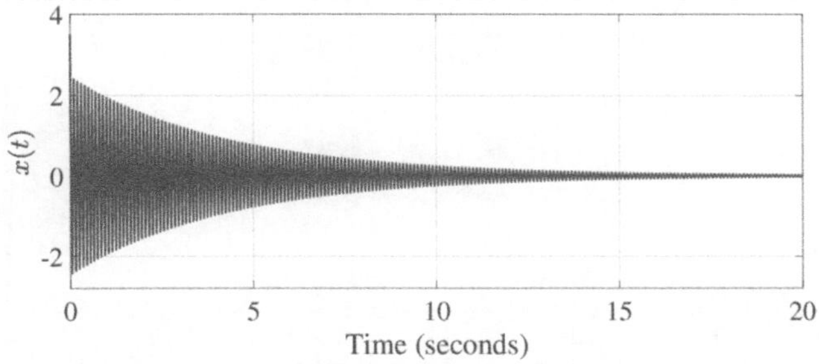

FIGURE 6.4: The system state x of (6.32) under the proposed control scheme with $\triangle B = 4.65$.

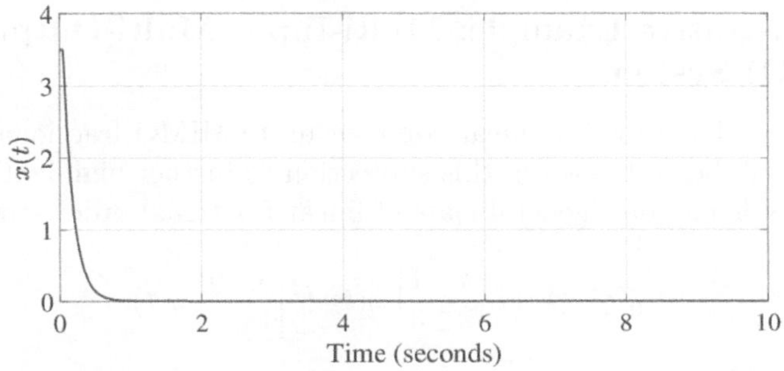

FIGURE 6.5: The system state x of (6.32) under the proposed control scheme with $\triangle B = -1.04$.

the system response with $\triangle A = 2$ and $\triangle B = 3.5$ is displayed in Fig. 6.6, from which it can be observed that the asymptotic stability of the closed-loop system is still guaranteed under the investigated control scheme. These results show that the proposed control scheme has a certain degree of robustness (or insensitivity) to parametric perturbations in the sense of ensuring the result in Theorem 6.1, even though the magnitudes of the perturbations exceed 50 % of the nominal parameters used in the controller design.

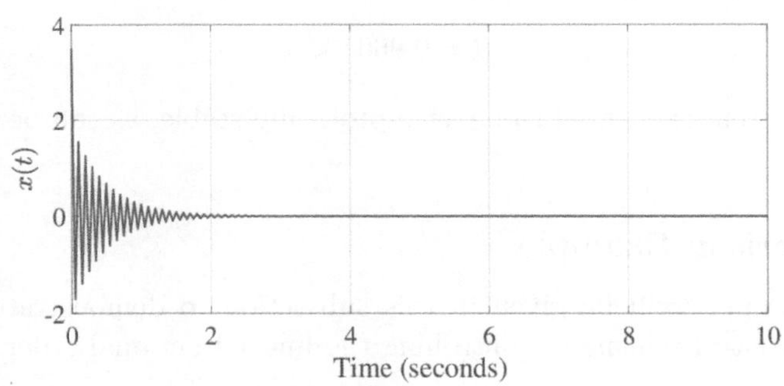

FIGURE 6.6: The system state x of (6.32) under the proposed control scheme with $\triangle A = 2$ and $\triangle B = 3.5$.

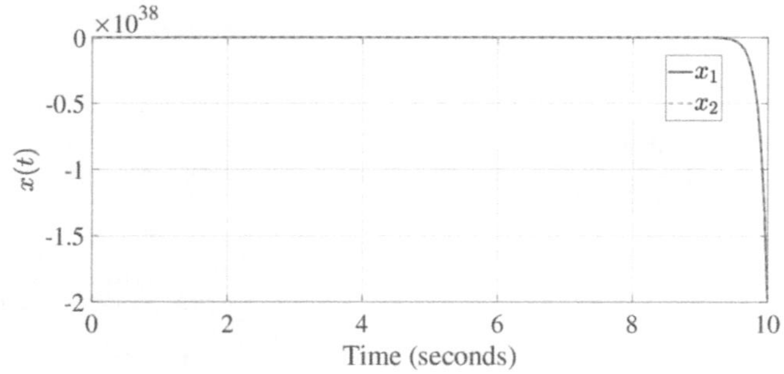

FIGURE 6.7: The open-loop system states of (6.36).

6.4.2 Demonstrative Example: Multi-Input-Multi-Output (MIMO) System

The stabilization problem for the linear commensurate MIMO fractional-order system described in (6.36) will be addressed in this subsection to further highlight the efficacy of the proposed control scheme for a general class of linear fractional-order systems.

$$ {}_0\mathcal{D}_t^{\alpha} x(t) = \begin{bmatrix} 4 & 1 \\ 2 & 3 \end{bmatrix} x(t) + \begin{bmatrix} 3 & 0 \\ 1 & 2 \end{bmatrix} u(t), \tag{6.36} $$

where $\alpha = 0.8$, $x(0) = [-3, 0.5]^T$, and states x_1 and x_2 are the outputs of the system. It can be noticed from (6.36) that Assumption 6.1 is satisfied with $rank\left(\begin{bmatrix} 3 & 0 \\ 1 & 2 \end{bmatrix}\right) = 2$. The solution of (6.36) without control signal is expressed as follows

$$ x(t) = \begin{bmatrix} \frac{\mathrm{E}_\alpha(2t^\alpha)}{3} + \frac{2\mathrm{E}_\alpha(5t^\alpha)}{3} & \frac{\mathrm{E}_\alpha(5t^\alpha)}{3} - \frac{\mathrm{E}_\alpha(2t^\alpha)}{3} \\ \frac{2\mathrm{E}_\alpha(5t^\alpha)}{3} - \frac{2\mathrm{E}_\alpha(2t^\alpha)}{3} & \frac{2\mathrm{E}_\alpha(2t^\alpha)}{3} + \frac{\mathrm{E}_\alpha(5t^\alpha)}{3} \end{bmatrix} x(0), \tag{6.37} $$

from which it can be observed that the open-loop system is unstable, as shown in Fig. 6.7. By designing the control matrix $\bar{K} = \begin{bmatrix} 2.5 & 1 \\ 1 & 2 \end{bmatrix}$ and sampling period $T_s = 0.05$ s that satisfy condition (6.18) as presented below

$$ \beta = 0.9665 < 1, \tag{6.38} $$

the closed-loop system is ensured to be asymptotically stable, as can be seen from Fig. 6.8 to Fig. 6.10.

6.4.3 Comparison Example

A comparison example will be given in this subsection to demonstrate the effectiveness of the proposed control scheme in controlling the linear fractional-order system studied in Section 5.3

$$ {}_0\mathcal{D}_t^{\alpha} x(t) = \frac{1}{C_\alpha} u(t) \tag{6.39} $$

with $x(t)$ and $u(t)$ respectively being the voltage and current of a fractional-order capacitor [203, 39, 205], capacitance $C_\alpha = 1$ F/(second)$^{1-\alpha}$, $\alpha = 0.5$, and the initial value of voltage $x(0) = 2.5$ V. From (6.39) it can be known that it is a special case of the considered class of systems described in (6.1) with $A = 0$. Fig. 6.11 displays a discharging circuit of

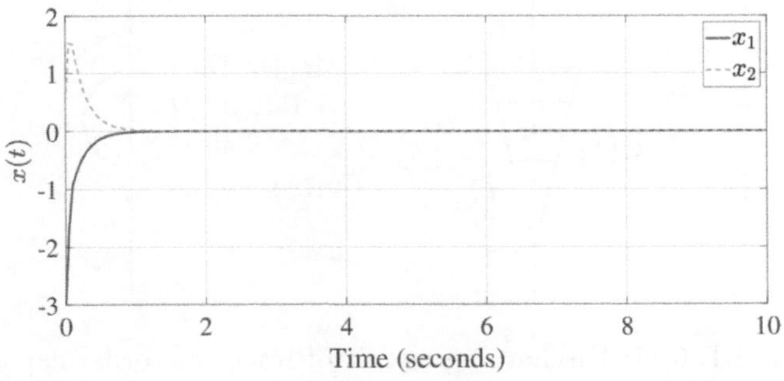

FIGURE 6.8: The system states of (6.36) under the proposed control scheme.

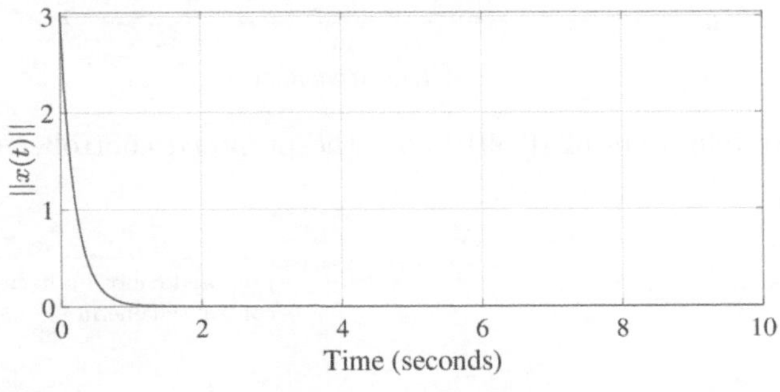

FIGURE 6.9: The value of $\|x(t)\|$ of (6.36) under the proposed control scheme.

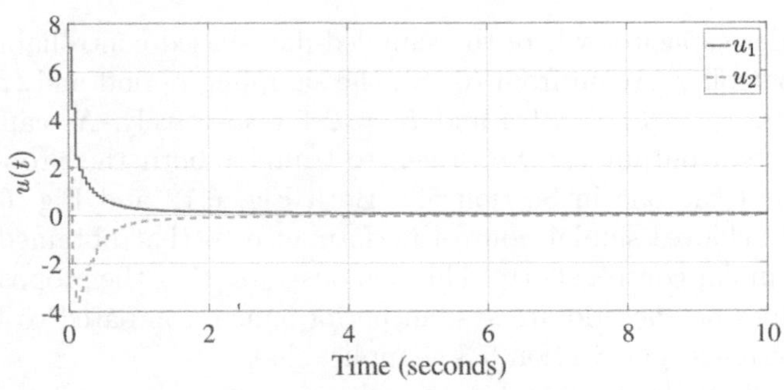

FIGURE 6.10: The control inputs of (6.36).

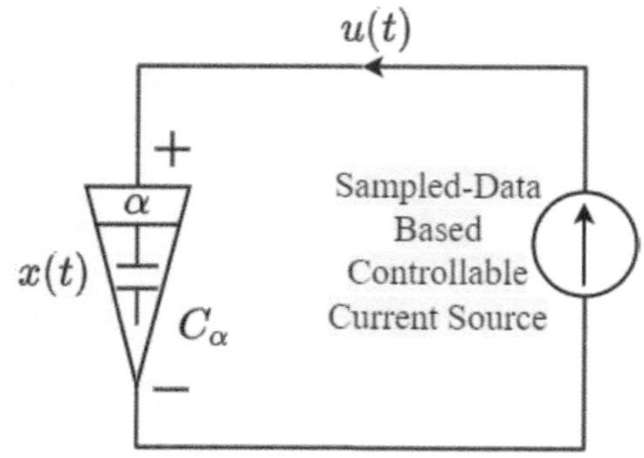

FIGURE 6.11: Discharging circuit of fractional-order capacitor.

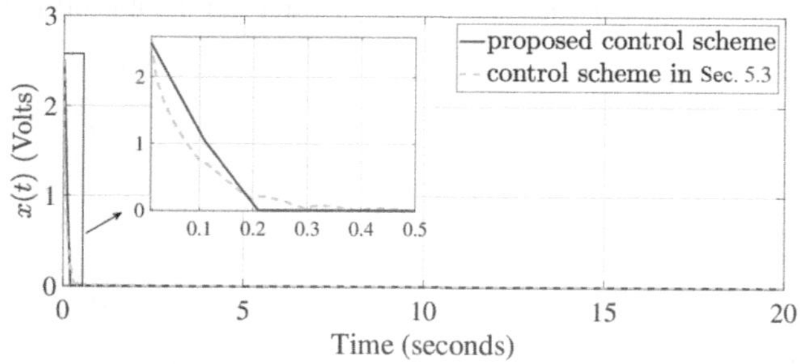

FIGURE 6.12: System state of (6.39) under the proposed controller and the one in Section 5.3.

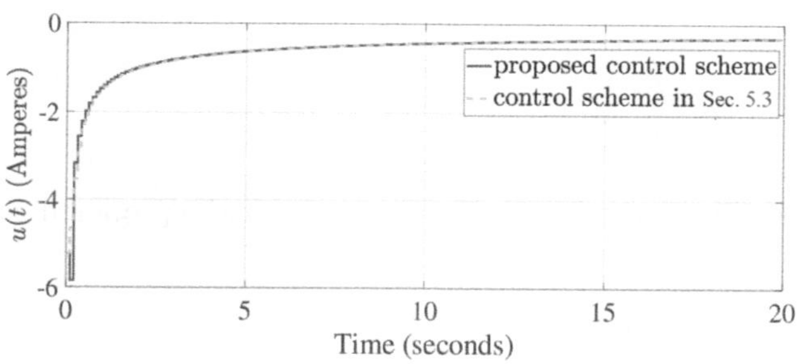

FIGURE 6.13: The proposed controller and controller in Section 5.3 of (6.39).

the fractional-order capacitor where the sampled-data-based controllable current source is treated as the controller. Again from (6.18), the sampling period and the designed control parameter are chosen as $T_s = 0.1$ s and $\bar{K} = 2.1$ respectively. As can be observed from Fig. 6.12, the system output can be driven to 0 under both the proposed sampled-data control scheme and the one in Section 5.3. Both Fig. 6.12 and Fig. 6.13 show that the proposed control achieves similar control performance to that obtained by the scheme in Section 5.3 with similar control efforts. This demonstrates that the proposed control strategy in this chapter does not deteriorate system performance, compared to that in Section 5.3 for the class of systems that Section 5.3 is applicable to.

Therefore based on these examples, the scheme in this chapter is more universal since the class of systems considered in Section 5.3 is only a subclass, while the same control objective as in Section 5.3 for this subclass can still be achieved under our proposed control input in this chapter without extra control costs.

Remark 6.5 *Fig. 6.1, Fig. 6.10, and Fig. 6.13 highlight that the bound of* $\|u(hT_s)\|$ $(h \in \mathbb{N}_{>2})$ *shown in (6.27) is generally conservative since from these figures it can be noticed that the upper bounds of the control signals are in practice much smaller. Therefore, reducing the conservativeness of the upper bound of the closed-loop signals will be an interesting and practically meaningful future work. In addition, even though according to the designed control law (6.9), (6.13), and (6.16) that the computation of the control input* $u(mT_s)$ *involves summing of series having* m *terms, due to the fast convergence of the system states, as can be seen from Fig. 6.2, Fig. 6.8, and Fig. 6.12, the computational burden of computing the control signals will not increase dramatically over the duration of the control application. Notwithstanding, to achieve the trade-off between getting better system performance and having less computational burden, future research on finite-dimensional sampled-data control for a general class of fractional-order linear systems is required.*

6.5 Conclusions

In this chapter, a novel sampled-data control strategy is presented for a general class of linear fractional-order systems. Criteria for asymptotic stability of the corresponding closed-loop system are derived for choosing design parameters and the established theoretical results are verified by simulation studies.

7

Finite-Dimensional Control for Linear Discrete-Time Fractional-Order Systems under DoS Attacks

In this chapter, the security control problem of discrete-time fractional-order networked systems under DoS attacks is considered. A practically applicable finite-dimensional control strategy will be developed for fractional-order systems that possess nonlocal characteristics. By employing the Lyapunov method, it is theoretically proved that under the proposed controller, the obtained closed-loop fractional system is globally ISS, even in the presence of DoS attacks. Finally, the effectiveness of the designed control method is demonstrated by the numerical example.

7.1 Introduction

The research and development of discrete-time fractional-order systems and their associated controls have recently been garnering increased interest, as described in the fourth paragraph of Subsection 1.3.2.

Attacks on communication links in a networked control system can be divided into two categories: DoS attacks and deception attacks. An interruption of communication between networks leads to a DoS attack [212], whereas deception attacks usually involve altering the data that are sent [213]. This chapter focuses on DoS attacks. As stated in the last paragraph of Subsection 1.3.2, nowadays, some controllers have been created to mitigate the impact of DoS attacks on integer-order systems. Despite this, there are a limited number of published security control studies for fractional-order systems in the literature, as mentioned in the fifth paragraph of Subsection 1.3.2. As of yet, the topic of DoS attacks and their effects on discrete-time fractional-order networked systems has not been explored in depth, which provides the impetus for this chapter.

In spite of the aforementioned discussion, in this chapter, we analyze the discrete-time fractional-order systems in which the plant and controller are connected via the network, while the attacker attempts to disrupt the control system's stability by hindering communication between the sensors and controller (measurement channel). The main contributions of this chapter are outlined in the following:

1. The development of a safety control protocol for discrete-time fractional-order systems subject to external disturbance and DoS attacks is investigated in this chapter, with the unique properties of fractional-order calculus being taken into account.

DOI: 10.1201/9781003620372-7

2. The controller proposed is finite-dimensional, which makes it possible to calculate the control input with only a limited number of prior system states, making it suitable for practical use.
3. A sufficient condition is provided to guarantee the global stability of the closed-loop system, resulting in the system output eventually settling at an ultimate bound around the origin.

The chapter is organized as follows: Section 7.2 presents the problem statement. The controller design procedure is given in Section 7.3. In Section 7.4, the proposed control strategy will be examined by simulation example, and finally, Section 7.5 provides the conclusions of this chapter.

7.2 Problem Formulation

Consider a discrete-time linear fractional system described as follows:

$$\begin{cases} {}_0\Delta_{k+1}^{\alpha} z(k+1) = Az(k) + Bu(k) + B_\omega \omega(k) \\ y(k) = C_2 z(k),\ z(0) = z_0 \end{cases} \tag{7.1}$$

where $z(k) \in \mathbb{R}^n$ is the state vector at time step $k \in \mathbb{N}_0$, ${}_0\Delta_{k+1}^{\alpha} z(k+1) = \left[{}_0\Delta_{k+1}^{\alpha_1} z_1(k+1), \cdots, {}_0\Delta_{k+1}^{\alpha_n} z_n(k+1)\right]^T$, $u(k) \in \mathbb{R}^m$ is the control input, $y(k) \in \mathbb{R}^p$ is the measurement output, $\omega(k) \in \mathbb{R}^r$ is the exogenous disturbance signal bounded as $\|\omega(k)\| \le q_\omega$ with $q_\omega > 0$ and A, B, B_ω, and C_2 are known real matrices with appropriate dimensions. In accordance with Remark 1.2 in Section 1.2, the state, control signal, and disturbance before the initial time are considered to be all equal to zero in this chapter, i.e., $z(q) = 0$, $u(q) = 0$ and $\omega(q) = 0$, $\forall q < 0$.

Remark 7.1 *For discrete-time fractional systems, several existing works [142, 143, 144, 145, 146, 147, 148] focus on the control problem. Nevertheless, none of these present studies concern the influence of attacks when designing control schemes, which highlights the advantages of this chapter where a safety control strategy is proposed for discrete-time fractional systems.*

According to [214], the GL fractional-order difference of a discrete-variable bounded function $f(k) : \mathbb{N}_0 \to \mathbb{R}$ is defined as

$$ {}_0\Delta_k^{\alpha} f(k) = \sum_{j=0}^{k} (-1)^j \binom{\alpha}{j} f(k-j),\ \alpha \in \mathbb{R}_0,\ k \in \mathbb{N}_0, \tag{7.2}$$

where

$$\binom{\alpha}{j} = \begin{cases} 1 & j = 0 \\ \frac{\alpha(\alpha-1)\cdots(\alpha-j+1)}{j!} & j > 0 \end{cases} \text{for all } j \in \mathbb{N}_0. \tag{7.3}$$

Define $c_j(\alpha) = (-1)^j \binom{\alpha}{j}$, and due to the fact that $|c_j(\alpha)| \le \frac{\alpha^j}{j!}$, for any $\alpha \in \mathbb{R}_0$, the sequence $\{c_j(\alpha)\}_{j \in \mathbb{N}_0}$ is absolutely summable.

For the ith state ($i = 1, 2, \cdots, n$), we have

$$ {}_0\Delta^{\alpha_i}_{k+1} z_i(k+1) = \sum_{j=0}^{k+1} c_j(\alpha_i) z_i(k+1-j), \tag{7.4} $$

in which α_i is the fractional order corresponding to z_i. Equation (7.4) can be rewritten as

$$ z_i(k+1) = {}_0\Delta^{\alpha_i}_{k+1} z_i(k+1) - \sum_{j=1}^{k+1} c_j(\alpha_i) z_i(k+1-j), \tag{7.5} $$

and hence, the evolution of $z(k+1)$ can be expressed as

$$ z(k+1) = Az(k) + Bu(k) + B_\omega \omega(k) - \sum_{j=1}^{k+1} F_j(\alpha) z(k+1-j) \tag{7.6} $$

where $F_j(\alpha) = \mathrm{diag}\big(c_j(\alpha_1),\, c_j(\alpha_2), \cdots, c_j(\alpha_n)\big) \in \mathbb{R}^n$. Alternatively, (7.6) can be presented as

$$ z(k+1) = \sum_{j=0}^{k} A_j z(k-j) + Bu(k) + B_\omega \omega(k) \tag{7.7} $$

in which $A_0 = A - F_1(\alpha)$ and $A_j = -F_{j+1}(\alpha)$ for $j \geq 1$. As a result, the linear discrete-time fractional-order system (7.1) can be described as

$$ \begin{cases} z(k+1) = \displaystyle\sum_{j=0}^{k} A_j z(k-j) + Bu(k) + B_\omega \omega(k) \\ y(k) = C_2 z(k), \ z(0) = z_0. \end{cases} \tag{7.8} $$

In this chapter, the identity function is denoted by Id, i.e., $Id(z) = z$. $\|z\|_{\text{-}\infty} = \sup\limits_{j\in\mathbb{Z}} \|z(j)\|$, and $\|z\|_{\text{-}[j_1,\, j_2]}$ is defined as:

$$ \|z\|_{\text{-}[j_1,\, j_2]} = \begin{cases} \max\limits_{j_1 \leq j \leq j_2} \|z(j)\|, & \text{if } j_1 \leq j_2 \\ 0, & \text{if } j_1 > j_2. \end{cases} $$

It can be seen from the expression of the GL fractional-order difference of function $f(k)$ given in (7.2) that the cumulative term used to implement the fractional-order difference will increase as k increases. Different from the controller proposed in [215], which is a linear weighted combination of all the past states of the observer, which will consequently result in the computational explosion as time goes by, a finite-dimensional controller that contains finite steps of recent states is considered in this chapter.

Reformulate (7.8) as

$$ \begin{cases} \hat{z}(k+1) = \hat{A}_L \hat{z}(k) + \hat{B}u(k) + \hat{B}_\omega \hat{\omega}(k) \\ y(k) = \hat{C}_2 \hat{z}(k), \end{cases} \tag{7.9} $$

where $L \in \mathbb{N}_0$, $\hat{z}(k) = \big[z^{\mathrm{T}}(k),\, z^{\mathrm{T}}(k-1), \cdots, z^{\mathrm{T}}(k-L)\big]^{\mathrm{T}} \in \mathbb{R}^{(L+1)n}$ is the recent-finite-steps state vector, $\hat{\omega}(k) = \big[\omega^{\mathrm{T}}(k),\, z^{\mathrm{T}}(k-L-1), \cdots, z^{\mathrm{T}}(0)\big]^{\mathrm{T}} \in \mathbb{R}^{r+(k-L)n}$, and $\hat{A}_L$, $\hat{B}$, $\hat{B}_\omega$

and $\hat{C}_2$ are defined as

$$\hat{A}_L = \begin{bmatrix} A_0 & A_1 & \cdots & A_{L-1} & A_L \\ I_n & 0 & \cdots & 0 & 0 \\ \vdots & \vdots & \ddots & \vdots & \vdots \\ 0 & 0 & \cdots & I_n & 0 \end{bmatrix}, \hat{B} = \begin{bmatrix} B \\ 0 \\ \vdots \\ 0 \end{bmatrix},$$

$$\hat{B}_\omega = \begin{bmatrix} B_\omega & A_{L+1} & \cdots & A_{k-1} & A_k \\ 0 & 0 & \cdots & 0 & 0 \\ \vdots & \vdots & \ddots & \vdots & \vdots \\ 0 & 0 & \cdots & 0 & 0 \end{bmatrix}, \hat{C}_2^T = \begin{bmatrix} C_2 \\ 0 \\ \vdots \\ 0 \end{bmatrix}.$$

Assumption 7.1 *The pair $(\hat{A}_L, \hat{B})$ is stabilizable.*

Assumption 7.2 *No DoS attacks occur at initial time $k = 0$.*

Remark 7.2 *Assumption 7.1 is needed for the existence of the positive matrices P and Q that will be applied during the controller design in Section 7.3. As will be mentioned in Section 7.3, under the influence of DoS attacks, in this chapter, the latest received state will be utilized for control until receiving the next successfully transmitted state, and thus, Assumption 7.2 is required for avoiding the whole control system becoming open-loop at the beginning of the control stage.*

The openness of network communication in networked control makes it susceptible to cyberattacks. This chapter investigates the effects of aperiodic DoS attacks on the measurement channel.

Suppose that a DoS attack occurs in the measurement channel at the instant $a_i \in [k - L, k)$, where i denotes the ith attack event. The control objective of this chapter is to design a finite-dimensional controller that the corresponding closed-loop fractional-order system (7.1) is globally ISS under DoS attacks, i.e., for arbitrary initial condition $z_0 \in \mathbb{R}^n$, the closed-loop state satisfies

$$\|z(k)\| \leq \beta_a(\|z_0\|, k) + \gamma_a(q_\omega), \; k \in \mathbb{N}_0, \tag{7.10}$$

where function β_a is a $\mathcal{KL}$-function and γ_a is a $\mathcal{K}$-function. Detailed definitions of the $\mathcal{KL}$-function, $\mathcal{K}_\infty$-function, and $\mathcal{K}$-function can be found in Section 5.3.

7.3 Controller Design

Define the difference vector as

$$\hat{e}(k) = \hat{z}(k)_a - \hat{z}(k) \tag{7.11}$$

where $\hat{z}(k)_a$ stands for the recent-finite-steps state vector under DoS attacks that is utilized for controller design. In this chapter, if a DoS attack occurs at the current time instant, or in other words, no information is available for the controller currently, then the latest received state will be used for controlling until the next successful state information transmission. The controller is designed as

$$u(k) = K_L \hat{z}(k)_a \tag{7.12}$$

with control matrix $K_L \in \mathbb{R}^{m \times (L+1)n}$.

Theorem 7.1 *Consider the closed-loop discrete-time fractional-order system consisting of the system (7.1) and controller (7.12). Under Assumption 7.1 and Assumption 7.2, for* $L \in \mathbb{N}_0$ *that satisfies* $0 < h_\Psi \max_i \phi_{\alpha_i}(L)_k < 1\,(i = 1, \cdots, n)$, *where* h_Ψ *and* $\phi_{\alpha_i}(L)_k$ *will be defined later, then the obtained closed-loop fractional system under DoS attacks is ensured to be globally ISS, with (7.10) being satisfied.*

Proof: With the controller designed in (7.12), the evolution of the recent-finite-steps state in the closed-loop system can be expressed as

$$\begin{aligned}\hat{z}(k+1) =&\hat{A}_L\hat{z}(k) + \hat{B}K_L\hat{z}(k)_a + \hat{B}_\omega\hat{\omega}(k)_a\\ =&\hat{A}_L\hat{z}(k) + \hat{B}K_L\hat{z}(k) + \hat{B}K_L\hat{e}(k) + \bar{\omega}(k)_a\\ =&\Phi\hat{z}(k) + \hat{B}K_L\hat{e}(k) + \bar{\omega}(k)_a\end{aligned} \tag{7.13}$$

where $\hat{\omega}(k)_a$ is the $\hat{\omega}(k)$ under DoS attacks, $\bar{\omega}(k)_a = \hat{B}_\omega\hat{\omega}(k)_a$ and $\Phi = \hat{A}_L + \hat{B}K_L$. According to Assumption 7.1, for any positive definite matrix Q, $\Phi^{\mathrm{T}}P\Phi - P + Q = 0$ holds for a positive definite matrix P.

Let the Lyapunov function be $W(k) = \hat{z}^{\mathrm{T}}(k)P\hat{z}(k)$, then

$$\begin{aligned}W(k+1) =&\hat{z}^{\mathrm{T}}(k+1)P\hat{z}(k+1)\\ =&\hat{z}^{\mathrm{T}}(k)(P-Q)\hat{z}(k) + 2\hat{z}^{\mathrm{T}}(k)\Phi^{\mathrm{T}}P\hat{B}K_L\hat{e}(k) + 2\hat{z}^{\mathrm{T}}(k)\Phi^{\mathrm{T}}P\bar{\omega}(k)_a\\ &+ \hat{e}^{\mathrm{T}}(k){K_L}^{\mathrm{T}}\hat{B}^{\mathrm{T}}P\hat{B}K_L\hat{e}(k) + 2\hat{e}^{\mathrm{T}}(k){K_L}^{\mathrm{T}}\hat{B}^{\mathrm{T}}P\bar{\omega}(k)_a + \bar{\omega}(k)_a^{\mathrm{T}}P\bar{\omega}(k)_a\\ =&W(k) - \hat{z}^{\mathrm{T}}(k)Q\hat{z}(k) + 2\hat{z}^{\mathrm{T}}(k)\Phi^{\mathrm{T}}P\hat{B}k_L\hat{e}(k) + 2\hat{z}^{\mathrm{T}}(k)\Phi^{\mathrm{T}}P\bar{\omega}(k)_a\\ &+ \hat{e}^{\mathrm{T}}(k){K_L}^{\mathrm{T}}\hat{B}^{\mathrm{T}}P\hat{B}K_L\hat{e}(k) + 2\hat{e}^{\mathrm{T}}(k){K_L}^{\mathrm{T}}\hat{B}^{\mathrm{T}}P\bar{\omega}(k)_a + \bar{\omega}(k)_a^{\mathrm{T}}P\bar{\omega}(k)_a.\end{aligned} \tag{7.14}$$

From (7.14), we obtain

$$\begin{aligned}W(k+1) - W(k) \leq& -\lambda_{\min}(Q)\|\hat{z}(k)\|^2 + \|2\Phi^{\mathrm{T}}P\hat{B}K_L\|\|\hat{z}(k)\|\|\hat{e}(k)\|\\ &+ \|2\Phi^{\mathrm{T}}P\|\|\hat{z}(k)\|\|\bar{\omega}(k)_a\| + [{K_L}^{\mathrm{T}}\hat{B}^{\mathrm{T}}P\hat{B}K_L]\|\hat{e}(k)\|^2\\ &+ \|2P\hat{B}K_L\|\|\hat{e}(k)\|\|\bar{\omega}(k)_a\| + P\|\bar{\omega}(k)_a\|^2.\end{aligned} \tag{7.15}$$

The following inequality should be guaranteed to ensure closed-loop system stability:

$$\|\hat{e}(k)\| \leq \sigma\|\hat{z}(k)\| + \sigma\|\bar{\omega}(k)_a\|, \tag{7.16}$$

where $\sigma \in \mathbb{R}_0$ is a suitable designed parameter. Substituting (7.16) into (7.15), we have

$$\begin{aligned}W(k+1) - W(k) \leq& -\lambda_{\min}(Q)\|\hat{z}(k)\|^2 + \|2\Phi^{\mathrm{T}}P\hat{B}K_L\|\sigma\|\hat{z}(k)\|^2 + \|2\Phi^{\mathrm{T}}P\|\|\hat{z}(k)\|\|\bar{\omega}(k)_a\|\\ &+ \|2\Phi^{\mathrm{T}}P\hat{B}K_L\|\sigma\|\hat{z}(k)\|\|\bar{\omega}(k)_a\| + \|2P\hat{B}K_L\|\sigma\|\bar{\omega}(k)_a\|^2\\ &+ [{K_L}^{\mathrm{T}}\hat{B}^{\mathrm{T}}P\hat{B}K_L]\sigma^2\left[\|\hat{z}(k)\|^2 + \|\bar{\omega}(k)_a\|^2 + 2\|\hat{z}(k)\|\|\bar{\omega}(k)_a\|\right]\\ &+ \|2P\hat{B}K_L\|\sigma\|\hat{z}(k)\|\|\bar{\omega}(k)_a\| + P\|\bar{\omega}(k)_a\|^2.\end{aligned} \tag{7.17}$$

Let $\zeta_1 = \lambda_{\min}(Q)$, $\zeta_2 = \|2\Phi^{\mathrm{T}}P\hat{B}K_L\|$, $\zeta_3 = \|2\Phi^{\mathrm{T}}P\|$, $\zeta_4 = \lambda_{\max}({K_L}^{\mathrm{T}}\hat{B}^{\mathrm{T}}P\hat{B}K_L)$, $\zeta_5 = \|2P\hat{B}K_L\|$, $\eta_2 = \lambda_{\max}(P)$, then

$$\begin{aligned}W(k+1) - W(k) \leq& -\zeta_1\|\hat{z}(k)\|^2 + \zeta_2\sigma\|\hat{z}(k)\|^2 + \zeta_2\sigma\|\hat{z}(k)\|\|\bar{\omega}(k)_a\| + \zeta_3\|\hat{z}(k)\|\|\bar{\omega}(k)_a\|\\ &+ \zeta_4\sigma^2\|\hat{z}(k)\|^2 + \zeta_4\sigma^2\|\bar{\omega}(k)_a\|^2 + \zeta_5\sigma\|\hat{z}(k)\|\|\bar{\omega}(k)_a\| + \zeta_5\sigma\|\bar{\omega}(k)_a\|^2\\ &+ \eta_2\|\bar{\omega}(k)_a\|^2 + 2\zeta_4\sigma^2\|\hat{z}(k)\|\|\bar{\omega}(k)_a\|\\ =& -(\zeta_1 - \zeta_2\sigma - \zeta_4\sigma^2)\|\hat{z}(k)\|^2 + (\zeta_2\sigma + \zeta_3 + \zeta_5\sigma + 2\zeta_4\sigma^2)\|\hat{z}(k)\|\|\bar{\omega}(k)_a\|\\ &+ (\zeta_4\sigma^2 + \zeta_5\sigma + \eta_2)\|\bar{\omega}(k)_a\|^2.\end{aligned} \tag{7.18}$$

Choosing σ that satisfies $\zeta_1 - \zeta_2\sigma - \zeta_4\sigma^2 > 0$ and letting $\zeta_6 = \zeta_1 - \zeta_2\sigma - \zeta_4\sigma^2$, $\zeta_7 = \zeta_2\sigma + \zeta_3 + \zeta_5\sigma + 2\zeta_4\sigma^2$, $\zeta_8 = \zeta_4\sigma^2 + \zeta_5\sigma + \eta_2$, $\zeta_9 = \frac{\zeta_7^2}{2\zeta_6} + \zeta_8$, then

$$W(k+1) - W(k) \leq -\zeta_6\|\hat{z}(k)\|^2 + \zeta_7\|\hat{z}(k)\|\|\bar{\omega}(k)_a\| + \zeta_8\|\bar{\omega}(k)_a\|^2. \tag{7.19}$$

Since for any positive real scalar δ, the following inequality holds:

$$\|\hat{z}(k)\|\|\bar{\omega}(k)_a\| \leq \frac{\|\hat{z}(k)\|^2}{2\delta} + \frac{\delta\|\bar{\omega}(k)_a\|^2}{2}, \tag{7.20}$$

thus, (7.19) becomes

$$W(k+1) - W(k) \leq -\zeta_6\|\hat{z}(k)\|^2 + \frac{\zeta_7}{2\delta}\|\hat{z}(k)\|^2 + \frac{\zeta_7\delta}{2}\|\bar{\omega}(k)_a\|^2 + \zeta_8\|\bar{\omega}(k)_a\|^2. \tag{7.21}$$

Let $\delta = \frac{\zeta_7}{\zeta_6}$, then we obtain

$$\begin{aligned} W(k+1) - W(k) \leq & -\zeta_6\|\hat{z}(k)\|^2 + \frac{\zeta_6}{2}\|\hat{z}(k)\|^2 + \frac{\zeta_7^2}{2\zeta_6}\|\bar{\omega}(k)_a\|^2 + \zeta_8\|\bar{\omega}(k)_a\|^2 \\ = & -\frac{\zeta_6}{2}\|\hat{z}(k)\|^2 + \zeta_9\|\bar{\omega}(k)_a\|^2. \end{aligned} \tag{7.22}$$

Define $\mu_1(\|\hat{z}(k)\|) = \lambda_{\min}(P)\|\hat{z}(k)\|^2$, $\mu_2(\|\hat{z}(k)\|) = \lambda_{\max}(P)\|\hat{z}(k)\|^2$, $\mu_3(\|\hat{z}(k)\|) = \frac{\zeta_6}{2}\|\hat{z}(k)\|^2$, $\mu_4(W(k)) = \mu_3 \circ \mu_2^{-1}(W(k)) = \frac{\zeta_6}{2\eta_2}W(k)$, $\mu_5(\|\bar{\omega}(k)_a\|) = \zeta_9\|\bar{\omega}(k)_a\|^2$, therefore we obtain

$$\mu_1(\|\hat{z}(k)\|) \leq W(k) \leq \mu_2(\|\hat{z}(k)\|), \tag{7.23}$$

and

$$\begin{aligned} W(k+1) - W(k) &\leq -\mu_3(\|\hat{z}(k)\|) + \mu_5(\|\bar{\omega}(k)_a\|) \\ &\leq -\mu_4(W(k)) + \mu_5(\|\bar{\omega}(k)_a\|). \end{aligned} \tag{7.24}$$

Given that μ_1, μ_2, μ_3, $\mu_5 \in \mathcal{K}_\infty$, it is evident that the function $W(k)$ is an ISS-Lyapunov function, implying the existence of a $\mathcal{KL}$-function $\beta_a : \mathbb{R}_0 \times \mathbb{R}_0 \to \mathbb{R}_0$ and a $\mathcal{K}$-function $\gamma_a : \mathbb{R}_0 \times \mathbb{R}_0$ such that

$$\|\hat{z}(k_0 + k)\| \leq \beta_a(\|\hat{z}(k_0)\|, k) + \gamma_a(\|\bar{\omega}_a\|_{\text{-}[k_0, k_0+k-1]}), \tag{7.25}$$

where $k_0, k \geq 0$. Let $\hat{\mu}_4(\|\bar{\omega}_a\|)$ be any $\mathcal{K}_\infty$-lower bound of $\mu_4 \in \mathcal{K}_\infty$ such that $Id - \hat{\mu}_4 \in \mathcal{K}$. Hence, we can have $\hat{\mu}_4(\|\bar{\omega}_a\|) = \hat{h}_4\|\bar{\omega}_a\|$, where $\hat{h}_4 = \min(\frac{\zeta_6}{2\eta_2}, \hat{\theta}_a)$ with $\hat{\theta}_a \in (0,1)$. Let $\rho_1(\|\bar{\omega}_a\|) = h_\rho\|\bar{\omega}_a\|$ with $h_\rho \in (0,1)$, then according to [216], (7.25) holds with $\gamma_a = \mu_1^{-1} \circ \hat{\gamma}_a$ where $\hat{\gamma}_a(\|\bar{\omega}_a\|) = \frac{\zeta_9}{\hat{h}_4 h_\rho}\|\bar{\omega}_a\|^2$ and $\beta_a(s,t) = \mu_1^{-1}(\hat{\beta}_a(\mu_2(s)), t)$ for a $\mathcal{KL}$-function $\hat{\beta}_a : \mathbb{R}_0 \times \mathbb{R}_0 \to \mathbb{R}_0$, therefore, (7.25) can be written as

$$\begin{aligned} \|\hat{z}(k_0 + k)\| \leq & \beta_a(\|\hat{z}(k_0)\|, k) + \sqrt{\frac{\zeta_9}{\hat{h}_4 h_\rho \lambda_{\min}(P)}}\|\bar{\omega}_a\|_{\text{-}[k_0, k_0+k-1]} \\ = & \beta_a(\|\hat{z}(k_0)\|, k) + h_\Psi\|\bar{\omega}_a\|_{\text{-}[k_0, k_0+k-1]} \end{aligned} \tag{7.26}$$

with $h_\Psi = \sqrt{\frac{\zeta_9}{\bar{h}_4 h_\rho \lambda_{\min}(P)}}$.

Since $F_{j+1}(\alpha) = \operatorname{diag}\big(c_{j+1}(\alpha_1), c_{j+1}(\alpha_2), \cdots, c_{j+1}(\alpha_n)\big)$, its maximum norm can be obtained as

$$\|F_{j+1}(\alpha)\|_\infty = \max_i |c_{j+1}(\alpha_i)| \le \max_i \frac{\alpha_i^{j+1}}{(j+1)!}. \tag{7.27}$$

For simplicity, the maximum norm of matrix F will be expressed as $\|F\|$. Thus, the following inequality can be derived as

$$\begin{aligned}
\|\bar{\omega}(k)_a\| \le & \|B_\omega\| q_\omega + \sum_{j=L+1}^{k} \|A_j\| \|z(k-j)_a\| \\
\le & \|B_\omega\| q_\omega + \sum_{j=L+1}^{k} \|A_j\| \|z\|_{\text{-}\infty} \\
= & \|B_\omega\| q_\omega + \sum_{j=L+1}^{k} \|F_{j+1}(\alpha)\| \|z\|_{\text{-}\infty} \\
\le & \|B_\omega\| q_\omega + \sum_{j=L+1}^{k} \max_i \frac{\alpha_i^{j+1}}{(j+1)!} \|z\|_{\text{-}\infty} \\
= & \|B_\omega\| q_\omega + \max_i \phi_{\alpha_i}(L)_k \|z\|_{\text{-}\infty}
\end{aligned} \tag{7.28}$$

where $\phi_{\alpha_i}(L)_k = \sum_{j=L+1}^{k} \frac{\alpha_i^{j+1}}{(j+1)!}$. Combining (7.26) with (7.28) results in

$$\|\hat{z}\|_{\text{-}\infty} \le \beta_a(\|\hat{z}(0)\|, 0) + h_\Psi \|B_\omega\| q_\omega + h_\Psi \max_i \phi_{\alpha_i}(L)_k \|z\|_{\text{-}\infty}. \tag{7.29}$$

Due to the fact that $\|z\|_{\text{-}\infty} \le \|\hat{z}\|_{\text{-}\infty}$, from (7.29), we obtain

$$\|\hat{z}\|_{\text{-}\infty} \le \beta_a(\|\hat{z}(0)\|, 0) + h_\Psi \|B_\omega\| q_\omega + h_\Psi \max_i \phi_{\alpha_i}(L)_k \|\hat{z}\|_{\text{-}\infty}. \tag{7.30}$$

Consider $L \in \mathbb{N}_0$ that satisfies $0 < h_\Psi \max_i \phi_{\alpha_i}(L)_k < 1$ and define $d_z = \big[1 - h_\Psi \max_i \phi_{\alpha_i}(L)_k\big]^{-1} \big[\beta_a(\|\hat{z}(0)\|, 0) + h_\Psi \|B_\omega\| q_\omega\big]$, it then further implies $\|\hat{z}\|_{\text{-}\infty} \le d_z$. Hence, (7.28) becomes

$$\|\bar{\omega}(k)_a\| \le \|B_\omega\| q_\omega + \max_i \phi_{\alpha_i}(L)_k d_z. \tag{7.31}$$

Furthermore, (7.26) turns into

$$\begin{aligned}
\|\hat{z}(k)\| \le & \beta_a(\|\hat{z}(0)\|, k) + h_\Psi \big[\|B_\omega\| q_\omega + \max_i \phi_{\alpha_i}(L)_k d_z\big] \\
= & \beta_a(\|\hat{z}(0)\|, k) + p_z
\end{aligned} \tag{7.32}$$

where $p_z = h_\Psi \big[\|B_\omega\| q_\omega + \max_i \phi_{\alpha_i}(L)_k d_z\big]$ and $\beta_a : \mathbb{R}_0 \times \mathbb{R}_0 \to \mathbb{R}_0$ is a $\mathcal{KL}$-function.

For any scalar $M \in \mathbb{N}$, the boundedness inequality of $\bar{\omega}_a$ can be written as

$$\begin{aligned}
\|\bar{\omega}(k)_a\| \leq & \|B_\omega\|q_\omega + \sum_{j=L+1}^{L+1+M} \|A_j\|\|z(k-j)_a\| + \sum_{j=L+2+M}^{k} \|A_j\|\|z(k-j)_a\| \\
\leq & \|B_\omega\|q_\omega + \sum_{j=L+1}^{L+1+M} \|A_j\|\|z\|_{\text{-}[k-L-1-M,\, k-L-1]} + \sum_{j=L+2+M}^{k} \|A_j\|\|z\|_{\text{-}[0,\, k-L-2-M]} \\
\leq & \|B_\omega\|q_\omega + \sum_{j=L+1}^{L+1+M} \|A_j\|\|z\|_{\text{-}[k-L-1-M,\, k-L]} + \max_i \phi_{\alpha_i}(L+1+M)_k d_z \\
\leq & \|B_\omega\|q_\omega + \sum_{j=L+1}^{L+1+M} \max_i \frac{\alpha_i^{j+1}}{(j+1)!}\|z\|_{\text{-}[k-L-1-M,\, k-L]} + \max_i \phi_{\alpha_i}(L+1+M)_k d_z \\
\leq & \|B_\omega\|q_\omega + \max_i \psi_{L\alpha_i}(M)\|z\|_{\text{-}[k-L-1-M,\, k-L]} + \max_i \phi_{\alpha_i}(L+1+M)_k d_z \qquad (7.33)
\end{aligned}$$

where $\psi_{L\alpha_i}(M) = \sum_{j=L+1}^{L+1+M} \frac{\alpha_i^{j+1}}{(j+1)!}$. Consider for constant $p_l \geq 0$ and function $\beta_{al} : \mathbb{R}_0 \times \mathbb{R}_0 \to \mathbb{R}_0$, $l \in \mathbb{N}_0$ such that for any $r,\, s \in \mathbb{R}_0$, $\beta_{al}(r,\, s)$ is bounded and $\beta_{al}(r,\, s) \to 0$ as $s \to \infty$, the following inequality holds:

$$\|\hat{z}(k)\| \leq \beta_{al}(\|\hat{z}(0)\|, k) + p_l. \qquad (7.34)$$

Step 1 ($l = 0$): Let $\beta_{a0}(r,\, s) = \beta_a(r,\, s)$ and $p_0 = p_z$, from (7.32) we then obtain

$$\|\hat{z}(k)\| \leq \beta_{a0}(\|\hat{z}(0)\|, k) + p_0. \qquad (7.35)$$

Step 2 ($l > 0$): For a $v_l \in \mathbb{R}$ that satisfies $v_l \in \left(0,\, \min\{1,\, \frac{\kappa-\chi}{\chi}p_l\}\right)$, where $\kappa \in (\chi,\, 1)$ and $\chi = h_\Psi \max_i \phi_{\alpha_i}(L)_k$, it then indicates $\chi(p_l + v_l) \leq \kappa p_l$. Choosing $v_{l+1} \in (0,\, \kappa v_l)$, there then exists a $k_l = k_l(v_{l+1})$ that for $k \geq k_l$, $\|z\|_{\text{-}[k-L,\, k]} \leq \|\hat{z}(k)\| \leq p_l + \frac{v_{l+1}}{2}$. This further implies that for any $M_l \in \mathbb{N}$, there exists a $k \geq \bar{k}_l = k_l + M_l + 2$ such that $\|z\|_{\text{-}[k-L-1-M_l,\, k-L]} \leq p_l + \frac{v_{l+1}}{2}$ holds. After combining with (7.33), we obtain

$$\|\bar{\omega}(k)_a\| \leq \|B_\omega\|q_\omega + \max_i \psi_{L\alpha_i}(M_l)(p_l + \frac{v_{l+1}}{2}) + \max_i \phi_{\alpha_i}(L+1+M_l)_k d_z, \text{ for } k \geq \bar{k}_l. \qquad (7.36)$$

By selecting M_l that satisfies $h_\Psi \max_i \phi_{\alpha_i}(L+1+M_l)_k d_z \leq \chi\frac{v_{l+1}}{2}$, from (7.26), we obtain

$$\begin{aligned}
\|\hat{z}(\bar{k}_l + k)\| \leq & \beta_a(\|\hat{z}(\bar{k}_l)\|,\, k) + h_\Psi\Big[\max_i \psi_{L\alpha_i}(M_l)(p_l + \frac{v_{l+1}}{2}) \\
& + \|B_\omega\|q_\omega + \max_i \phi_{\alpha_i}(L+1+M_l)_k d_z\Big]. \qquad (7.37)
\end{aligned}$$

Since for $L,\, M_l \in \mathbb{N}_0$, $\psi_{L\alpha_i}(M_l) \leq \phi_{\alpha_i}(L)_k$, hence,

$$\begin{aligned}
\|\hat{z}(\bar{k}_l + k)\| \leq & \beta_a(\|\hat{z}(\bar{k}_l)\|,\, k) + h_\Psi\|B_\omega\|q_\omega + h_\Psi \max_i \phi_{\alpha_i}(L)_k(p_l + \frac{v_{l+1}}{2}) + \chi\frac{v_{l+1}}{2} \\
\leq & \beta_a(\|\hat{z}(\bar{k}_l)\|,\, k) + h_\Psi\|B_\omega\|q_\omega + \chi(p_l + \frac{v_{l+1}}{2}) + \chi\frac{v_{l+1}}{2} \\
\leq & \beta_a(\|\hat{z}(\bar{k}_l)\|,\, k) + h_\Psi\|B_\omega\|q_\omega + \chi(p_l + v_l) \\
\leq & \beta_a(\|\hat{z}(\bar{k}_l)\|,\, k) + h_\Psi\|B_\omega\|q_\omega + \kappa p_l. \qquad (7.38)
\end{aligned}$$

Thus, combining with (7.34) gives

$$\|\hat{z}(k)\| \leq \beta_{a(l+1)}(\|\hat{z}(0)\|,\, k) + \kappa p_l + h_\Psi \|B_\omega\| q_\omega \tag{7.39}$$

where

$$\beta_{a(l+1)}(r,\, k) = \begin{cases} \beta_{al}(r,\, k) + (1-\kappa)p_l - h_\Psi \|B_\omega\| q_\omega, & k \in [0,\, \bar{k}_l - 1] \\ \min\left\{\beta_{al}(r,\, k) + (1-\kappa)p_l - h_\Psi \|B_\omega\| q_\omega,\, \beta_a(\|\hat{z}(\bar{k}_l)\|,\, k - \bar{k}_l)\right\}, & k \in [\bar{k}_l,\, +\infty) \end{cases} \tag{7.40}$$

which recursively satisfies that $\beta_{a(l+1)}(r,\, k) \to 0$ as $k \to \infty$, and it can be noticed from (7.34) that the state will finally converge to p_l. Also, (7.39) can be written as $\|\hat{z}(k)\| \leq \beta_{a(l+1)}(\|\hat{z}(0)\|,\, k) + p_{l+1}$, where

$$\begin{aligned} p_{l+1} &= \kappa p_l + h_\Psi \|B_\omega\| q_\omega \\ &= \kappa\big(\kappa p_{l-1} + h_\Psi \|B_\omega\| q_\omega\big) + h_\Psi \|B_\omega\| q_\omega \\ &= \kappa\big(\kappa(\kappa p_{l-2} + h_\Psi \|B_\omega\| q_\omega) + h_\Psi \|B_\omega\| q_\omega\big) + h_\Psi \|B_\omega\| q_\omega \\ &= \cdots \\ &= \kappa^{l+1} p_0 + h_\Psi \|B_\omega\| q_\omega \Big(\sum_{j=0}^{l} \kappa^j\Big). \end{aligned} \tag{7.41}$$

As a result, (7.34) indicates

$$\|\hat{z}(k)\| \leq \beta_{a\infty}(\|\hat{z}(0)\|,\, k) + p_\infty \tag{7.42}$$

where $\beta_{a\infty}(\|\hat{z}(0)\|,\, k) \to 0$ as $k \to \infty$ and

$$p_\infty = \lim_{l \to \infty} p_l = h_\Psi \|B_\omega\| q_\omega \frac{\kappa}{1-\kappa}. \tag{7.43}$$

In light of Assumption 7.2, $\|\hat{z}(0)\| = \|z(0)\|$ and $\|z(k)\| \leq \|\hat{z}(k)\|$, thus,

$$\begin{aligned} \|z(k)\| \leq& \beta_a(\|z(0)\|,\, k) + h_\Psi \|B_\omega\| q_\omega \frac{\kappa}{1-\kappa} \\ =& \beta_a(\|z(0)\|,\, k) + \gamma_a(q_\omega) \end{aligned} \tag{7.44}$$

where $\gamma_a(r) = c_{\gamma a} r$ with $c_{\gamma a} = h_\Psi \|B_\omega\| \frac{\kappa}{1-\kappa}$. Therefore, under DoS attacks, the closed-loop fractional-order system (7.1) with controller (7.12) is globally ISS, and the control objective is achieved. ■

Remark 7.3 *Although there has been considerable research into secure control for integer-order systems [152, 154, 155, 212], comparatively little attention has been given to systems in fractional-order, and even less to those in the discrete-time domain. The research in this chapter is the first to address the security control of discrete-time fractional-order systems under DoS attacks while taking into account the memory and heredity effects of fractional calculus.*

Remark 7.4 *[156] and [157] respectively investigate the control problem for fractional-order MASs and complex networks which are vulnerable to DoS attacks in the continuous-time domain. The control issue of discrete-time fractional-order MASs under DoS attacks is studied in [158], yet the non-local characteristics of fractional-order calculus are ignored when designing the control scheme. Different from such mentioned works, the control for discrete-time fractional systems under the effect of DoS attacks that rigorously consider the unique hereditary and infinite memory properties of fractional calculus is addressed in this chapter for the first time.*

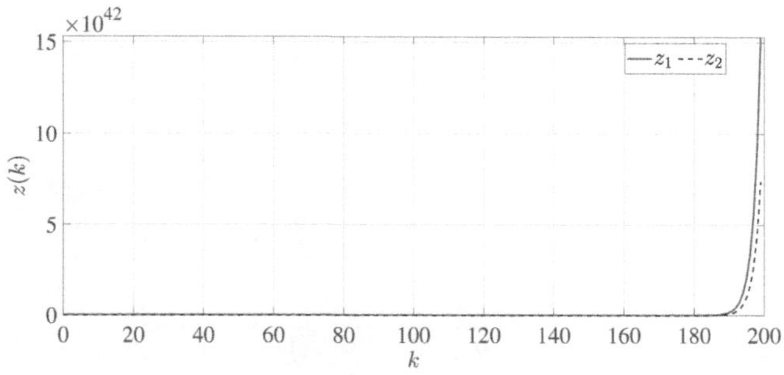

FIGURE 7.1: System output z_1 and z_2 without control effort.

7.4 Numerical Example

Consider the fractional-order discrete-time system shown in (7.1) with

$$A = \begin{bmatrix} 1 & 1 \\ 1 & -1 \end{bmatrix}, B = \begin{bmatrix} 1 \\ 1 \end{bmatrix}, B_\omega = \begin{bmatrix} 0 & 0 \\ 0 & 1 \end{bmatrix}, C_2 = \begin{bmatrix} 1 & 1 \end{bmatrix},$$

${}_0\Delta^{\alpha}_{k+1} z(k+1) = \left[{}_0\Delta^{0.1}_{k+1} z_1(k+1), {}_0\Delta^{0.4}_{k+1} z_n(k+1)\right]^T$ $(k \in \mathbb{N}_0)$, $z_0 = [0.1, 0.5]^T$, and the external disturbance $\omega(k)$ is a uniformly distributed random signal with $q_\omega = 0.02$. The numerical example and its visualized results are both implemented in MATLAB.

The results of the open-loop fractional-order system without control input, depicted in Fig. 7.1, demonstrate its instability, as evidenced by the outputs z_1 and z_2. By choosing $L = 1, \sigma = 0.1$, and $K_L = [-1, 0.01, 0.01, 0.05]^T$, we can notice that even under the influence of DoS attacks, the closed-loop system is ISS under our proposed controller, as presented in Fig. 7.2 and Fig. 7.3. Furthermore, it is presented in Fig. 7.2 that under the proposed control input, both system outputs can be driven to zero instead of just being bounded, as illustrated in the proof of Theorem 7.1, which further indicates an interesting future work to explore the design of security control law for fractional-order systems to achieve asymptotic stability. Moreover, as displayed in Fig. 7.4 and Fig. 7.5 where $K_L = [-1, 0.01]^T$ and $L = 0$, it is worth noting that with a proper selection of design parameters, the global stability of the closed-loop system in the presence of DoS attacks can be ensured under our proposed controller, even if only the current state is considered in designing the control strategy. Consequently, the effectiveness of the investigated control strategy is verified by the simulation results.

7.5 Conclusions

The security control issue for discrete-time fractional-order linear systems under DoS attacks is addressed in this chapter. A finite-dimensional controller, which requires merely the information of a finite number of the previous state, and hence is practically useful, is proposed in this chapter. Our proposed control law is verified by numerical simulation, ensuring the global input-to-state stability of the associated closed-loop system.

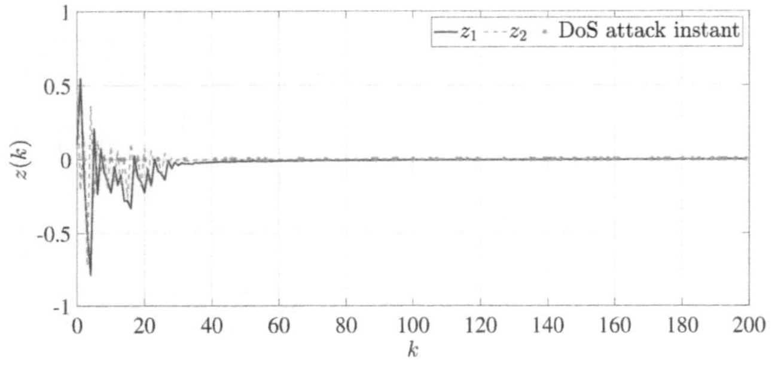

FIGURE 7.2: System output z_1 and z_2 with DoS attacks under the proposed control law where $L = 1$.

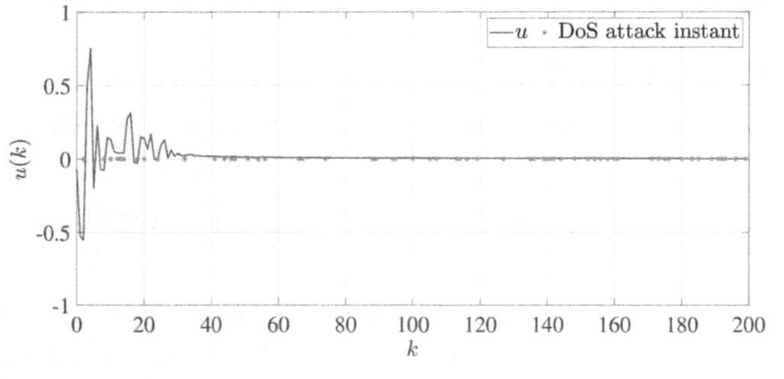

FIGURE 7.3: Control input u with $L = 1$.

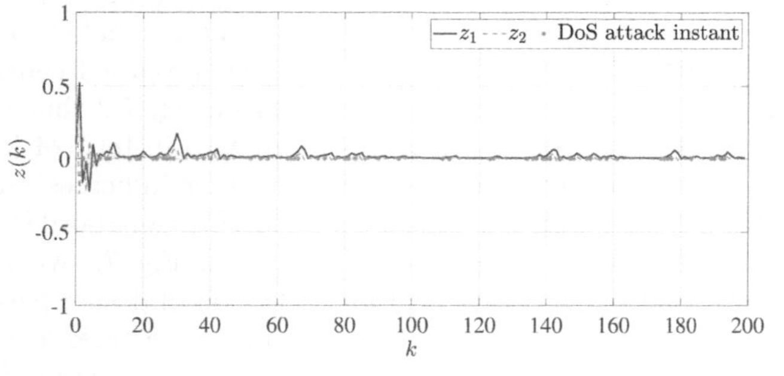

FIGURE 7.4: System output z_1 and z_2 with DoS attacks under the proposed control law where $L = 0$.

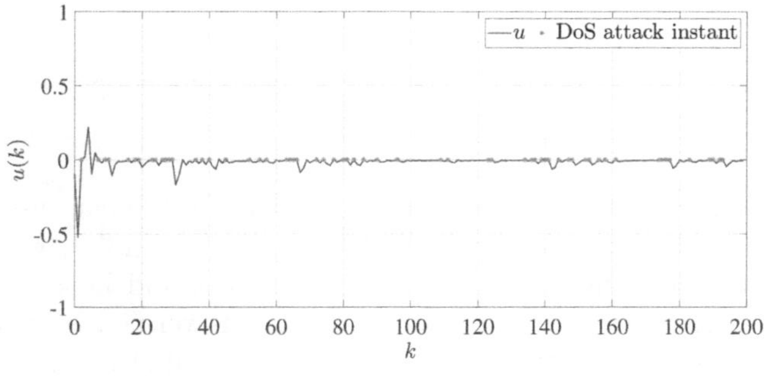

FIGURE 7.5: Control input u with $L = 0$.

8

Adaptive Fractional-Order Backstepping Control for a General Class of Nonlinear Uncertain Integer-Order Systems with External Disturbance

In this chapter, we consider the adaptive fractional-order backstepping control problem for a class of high-order integer-order systems with uncertainties and unknown external disturbance. To obtain enhanced control performance, fractional-order calculus is integrated within the conventional backstepping controller design procedure. Theoretical proof is provided based on the Lyapunov stability theorem to ensure the global stability of the closed-loop system in the sense that all the closed-loop signals are uniformly ultimately bounded and the output tracking error to a reference signal converges to an adjustable arbitrarily small range by tuning certain design parameters. Both numerical simulation and experimental results verify the efficacy of the proposed adaptive fractional-order backstepping control scheme.

8.1 Introduction

In the previous chapters, fractional-order plants with integer-order controllers and fractional-order plants with fractional-order controllers have been investigated. Note that even though integer-order plants with integer-order controllers have been widely studied, it is not the focus of this book. For completeness, from this chapter onwards, we will discuss designing fractional-order controllers for integer-order systems.

In this chapter, the adaptive fractional-order backstepping control problem for high-order nonlinear uncertain integer-order systems with unknown time-varying disturbance will be addressed. This chapter can be treated as a generalization of the traditional backstepping control design method for integer-order systems, and the fractional order in the proposed fractional-order controller for integer-order systems can be designed independently, which introduces additional degrees of freedom and flexibility in regulating the controller and hence can further realize enhanced control performance. The class of integer-order systems studied in this chapter can be of arbitrary order, unlike existing related works that are only for second-order integer systems, as stated in Subsection 1.3.3. No non-causal differential operation is involved in designing the fractional-order controller, which consequently will not cause practical issues in implementation. Furthermore, in light of the Lyapunov stability analysis approach, it is theoretically shown that in this chapter, the corresponding

DOI: 10.1201/9781003620372-8

closed-loop system under the proposed fractional-order controller is globally stable in the sense that all its signals are uniformly ultimately bounded and the system output error converges to an arbitrarily small tunable bound by adjusting certain design parameters.

The main contributions of this chapter are summarized as follows.

1. Different from the backstepping-based adaptive fractional-order control schemes proposed in [20, 21, 22, 23, 24] that are only applicable for second-order integer systems, the class of integer-order uncertain systems considered is allowed to be arbitrary order and is subject to external time-varying disturbance.
2. In light of the Lyapunov stability analysis approach, it is theoretically shown that the corresponding closed-loop system under the proposed fractional-order controller is globally stable in the sense that all its signals are uniformly ultimately bounded and the system output error converges to an arbitrarily small tunable bound by adjusting certain design parameters. Consequently, compared with the controllers in [21] and [23], where assumption about the bound of system output is required, the application potential and field of the proposed control method are significantly broadened.
3. The obtained controller does not rely on non-causal differential operations and thus has no practical issues in implementation, in contrast to the control scheme in [20, 21, 22].
4. This chapter is a generalization of the traditional backstepping control design method for integer-order systems. Distinct from Chapter 2 to Chapter 4, where a class of fractional-order uncertain nonlinear systems is investigated, integer-order nonlinear systems subject to parametric uncertainties and external disturbances are studied in this chapter. Additionally, compared to the controller in Chapter 4, whose fractional order is determined by the characteristic of the fractional-order systems under study, the fractional order in our proposed fractional-order controller for integer-order systems can be designed independently, which introduces flexibility in regulating the controller and hence can further realize enhanced control performance.

The outline of this chapter is as follows. In Section 8.2, the mathematical preliminaries are formulated. An adaptive fractional-order backstepping controller is designed in Section 8.3 for high-order nonlinear integer-order systems, followed by the stability analysis for the resulting closed-loop system. Section 8.4 carries out the simulation and experimental verification. Finally, the conclusions are given in Section 8.5.

8.2 Problem Formulation

Consider the following class of nonlinear nth-order systems with parametric uncertainties and time-varying external disturbance:

$$\begin{cases} \dot{x}_1 = x_2 + \psi_1^T(x_1)\theta_1 \\ \dot{x}_2 = x_3 + \psi_2^T(x_1, x_2)\theta_2 \\ \quad \vdots \\ \dot{x}_{n-1} = x_n + \psi_{n-1}^T(x_1, x_2, \cdots, x_{n-1})\theta_{n-1} \\ \dot{x}_n = bh(x)u + \psi_n^T(x)\theta_n + d \\ y = x_1 \end{cases} \tag{8.1}$$

where n is an arbitrary positive integer, $u \in \mathbb{R}$ and $y \in \mathbb{R}$ respectively represent the input and output of the system, $x = [x_1, x_2, \cdots, x_n]^T \in \mathbb{R}^n$ is the system state vector, $\psi_i \in \mathbb{R}^q$ $(i = 1, \cdots, n)$ denote known smooth nonlinear functions, $\theta_i \in \mathbb{R}^q$ are unknown constants, the control coefficient $b \in \mathbb{R}$ is a non-zero unknown constant, $h(x)$ is a known nonlinear smooth function, and d is an unknown time-varying external disturbance upper bounded by an unknown constant D.

Note that different from the classes of systems studied in Chapter 2 to Chapter 7 that are of fractional-order, integer-order systems are investigated in this chapter using a fractional-order control scheme for improving control performance.

Remark 8.1 *In this chapter, both fractional integral and fractional derivative relevant terms will be contained in the backstepping-based designed control input, accordingly, under a similar motive that is well-known in linear control theory, the closed-loop system transient and steady performance, compared with that under traditional integer backstepping control scheme, can be further improved via adjusting the additional design flexibility introduced by fractional integration and differentiation.*

Assumption 8.1 *The reference signal for the high-order systems x_r and its derivatives $x_r^{(i)}$ $(i = 1, \cdots, n)$ are all smooth, bounded, and known.*

Assumption 8.2 *The sign of b is available for controller construction, and $h(x) \neq 0$.*

Remark 8.2 *Without loss of generality, Assumption 8.1 and Assumption 8.2 are required for standard backstepping-based adaptive tracking control [86].*

The control objective in this chapter is to design a fractional-order adaptive backstepping controller for the systems depicted in (8.1) to ensure that the resulting closed-loop system is globally stable in the sense that all the closed-loop signals are uniformly ultimately bounded for arbitrary initial conditions, and the system output tracking error can converge to an arbitrarily small tunable bound by adjusting certain design parameters.

In the rest of this chapter, the fractional-order integral and Caputo-type derivative of a function $g(t)$, i.e., ${}_0I_t^\alpha g(t)$ and ${}_0\mathcal{D}_t^\alpha g(t)$, will be expressed as $I^\alpha g(t)$ and $\mathcal{D}^\alpha g(t)$ for simplification.

The following lemma is first established for the development and analysis of the proposed control scheme.

Lemma 8.1 ([191]) *For any $y \in \mathbb{R}$ and $\lambda > 0$, the following inequality holds:*

$$0 \leq |y| - y \tanh\left(\frac{y}{\lambda}\right) \leq 0.2785\lambda. \tag{8.2}$$

8.3 Design of Fractional-Order Adaptive Backstepping Controller and Stability Analysis

8.3.1 Adaptive Fractional-Order Backstepping Controller Design

To achieve the above control objective, a fractional-order adaptive controller will be designed, by following the iterative backstepping procedure.

Step 1: Define a virtual control signal and virtual error respectively as α_1 and $z_2 = x_2 - \alpha_1$, and the derivative of a Lyapunov function $\overline{U}_1 = \frac{1}{2}z_1^2$ can be calculated as

$$\dot{\overline{U}}_1 = z_1\dot{z}_1 = z_1z_2 + z_1\alpha_1 + z_1\psi_1^T\theta_1 - z_1\dot{x}_r, \tag{8.3}$$

where $z_1 = y - x_r$ is the system output tracking error.

After designing the virtual control signal as

$$\alpha_1 = -k_1 z_1 + \dot{x}_r - \psi_1^T \hat{\theta}_1 \tag{8.4}$$

where $k_1 > 0$ and $\hat{\theta}_1$ is the estimate of the unknown θ_1, then $\dot{\overline{U}}_1$ becomes

$$\dot{\overline{U}}_1 = z_1 z_2 - k_1 z_1^2 - z_1 \psi_1^T \tilde{\theta}_1 \tag{8.5}$$

in which $\tilde{\theta}_1 = \hat{\theta}_1 - \theta_1$.

By considering the Lyapunov function $U_1 = \overline{U}_1 + \frac{1}{2}\tilde{\theta}_1^T \Gamma_1^{-1} \tilde{\theta}_1$ where Γ_1 is a positive definite matrix, we design the following adaptive law

$$\dot{\hat{\theta}}_1 = z_1 \Gamma_1 \psi_1 - \nu \Gamma_1 \hat{\theta}_1, \text{ where } \nu > \frac{1}{2}, \tag{8.6}$$

so that

$$\dot{U}_1 = z_1 z_2 - k_1 z_1^2 - \nu \tilde{\theta}_1^T \hat{\theta}_1. \tag{8.7}$$

Step 2: Defining a virtual error as $z_3 = x_3 - \alpha_2$ where the virtual control signal α_2 is designed as

$$\alpha_2 = -z_1 - k_2 z_2 - \psi_{v,2}^T \hat{\theta}_{v,2} - \zeta_2, \tag{8.8}$$

in which $k_2 > 0$, $\hat{\theta}_{v,2}$ represents the estimate of $\theta_{v,2}$, and

$$\theta_{v,2} = \left[\theta_2^T, \theta_1^T\right]^T, \ \psi_{v,2} = \left[\psi_2^T, \left(-\frac{\partial \alpha_1}{\partial x_1}\right)\psi_1^T\right]^T,$$
$$\zeta_2 = \left(-\frac{\partial \alpha_1}{\partial x_1}\right)x_2 + \left(-\frac{\partial \alpha_1}{\partial x_r}\right)\dot{x}_r + \left(-\frac{\partial \alpha_1}{\partial \dot{x}_r}\right)x_r^{(2)} + \left(-\frac{\partial \alpha_1}{\partial \hat{\theta}_1}\right)\dot{\hat{\theta}}_1,$$

resulting in the time derivative of Lyapunov function $\overline{U}_2 = U_1 + \frac{1}{2}z_2^2$ as

$$\dot{\overline{U}}_2 = - k_1 z_1^2 - k_2 z_2^2 + z_2 z_3 - \nu \tilde{\theta}_1^T \hat{\theta}_1 - z_2 \psi_{v,2}^T \tilde{\theta}_{v,2} \tag{8.9}$$

where $\tilde{\theta}_{v,2} = \hat{\theta}_{v,2} - \theta_{v,2}$. By designing adaptive law as

$$\dot{\hat{\theta}}_{v,2} = z_2 \Gamma_2 \psi_{v,2} - \nu \Gamma_2 \hat{\theta}_{v,2} \tag{8.10}$$

where matrix Γ_2 is positive definite, the time derivative of Lyapunov function $U_2 = \overline{U}_2 + \frac{1}{2}\tilde{\theta}_{v,2}^T \Gamma_2^{-1} \tilde{\theta}_{v,2}$ is obtained as follows

$$\dot{U}_2 = - k_1 z_1^2 - k_2 z_2^2 + z_2 z_3 - \nu \tilde{\theta}_1^T \hat{\theta}_1 - \nu \tilde{\theta}_{v,2}^T \hat{\theta}_{v,2}. \tag{8.11}$$

Step i $(i = 3, \cdots, n-2)$*:* The design process from Step 3 to Step $n-2$ is similar to Step 2, hence detailed derivation is omitted here for simplification, with design results summarized as follows.

Virtual errors:

$$z_{i+1} = x_{i+1} - \alpha_i. \tag{8.12}$$

Virtual control signals:

$$\alpha_i = -z_{i-1} - k_i z_i - \psi_{v,i}^T \hat{\theta}_{v,i} - \zeta_i \tag{8.13}$$

with $k_i > 0$, $\hat{\theta}_{v,i}$ being the estimate of $\theta_{v,i}$, and

$$\begin{aligned}
\theta_{v,i} &= \left[\theta_i^T, \theta_{i-1}^T, \cdots, \theta_1^T\right]^T, \\
\psi_{v,i} &= \left[\psi_i^T, \left(-\frac{\partial \alpha_{i-1}}{\partial x_{i-1}}\right)\psi_{i-1}^T, \cdots, \left(-\frac{\partial \alpha_{i-1}}{\partial x_1}\right)\psi_1^T\right]^T, \\
\zeta_i &= \sum_{j=2}^{i}\left(-\frac{\partial \alpha_{i-1}}{\partial x_{j-1}}\right)x_j + \sum_{j=1}^{i}\left(-\frac{\partial \alpha_{i-1}}{\partial x_r^{(j-1)}}\right)x_r^{(j)} + \sum_{j=1}^{i-1}\left(-\frac{\partial \alpha_{i-1}}{\partial \hat{\theta}_{v,j}}\right)\dot{\hat{\theta}}_{v,j}.
\end{aligned}$$

Adaptive laws:

$$\dot{\hat{\theta}}_{v,i} = z_i \Gamma_i \psi_{v,i} - \nu \Gamma_i \hat{\theta}_{v,i}, \tag{8.14}$$

where Γ_i denotes a positive definite matrix.

By representing θ_1 as $\theta_{v,1}$, the time derivative of Lyapunov function $U_{n-2} = \frac{1}{2}\sum_{j=1}^{n-2} z_j^2 + \frac{1}{2}\sum_{j=1}^{n-2} \tilde{\theta}_{v,j}^T \Gamma_j^{-1} \tilde{\theta}_{v,j}$ is obtained at the end of Step $n-2$:

$$\dot{U}_{n-2} = -\sum_{j=1}^{n-2} k_j z_j^2 + z_{n-2} z_{n-1} - \nu \sum_{j=1}^{n-2} \tilde{\theta}_{v,j}^T \hat{\theta}_{v,j}. \tag{8.15}$$

Step $n-1$: By defining virtual error $z_n = x_n - \alpha_{n-1}$ and designing the virtual control signal α_{n-1} as

$$\alpha_{n-1} = -z_{n-2} - k_{n-1} z_{n-1} - \psi_{v,n-1}^T \hat{\theta}_{v,n-1} - \zeta_{n-1} - c_1 I^{\beta} z_{n-1}, \tag{8.16}$$

where $k_{n-1} > 0$, $c_1 > 0$, $\beta \in [0,1]$, $\hat{\theta}_{v,n-1}$ is the estimate of $\theta_{v,n-1}$, and

$$\begin{aligned}
\theta_{v,n-1} &= \left[\theta_{n-1}^T, \theta_{n-2}^T, \cdots, \theta_1^T\right]^T, \\
\psi_{v,n-1} &= \left[\psi_{n-1}^T, \left(-\frac{\partial \alpha_{n-2}}{\partial x_{n-2}}\right)\psi_{n-2}^T, \cdots, \left(-\frac{\partial \alpha_{n-2}}{\partial x_1}\right)\psi_1^T\right]^T, \\
\zeta_{n-1} &= \sum_{j=2}^{n-1}\left(-\frac{\partial \alpha_{n-2}}{\partial x_{j-1}}\right)x_j + \sum_{j=1}^{n-1}\left(-\frac{\partial \alpha_{n-2}}{\partial x_r^{(j-1)}}\right)x_r^{(j)} + \sum_{j=1}^{n-2}\left(-\frac{\partial \alpha_{n-2}}{\partial \hat{\theta}_{v,j}}\right)\dot{\hat{\theta}}_{v,j},
\end{aligned}$$

then we can compute the derivative of Lyapunov function $\overline{U}_{n-1} = U_{n-2} + \frac{1}{2}z_{n-1}^2$ as follows

$$\dot{\overline{U}}_{n-1} = -\sum_{j=1}^{n-1} k_j z_j^2 + z_{n-1} z_n - \nu \sum_{j=1}^{n-2} \tilde{\theta}_{v,j}^T \hat{\theta}_{v,j} - c_1 z_{n-1} I^{\beta} z_{n-1} - z_{n-1} \psi_{v,n-1}^T \tilde{\theta}_{v,n-1} \tag{8.17}$$

in which $\tilde{\theta}_{v,n-1} = \hat{\theta}_{v,n-1} - \theta_{v,n-1}$.

By designing the adaptive law in this step as

$$\dot{\hat{\theta}}_{v,n-1} = z_{n-1} \Gamma_{n-1} \psi_{v,n-1} - \nu \Gamma_{n-1} \hat{\theta}_{v,n-1} \tag{8.18}$$

where Γ_{n-1} is a positive definite matrix, the time derivative of the Lyapunov function $U_{n-1} = \overline{U}_{n-1} + \frac{1}{2}\tilde{\theta}_{v,n-1}^T \Gamma_{n-1}^{-1} \tilde{\theta}_{v,n-1}$ is

$$\dot{U}_{n-1} = -\sum_{j=1}^{n-1} k_j z_j^2 - \nu \sum_{j=1}^{n-1} \tilde{\theta}_{v,j}^T \hat{\theta}_{v,j} + z_{n-1} z_n - c_1 z_{n-1} I^{\beta} z_{n-1}. \tag{8.19}$$

Step n: Firstly, the virtual control signal α_{n-1} is rewritten as

$$\alpha_{n-1} = \alpha_{(n-1)1}(x_1, \cdots, x_{n-1}, \hat{\theta}_{v,1}, \cdots, \hat{\theta}_{v,n-1}, x_r, \cdots, x_r^{(n-1)}) - c_1 I^\beta z_{n-1}, \tag{8.20}$$

then its derivative is derived as

$$\begin{aligned}\dot{\alpha}_{n-1} =& \frac{\partial \alpha_{(n-1)1}}{\partial x_1}(x_2 + \psi_1^T \theta_1) + \frac{\partial \alpha_{(n-1)1}}{\partial x_2}(x_3 + \psi_2^T \theta_2) + \cdots + \frac{\partial \alpha_{(n-1)1}}{\partial x_{n-1}}(x_n \\ &+ \psi_{n-1}^T \theta_{n-1}) + \frac{\partial \alpha_{(n-1)1}}{\partial \hat{\theta}_{v,1}} \dot{\hat{\theta}}_{v,1} + \cdots + \frac{\partial \alpha_{(n-1)1}}{\partial \hat{\theta}_{v,n-1}} \dot{\hat{\theta}}_{v,n-1} + \frac{\partial \alpha_{(n-1)1}}{\partial x_r} \dot{x}_r \\ &+ \cdots + \frac{\partial \alpha_{(n-1)1}}{\partial x_r^{(n-1)}} x_r^{(n)} - c_1 \mathcal{D}^{1-\beta} z_{n-1}.\end{aligned} \tag{8.21}$$

According to Property 1.2 and Definition 1.11, $\mathcal{D}^{1-\beta} z_{n-1}$ can be expressed as follows:

$$\begin{aligned}\mathcal{D}^{1-\beta} z_{n-1} =& \mathcal{D}^{1-\beta} x_{n-1} - \mathcal{D}^{1-\beta} \alpha_{n-2} \\ =& \frac{1}{\Gamma(\beta)} \int_0^t \frac{\dot{x}_{n-1}}{(t-\tau)^{1-\beta}} \mathrm{d}\tau - \frac{1}{\Gamma(\beta)} \int_0^t \frac{\dot{\alpha}_{n-2}}{(t-\tau)^{1-\beta}} \mathrm{d}\tau \\ =& \frac{1}{\Gamma(\beta)} \int_0^t \frac{x_n}{(t-\tau)^{1-\beta}} \mathrm{d}\tau + \frac{1}{\Gamma(\beta)} \int_0^t \frac{\psi_{n-1}^T \theta_{n-1}}{(t-\tau)^{1-\beta}} \mathrm{d}\tau - \frac{1}{\Gamma(\beta)} \int_0^t \Big[\frac{\partial \alpha_{n-2}}{\partial x_1}(x_2 \\ &+ \psi_1^T \theta_1) + \cdots + \frac{\partial \alpha_{n-2}}{\partial x_{n-2}}(x_{n-1} + \psi_{n-2}^T \theta_{n-2}) + \frac{\partial \alpha_{n-2}}{\partial \hat{\theta}_{v,1}} \dot{\hat{\theta}}_{v,1} + \cdots \\ &+ \frac{\partial \alpha_{n-2}}{\partial \hat{\theta}_{v,n-2}} \dot{\hat{\theta}}_{v,n-2} + \frac{\partial \alpha_{n-2}}{\partial x_r} \dot{x}_r + \cdots + \frac{\partial \alpha_{n-2}}{\partial x_r^{(n-2)}} x_r^{(n-1)} \Big] (t-\tau)^{\beta-1} \mathrm{d}\tau \\ =& I^\beta x_n + I^\beta \psi_{n-1}^T \theta_{n-1} - I^\beta \Big(\frac{\partial \alpha_{n-2}}{\partial x_1} x_2 \Big) - I^\beta \Big(\frac{\partial \alpha_{n-2}}{\partial x_1} \psi_1^T \Big) \theta_1 - \cdots \\ &- I^\beta \Big(\frac{\partial \alpha_{n-2}}{\partial x_{n-2}} x_{n-1} \Big) - I^\beta \Big(\frac{\partial \alpha_{n-2}}{\partial x_{n-2}} \psi_{n-2}^T \Big) \theta_{n-2} - I^\beta \Big(\frac{\partial \alpha_{n-2}}{\partial \hat{\theta}_{v,1}} \dot{\hat{\theta}}_{v,1} \Big) \\ &- \cdots - I^\beta \Big(\frac{\partial \alpha_{n-2}}{\partial \hat{\theta}_{v,n-2}} \dot{\hat{\theta}}_{v,n-2} \Big) - I^\beta \Big(\frac{\partial \alpha_{n-2}}{\partial x_r} \dot{x}_r \Big) - \cdots - I^\beta \Big(\frac{\partial \alpha_{n-2}}{\partial x_r^{(n-2)}} x_r^{(n-1)} \Big).\end{aligned} \tag{8.22}$$

Considering the Lyapunov function $\overline{U}_n = U_{n-1} + \frac{1}{2} z_n^2$, its derivative is

$$\dot{\overline{U}}_n = \dot{U}_{n-1} + z_n(\dot{x}_n - \dot{\alpha}_{n-1}) = \dot{U}_{n-1} + z_n(bh(x)u + \psi_n^T \theta_n + d - \dot{\alpha}_{n-1}). \tag{8.23}$$

By using the property that $z_n d \leq |z_n| D \leq 0.2785\lambda D + z_n \tanh(z_n/\lambda) D$ with $\lambda > 0$ obtained from Lemma 8.1 and substituting (8.19), (8.21) and (8.22) into (8.23), we get

$$\begin{aligned}\dot{\overline{U}}_n \leq & -\sum_{j=1}^{n-1} k_j z_j^2 - \nu \sum_{j=1}^{n-1} \tilde{\theta}_{v,j}^T \hat{\theta}_{v,j} + z_n bh(x)u + z_{n-1} z_n + 0.2785\lambda D \\ &- c_1 z_{n-1} I^\beta z_{n-1} + z_n \tanh\Big(\frac{z_n}{\lambda}\Big) D + z_n \psi_{v,n}^T \theta_{v,n} + z_n \zeta_n,\end{aligned} \tag{8.24}$$

where

$$\psi_{v,n} =\Big[\psi_n^T, \Big(-\frac{\partial\alpha_{(n-1)1}}{\partial x_{n-1}}\Big)\psi_{n-1}^T + c_1 I^\beta \psi_{n-1}^T, \Big(-\frac{\partial\alpha_{(n-1)1}}{\partial x_{n-2}}\Big)\psi_{n-2}^T$$
$$+\Big[-c_1 I^\beta\Big(\frac{\partial\alpha_{n-2}}{\partial x_{n-2}}\psi_{n-2}^T\Big)\Big], \cdots, \Big(-\frac{\partial\alpha_{(n-1)1}}{\partial x_1}\Big)\psi_1^T + \Big[-c_1 I^\beta\Big(\frac{\partial\alpha_{n-2}}{\partial x_1}\psi_1^T\Big)\Big]\Big]^T,$$
$$\theta_{v,n} =\big[\theta_n^T, \theta_{n-1}^T, \cdots, \theta_1^T\big]^T,$$
$$\zeta_n =\sum_{j=2}^{n}\Big(-\frac{\partial\alpha_{(n-1)1}}{\partial x_{j-1}}\Big)x_j + \sum_{j=1}^{n-1}\Big(-\frac{\partial\alpha_{(n-1)1}}{\partial\hat{\theta}_{v,j}}\Big)\dot{\hat{\theta}}_{v,j} + \sum_{j=1}^{n}\Big(-\frac{\partial\alpha_{(n-1)1}}{\partial x_r^{(j-1)}}\Big)x_r^{(j)}$$
$$+ c_1 I^\beta x_n + \Big[-c_1 I^\beta\Big(\sum_{j=2}^{n-1}\frac{\partial\alpha_{n-2}}{\partial x_{j-1}}x_j + \sum_{j=1}^{n-2}\frac{\partial\alpha_{n-2}}{\partial\hat{\theta}_{v,j}}\dot{\hat{\theta}}_{v,j} + \sum_{j=1}^{n-1}\frac{\partial\alpha_{n-2}}{\partial x_r^{(j-1)}}x_r^{(j)}\Big)\Big].$$

Define $\rho = \frac{1}{b}$. Based on Assumption 8.2, the control input u is designed as

$$u =\frac{\hat{\rho}}{h(x)}\alpha_n, \tag{8.25}$$

where

$$\alpha_n = - z_{n-1} - k_n z_n - \tanh\Big(\frac{z_n}{\lambda}\Big)\hat{D} - \zeta_n + c_1 z_{n-1} f_z(z_n,\sigma) I^\beta z_{n-1} - \psi_{v,n}^T\hat{\theta}_{v,n} \tag{8.26}$$

with $k_n > 0$, $\sigma \in (0,1)$,

$$f_z(z_n,\sigma) = \begin{cases} \frac{1}{z_n}, & |z_n| \geq \sigma \\ \frac{z_n}{z_n^2-\sigma^2+\sigma}, & |z_n| < \sigma, \end{cases} \tag{8.27}$$

and $\hat{\rho}$, $\hat{\theta}_{v,n}$ and $\hat{D}$ being the estimates of ρ, $\theta_{v,n}$ and D, respectively. Construct Lyapunov function $\overline{U}_{n1} = \overline{U}_n + \frac{|b|}{2\mu}\tilde{\rho}^2$ with $\mu > 0$ and $\tilde{\rho} = \hat{\rho} - \rho$, then by combining with (8.24) and (8.25), its derivative is computed as

$$\dot{\overline{U}}_{n1} =\dot{\overline{U}}_n + \frac{|b|}{\mu}\tilde{\rho}\dot{\hat{\rho}}$$
$$\leq -\sum_{j=1}^{n-1} k_j z_j^2 - \nu\sum_{j=1}^{n-1}\tilde{\theta}_{v,j}^T\hat{\theta}_{v,j} + z_{n-1}z_n - c_1 z_{n-1} I^\beta z_{n-1} + 0.2785\lambda D$$
$$+ z_n \tanh\Big(\frac{z_n}{\lambda}\Big)D + z_n\psi_{v,n}^T\theta_{v,n} + z_n\zeta_n + z_n b(\rho+\tilde{\rho})\alpha_n + \frac{|b|}{\mu}\tilde{\rho}\dot{\hat{\rho}}. \tag{8.28}$$

Design

$$\dot{\hat{\rho}} = -\mu\text{sign}(b) z_n\alpha_n - \iota\mu\hat{\rho}, \tag{8.29}$$

where $\iota > \frac{1}{2}$ and sign(b) denotes the sign of b, then it gives

$$\dot{\overline{U}}_{n1} \leq -\sum_{j=1}^{n} k_j z_j^2 - \nu\sum_{j=1}^{n-1}\tilde{\theta}_{v,j}^T\hat{\theta}_{v,j} + 0.2785\lambda D - z_n\psi_{v,n}^T\tilde{\theta}_{v,n} - |b|\iota\tilde{\rho}\hat{\rho}$$
$$- z_n\tanh\Big(\frac{z_n}{\lambda}\Big)\tilde{D} - c_1 z_{n-1} I^\beta z_{n-1} + c_1 z_{n-1} z_n f_z(z_n,\sigma) I^\beta z_{n-1}, \tag{8.30}$$

in which $\tilde{\theta}_{v,n} = \hat{\theta}_{v,n} - \theta_{v,n}$ and $\tilde{D} = \hat{D} - D$.

From the definition of f_z in (8.27), it is clear that when $|z_n| \geq \sigma$,

$$\dot{\overline{U}}_{n1} \leq -\sum_{j=1}^{n} k_j z_j^2 - \nu \sum_{j=1}^{n-1} \tilde{\theta}_{v,j}^T \hat{\theta}_{v,j} + 0.2785\lambda D - z_n \psi_{v,n}^T \tilde{\theta}_{v,n} - z_n \tanh\left(\frac{z_n}{\lambda}\right)\tilde{D} - |b|\iota\tilde{\rho}\hat{\rho}. \tag{8.31}$$

Then for $|z_n| < \sigma$, it gives

$$\begin{aligned}|z_n| =& |x_n - \alpha_{n-1}| \\ =& \left|x_n + z_{n-2} + k_{n-1}z_{n-1} + \psi_{v,n-1}^T \hat{\theta}_{v,n-1} + \zeta_{n-1} + c_1 I^\beta z_{n-1}\right| < \sigma,\end{aligned} \tag{8.32}$$

which implies that the sum of last two terms on the right-hand side of (8.30) is bounded by a positive constant H, whose value is related to σ. Thus overall, we have

$$\begin{aligned}\dot{\overline{U}}_{n1} \leq& -\sum_{j=1}^{n} k_j z_j^2 - \nu \sum_{j=1}^{n-1} \tilde{\theta}_{v,j}^T \hat{\theta}_{v,j} - z_n \psi_{v,n}^T \tilde{\theta}_{v,n} - z_n \tanh\left(\frac{z_n}{\lambda}\right)\tilde{D} \\ & + 0.2785\lambda D + H - |b|\iota\tilde{\rho}\hat{\rho}.\end{aligned} \tag{8.33}$$

Finally, with the adaptive laws designed as follows

$$\dot{\hat{\theta}}_{v,n} = z_n \Gamma_n \psi_{v,n} - \nu \Gamma_n \hat{\theta}_{v,n}, \tag{8.34}$$

$$\dot{\hat{D}} = \gamma z_n \tanh\left(\frac{z_n}{\lambda}\right) - \gamma\eta\hat{D}, \tag{8.35}$$

where Γ_n is a positive definite matrix, $\gamma > 0$, and $\eta > \frac{1}{2}$, the time derivative of Lyapunov function $U_n = \overline{U}_{n1} + \frac{1}{2}\tilde{\theta}_{v,n}^T \Gamma_n^{-1} \tilde{\theta}_{v,n} + \frac{1}{2\gamma}\tilde{D}^2$ satisfies that

$$\dot{U}_n \leq -\sum_{j=1}^{n} k_j z_j^2 - \nu \sum_{j=1}^{n} \tilde{\theta}_{v,j}^T \hat{\theta}_{v,j} - \eta\tilde{D}\hat{D} + 0.2785\lambda D + H - |b|\iota\tilde{\rho}\hat{\rho}. \tag{8.36}$$

8.3.2 Stability Analysis

We now present our main results in the following theorem.

Theorem 8.1 *Consider the closed-loop adaptive system consisting of nonlinear high-order system (8.1) and fractional-order adaptive backstepping controller with control law (8.25) and adaptive laws (8.6), (8.10), (8.14), (8.18), (8.29), (8.34) and (8.35). The boundedness of all the signals in the closed-loop system is globally ensured and the output error* z_1 *satisfies*

$$|z_1|_\infty \leq \sqrt{\max\left\{2U_n(0), \frac{2M_{h2}}{M_{h1}}\right\}} \tag{8.37}$$

and

$$\lim_{t\to\infty} |z_1| \leq \sqrt{\frac{2M_{h2}}{M_{h1}}}, \tag{8.38}$$

where

$$U_n(0) =\sum_{i=1}^{n}\frac{1}{2}z_i^2(0) + \sum_{i=1}^{n}\frac{1}{2}\tilde{\theta}_{v,i}^T(0)\Gamma_i^{-1}\tilde{\theta}_{v,i}(0) + \frac{\tilde{D}^2(0)}{2\gamma} + \frac{|b|}{2\mu}\tilde{\rho}^2(0),$$

$$M_{h1} =\min\left\{2k_i, \frac{1}{\lambda_{\max}(\Gamma_i^{-1})}, \gamma, \mu\right\} \ (i = 1, \cdots, n),$$

$$M_{h2} =\frac{\nu^2}{4\nu-2}\sum_{i=1}^{n}\|\theta_{v,i}\|^2 + \frac{\eta^2}{4\eta-2}D^2 + H + 0.2785\lambda D + \frac{\iota^2}{(4\iota-2)|b|}.$$

Proof: Firstly, according to Young's inequality, we have

$$\begin{aligned}
-\nu\tilde{\theta}_{v,i}^T\hat{\theta}_{v,i} =& -\nu\tilde{\theta}_{v,i}^T(\tilde{\theta}_{v,i} + \theta_{v,i}) \\
\leq& -\nu\|\tilde{\theta}_{v,i}\|^2 + \frac{\nu}{2a_1}\|\theta_{v,i}\|^2 + \frac{a_1\nu}{2}\|\tilde{\theta}_{v,i}\|^2 \\
=&\frac{\nu}{2a_1}\|\theta_{v,i}\|^2 - \frac{\nu(2-a_1)}{2}\|\tilde{\theta}_{v,i}\|^2
\end{aligned} \tag{8.39}$$

where $i = 1, \cdots, n$ and $a_1 > 0$. By letting $a_1 = \frac{2\nu-1}{\nu}$, then

$$\begin{aligned}
-\nu\tilde{\theta}_{v,i}^T\hat{\theta}_{v,i} \leq&\frac{\nu^2}{4\nu-2}\|\theta_{v,i}\|^2 - \frac{1}{2}\|\tilde{\theta}_{v,i}\|^2 \\
\leq&\frac{\nu^2}{4\nu-2}\|\theta_{v,i}\|^2 - \frac{1}{2\lambda_{\max}(\Gamma_i^{-1})}\tilde{\theta}_{v,i}^T\Gamma_i^{-1}\tilde{\theta}_{v,i}.
\end{aligned} \tag{8.40}$$

Secondly, by choosing a positive constant $a_2 = \frac{2\eta-1}{\eta}$, we have

$$\begin{aligned}
-\eta\tilde{D}\hat{D} =& -\eta\tilde{D}(\tilde{D} + D) \leq -\eta\tilde{D}^2 + \frac{\eta}{2a_2}D^2 + \frac{\eta a_2}{2}\tilde{D}^2 \\
=&\frac{\eta^2}{4\eta-2}D^2 - \frac{\tilde{D}^2}{2}.
\end{aligned} \tag{8.41}$$

Thirdly, design $a_3 = \frac{2\iota-1}{\iota}$ and it gives

$$\begin{aligned}
-|b|\iota\tilde{\rho}\hat{\rho} =& -\iota|b|\tilde{\rho}(\tilde{\rho} + \rho) \leq \iota|b|\tilde{\rho}^2 + \frac{\iota|b|}{2a_3}\rho^2 + \frac{\iota|b|a_3}{2}\tilde{\rho}^2 \\
=&\frac{\iota^2}{(4\iota-2)|b|} - \frac{|b|}{2}\tilde{\rho}^2.
\end{aligned} \tag{8.42}$$

Hence substituting (8.40) to (8.42) in (8.36) gives

$$\begin{aligned}
\dot{U}_n \leq& -\sum_{j=1}^{n}k_jz_j^2 - \sum_{j=1}^{n}\frac{1}{2\lambda_{\max}(\Gamma_j^{-1})}\tilde{\theta}_{v,j}^T\Gamma_j^{-1}\tilde{\theta}_{v,j} - \frac{\gamma}{2\gamma}\tilde{D}^2 + 0.2785\lambda D + H \\
&+ \frac{\eta^2}{4\eta-2}D^2 + \frac{\iota^2}{(4\iota-2)|b|} - \frac{|b|\mu}{2\mu}\tilde{\rho}^2 + \sum_{j=1}^{n}\frac{\nu^2}{4\nu-2}\|\theta_{v,j}\|^2 \\
\leq& -M_{h1}U_n + M_{h2}.
\end{aligned} \tag{8.43}$$

By taking integral on the differential inequality (8.43), we get

$$U_n(t) \leq U_n(0)e^{-M_{h1}t} + \frac{M_{h2}}{M_{h1}}(1 - e^{-M_{h1}t}), \tag{8.44}$$

which shows that U_n is uniformly bounded and hence $z_i\,(i = 1, 2, \cdots, n)$, $\tilde{\theta}_{v,i}$, $\tilde{\rho}$, and $\tilde{D}$ are bounded. From (8.4), (8.8), (8.13), and the boundedness of $x_r^{(i)}\ (i = 1, \cdots, n)$ according to Assumption 8.1, we obtain that $\alpha_p\,(p = 1, \cdots, n-2)$ are bounded. As a result, the boundedness of $x_j\,(j = 1, \cdots, n-1)$ can be ensured from the boundedness of z_j and α_p.

For α_{n-1} that contains the fractional-order integral of z_{n-1}, as $t^{\beta-1}$ will monotonically decrease to zero for $\beta \in (0, 1)$ and be equal to 1 for $\beta = 1$, the boundedness of $I^\beta z_{n-1}$ can be guaranteed by the boundedness of z_{n-1}, as seen below

$$I^\beta z_{n-1} = \frac{1}{\Gamma(\beta)} \int_0^t \frac{z_{n-1}(\tau)}{(t-\tau)^{1-\beta}} \mathrm{d}\tau = \frac{1}{\Gamma(\beta)} (z_{n-1} * t^{\beta-1}), \tag{8.45}$$

where $*$ denotes the convolution operator. Thus it can be noted that, together with the boundedness of $x_r^{(i)}$ in accordance with Assumption 8.1, α_{n-1} is bounded. The boundedness of $x_n = z_n + \alpha_{n-1}$ is therefore guaranteed. Analogously, together this with the control input u given in (8.25), it can be shown that u is bounded based on Assumption 8.1. Hence all the closed-loop signals are globally bounded and the output tracking error z_1 is continuously bounded by a tunable bound $\sqrt{\max\left\{2U_n(0), \frac{2M_{h2}}{M_{h1}}\right\}}$. Moreover, it can be noticed from (8.44) that as $t \to \infty$, we have $U_n(t) \le \frac{M_{h2}}{M_{h1}}$, which indicates $\lim_{t\to\infty}|z_1| \le \sqrt{\frac{2M_{h2}}{M_{h1}}}$, or in other words, the output tracking error can be driven to an adjustable arbitrarily small bound by properly selecting the design parameters $k_i\,(i = 1, \cdots, n)$, Γ_i, γ, η, ν, μ, ι, λ, and σ. ■

Remark 8.3 *In order to briefly illustrate the design idea of the proposed control scheme, the second-order system case is taken as an example. As can be observed from (8.16), because of the introduction of the fractional-order integral term $I^\beta z_1$ in the virtual control signal α_1, both $I^\beta z_1$ and $\mathcal{D}^{1-\beta} z_1$ will be contained in the real control input u. Consequently, as is well-known from linear control theory, the system transient and steady-state performances can be improved by considering the integration as well as the differentiation of the output error z_1 in the proposed control law. Furthermore, in comparison with including integer-order integral of z_1 in α_1, which will result in terms related to $\int z_1 \mathrm{d}t$ and z_1 in the control signal u, introducing fractional-order integral of z_1 inside α_1 provides one more degree of freedom for the control parameters to be designed. In addition, since the term*

$$\mathcal{D}^{1-\beta} z_1 = \frac{1}{\Gamma(\beta)} \int_0^t \frac{\dot{z}_1(\tau)}{(t-\tau)^{1-\beta}} \mathrm{d}\tau \tag{8.46}$$

in the control signal can be treated as weighted integral of the first-order derivative of z_1, which reveals the "memory" effect of fractional-order calculus, hence the variation trend of the output error is more comprehensively considered in the proposed controller when implementing control effort on the system. Therefore, the overall control performance can be further enhanced by properly choosing the fractional order β. Nevertheless, since realizing the differentiation of certain variables (e.g. the system output error z_1) is unrealistic in practice, the term $\mathcal{D}^{1-\beta} z_1$ is further handled as in (8.22) to avoid the differential operation contained in the control signal.

Remark 8.4 *In [21] and [23], the lower bound of the absolute value of system output error is required to be constantly larger than a specific positive parameter for $t \in (0, T)$, where $T > 0$ is a certain time instant, in order to guarantee the stability of the closed-loop system. Moreover, the values of the control parameters in [21] are related to the lower bound of the system output error as well as the value of T, which adds constraint on the designed* *controller, while in [23], the* sign *function is applied in the control law to compensate for the term related to the lower bound of the output error and T, yet it might induce chattering phenomenon in the control signal. As one of the contributions of this chapter, such limitations are removed in our designed control scheme.*

Remark 8.5 *For the adjustable bound of $\lim_{t\to\infty}|z_1|$ shown in (8.38), it can be tuned arbitrarily small by increasing k_i, Γ_i, γ, and μ arbitrarily and designing ν, η, λ, ι, and σ to be smaller. Nevertheless, larger k_i, Γ_i, γ, and μ may lead to a larger control effort, as can be seen from (8.4), (8.6), (8.8), (8.10), (8.13), (8.14), (8.16), (8.18), (8.25), (8.26), (8.29), (8.34), and (8.35). Thus, when choosing design parameters for a particular system, the trade-off between control performance and control force should be made. Besides, (8.37) shows that z_1 is continuously bounded by a tunable bound $\sqrt{\max\left\{2U_n(0), \frac{2M_{h2}}{M_{h1}}\right\}}$ for $\forall t > 0$, in which $U_n(0)$ depends on certain design parameters, initial conditions, and initial parameter estimates and can be made smaller with larger Γ_i, γ, and μ and smaller initial estimate errors. Thus it implies that not only steady-state but also transient tracking error performance can be improved by proper selection of parameter initial estimates and certain design parameters.*

8.4 Simulation and Experimental Verification

Oustaloup filter [29] is adopted in this chapter to implement fractional-order integrals with the interested frequency range (ω_b, ω_h) being $(10^{-4}\,\text{rad/s}, 10^{2}\,\text{rad/s})$ and the order of filter N being 5.

In the simulation and experimental studies, to emphasize the advantages of the fractional-order control strategy, the results obtained under the proposed adaptive fractional-order backstepping controller will be compared with those under the traditional integer-order one, which can be achieved by setting control parameters $c_1 = \beta = 0$.

8.4.1 Simulation Verification

In this subsection, the following third-order chaotic system will be studied to demonstrate the proposed control scheme for high-order integer-order systems:

$$\begin{cases} \dot{x}_1 = x_2 + \dfrac{10}{7}(x_1 - x_1^3) \\ \dot{x}_2 = x_3 + 10x_1 - x_2 \\ \dot{x}_3 = u - \dfrac{100}{7}x_2 + 0.5\sin(2t) + 3\underline{U}(t-5) \\ y = x_1, \end{cases} \tag{8.47}$$

where $\psi_1 = \frac{10}{7}(x_1 - x_1^3)$, $\psi_2 = 10x_1 - x_2$, $\psi_3 = -\frac{100}{7}x_2$, $b = 1$, $\theta_1 = \theta_2 = \theta_3 = 1$ and the external unknown disturbance $d = 0.5\text{sin}(2t) + 3\underline{U}(t-5)$ with $\underline{U}(t)$ standing for a unit step function. The designed control parameters are selected as $k_1 = k_2 = k_3 = 2$, $c_1 = 0.55$, $\Gamma_1 = \Gamma_2 = \Gamma_3 = 0.05I$, $\mu = 0.1$, $\gamma = 1$, $\iota = \nu = \eta = 0.51$, $\lambda = 0.15$, and $\sigma = 0.05$. With initial conditions $x(0) = [0,\ -2,\ 1]^T$, $\hat{\theta}_i(0) = 0.8\,(i = 1,2,3)$, $\hat{\rho}(0) = 0.8$, and $\hat{D}(0) = 0.6$, the control problem for this third-order system simulation study is to stabilize the output of the corresponding closed-loop system under our investigated fractional-order adaptive controller.

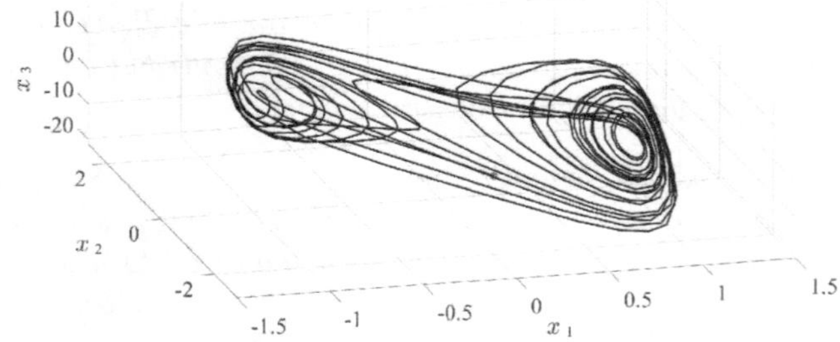

FIGURE 8.1: System states of (8.47) without control input, where *red dot* denotes the initial condition.

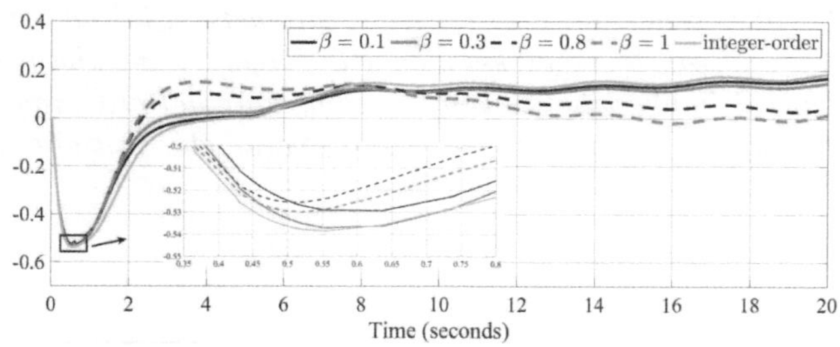

FIGURE 8.2: System output y of (8.47) under the proposed fractional-order controller with different β and traditional integer-order controller.

It can be seen from Fig. 8.1 that chaotic phenomenon exists when no control signal is implemented to (8.47). After applying our designed fractional-order adaptive backstepping-based controller with different values of β on such a third-order system, the system is stable with results presented in Fig. 8.2 and Fig. 8.4, where "integer-order" represents the case with traditional adaptive backstepping controller applied. It is observed from Fig. 8.2 that the system output $y = z_1$ converges to a small region around the origin, and the corresponding transient upper bound can be reduced with the investigated control scheme in this chapter. Fig. 8.3 presents the performance index $J_{z_1} = \int_0^t z_1(\tau)^2 \mathrm{d}\tau$ of the closed-loop system and it is observed that the integer-order controller yields significantly greater value than those under the designed fractional-order control strategy. In particular, as can be seen from Fig. 8.2 and Fig. 8.3, for the considered system (8.47), the minimum values of $|z_1|_\infty$ and J_{z_1} can be achieved in the case of $\beta = 0.8$. Furthermore, as shown in Fig. 8.4, we can notice that

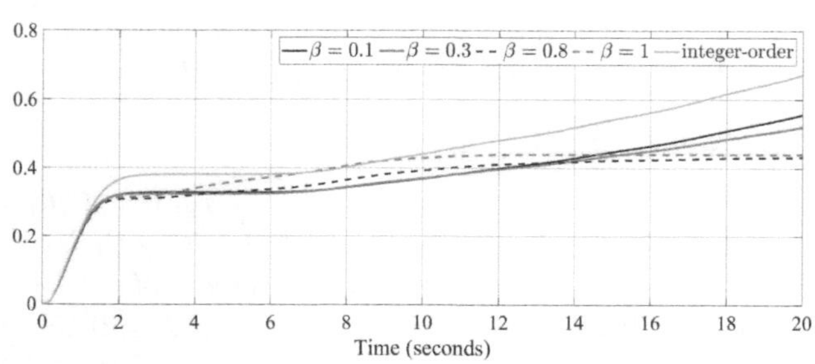

FIGURE 8.3: The performance index J_{z_1} of (8.47) under the proposed fractional-order controller with different β and traditional integer-order controller.

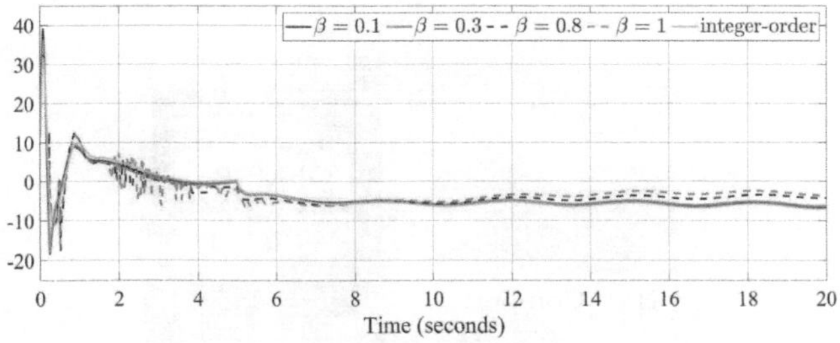

FIGURE 8.4: The control input u of (8.47) under the proposed fractional-order controller with different β and traditional integer-order controller.

the fractional-order control signals have similar magnitudes as the traditional integer-order one, implying that better performances are achieved without increasing control costs.

8.4.2 Experimental Results

An experimental test is to be presented in this subsection to illustrate the developed control scheme for second-order integer-order systems and also verify the theoretical results.

The dynamic model of a permanent magnet linear synchronous motor (PMLSM) in [20] is used and given below

$$\begin{cases} \dot{x}_1 = x_2 \\ \dot{x}_2 = 180.33u - 115.025x_2 + 0.01\big[\cos(t) + \sin(t) + \underline{U}(t-2)\big], \end{cases} \tag{8.48}$$

where states x_1 and x_2 respectively represent the position and velocity of the three-phase armature winding of the PMLSM with $y = x_1$ being the system output, and the winding current i_q is treated as the control signal u. From (8.48), it can be observed that $\psi_1 = 0$, $\psi_2 = -115.025x_2$, $b = 180.33$, $\theta_2 = 1$, and the unknown disturbance $d = 0.01\big[\cos(t) + \sin(t) + \underline{U}(t-2)\big]$. The experimental hardware setup is presented in Fig. 8.5, in which the PMLSM is implemented in OPAL-RT with a host computer providing the simulation and debug environment for it.

For experimental purposes, the initial values of the system states and parameter estimates are correspondingly set as $x(0) = [0.1 \text{ mm}, 0.5 \text{ mm/s}]^T$ and $\hat{\theta}_2(0) = 0.8$, $\hat{\rho}(0) = 0.005$, $\hat{D}(0) = 0.6$. The control parameters are selected as $k_1 = 1$, $k_2 = 0.03$, $c_1 = 5$, $\Gamma_2 = 10$, $\gamma = 10$, $\mu = 10$, $\iota = \nu = \eta = 2$, $\lambda = 0.02$, and $\sigma = 0.01$. The reference trajectory is $x_r = \sin(0.5t)$.

The experimental results are shown in Fig. 8.6 to Fig. 8.8. From Fig. 8.6, we can observe that the output tracking error $z_1 = y - x_r$ converges towards a small bound around zero, and its transient upper bound is lowered under our investigated fractional-order adaptive controller compared with the traditional integer-order ones, which highlights the advantage of the control scheme developed in this chapter. Fig. 8.7 reveals the proposed control signals under different values of β, which have similar magnitudes to the conventional one. Specifically, it is noticed from Fig. 8.6 and Fig. 8.8 that for this second-order experimental test, the values of J_{z_1} and $|z_1|_\infty$ will reduce with the decrease of β from 0.7 to 0.1, particularly when $\beta = 0.1$, the maximum overshoot of z_1 and the value of J_{z_1} at $t = 20\,\text{s}$ reduced respectively by half and 35 % compared to those under integer-order controller.

Consequently, all the results in this section illustrate the efficacy of the proposed control scheme and validate the established theoretical results.

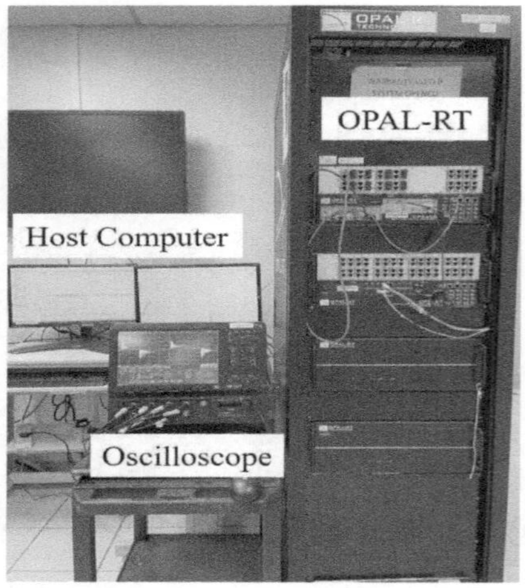

FIGURE 8.5: Experimental setup.

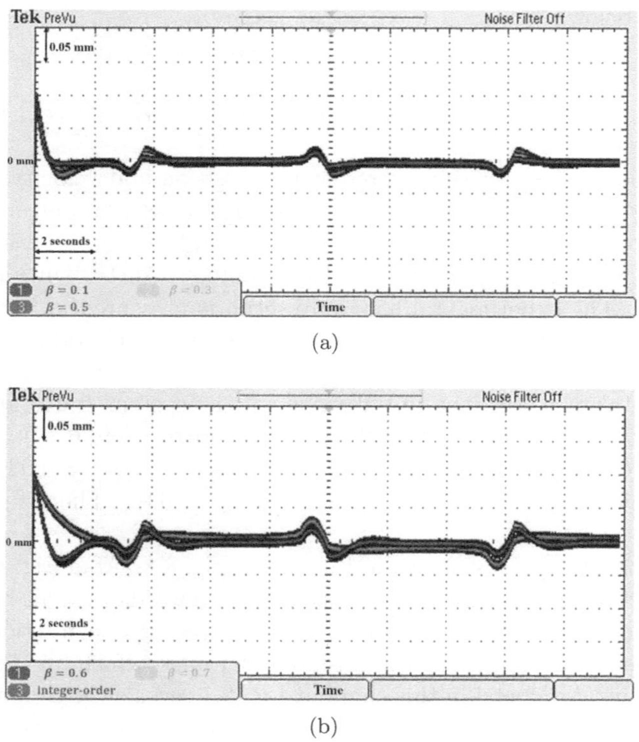

FIGURE 8.6: The experimental system output error z_1 of (8.48) under the proposed fractional-order controller and traditional integer-order controller. (a) Under the proposed fractional-order controller with $\beta = 0.1$, $\beta = 0.3$, and $\beta = 0.5$. (b) Under the proposed fractional-order controller with $\beta = 0.6$, $\beta = 0.7$, and the traditional integer-order controller.

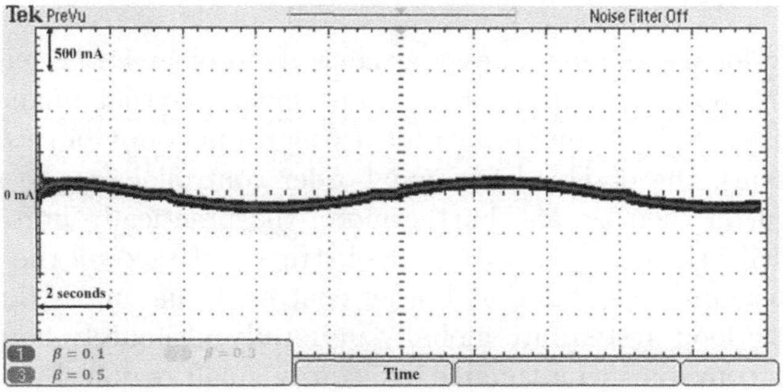

(a)

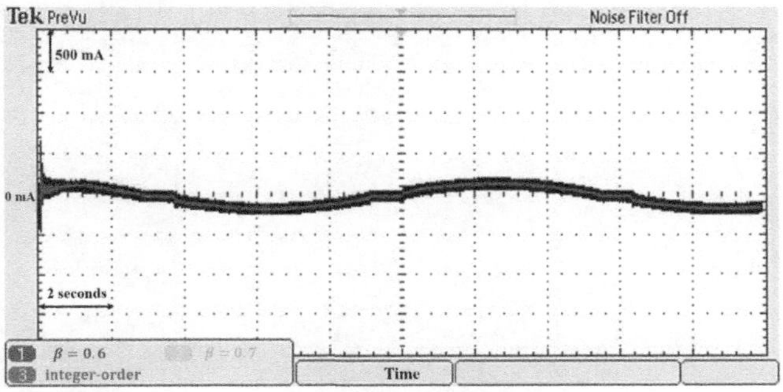

(b)

FIGURE 8.7: The experimental control signal u of (8.48) under the proposed fractional-order controller and traditional integer-order controller. (a) Under the proposed fractional-order controller with $\beta = 0.1$, $\beta = 0.3$, and $\beta = 0.5$. (b) Under the proposed fractional-order controller with $\beta = 0.6$, $\beta = 0.7$, and the traditional integer-order controller.

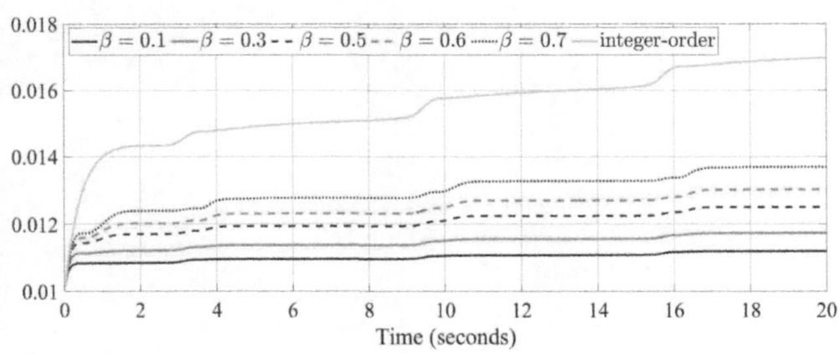

FIGURE 8.8: The performance index J_{z_1} of (8.48) under the proposed fractional-order controller and traditional integer-order controller.

8.5 Conclusions

The adaptive fractional-order backstepping control problem for a general class of high-order nonlinear integer-order uncertain systems under external unknown disturbance is investigated in this chapter. Compared with existing results only for second-order nonlinear integer-order systems, the designed fractional-order controller can be applied to systems without restriction on their orders. Furthermore, the practically unrealizable differential operation is avoided in our investigated control strategy. Based on the Lyapunov stability analysis method, the proposed fractional-order controller guarantees that all the signals in the resulting closed-loop system are globally uniformly ultimately bounded and the output tracking error converges to a tunable arbitrarily small region. Finally, simulation and experimental results also verify the effectiveness of the proposed control scheme.

9

Prescribed Performance-Based Adaptive Fractional Backstepping Control of Integer-Order Nonlinear Uncertain Systems with External Disturbance

In this chapter, an adaptive fractional backstepping controller is designed for integer-order nonlinear uncertain systems of arbitrary order with prescribed performance subject to unknown time-varying disturbances. The integration of fractional adaptive backstepping control procedure and prescribed performance bound technique, which describes the convergence rate and largest overshoot of the output error, into a unitary framework can result in an enhanced control scheme with high precision. Based on Lyapunov analysis, it is theoretically demonstrated that under the proposed control scheme, all the closed-loop signals remain bounded and the output tracking error with respect to the reference signal, which is strictly confined within the specified performance bound for all time, will asymptotically converge to zero. Simulation studies are presented to show the effectiveness of the proposed fractional adaptive PPB control strategy. Additionally, this chapter presents a scheme that assures perfect output tracking and provides a transient output error trajectory that can be confined to a range with predefined convergence rate and maximum overshoot, which shows a significant improvement from Chapter 8 where fractional adaptive backstepping control is developed for high-order nonlinear integer-order uncertain systems, yet could only maintain the maximum value of the transient system output error within a tunable limit.

9.1 Introduction

Motivated by the statement in the third paragraph of Subsection 1.3.3, in this chapter, we propose an adaptive fractional-order prescribed performance backstepping control scheme for achieving asymptotic output tracking for a general class of integer-order nonlinear systems of arbitrary order in the presence of parametric uncertainties and external time-varying unknown disturbance. The main approaches and contributions of this chapter can be summarized as follows:

1. Fractional adaptive backstepping control combined with the prescribed performance bound technique for nonlinear integer-order systems is proposed in this chapter, resulting in greater flexibility in adjusting control parameters and an enhanced control capability compared to the counterpart standard integer control.

DOI: 10.1201/9781003620372-9

Additionally, this chapter presents a scheme that assures perfect output tracking and provides a transient output error trajectory that can be confined to a range with predefined convergence rate and maximum overshoot, which shows a major improvement from Chapter 8 where fractional adaptive backstepping control is developed for high-order nonlinear integer-order uncertain systems, yet could only maintain the maximum value of the transient system output error within a tunable limit. In particular, an additional term u_c, as will be presented later in the chapter, is introduced in the studied control scheme together with the designed adaptive laws to address the limitation in Chapter 8 that cannot achieve asymptotic output tracking.

2. In contrast to existing works [163, 164, 165, 166, 167, 168] that can only be applicable to second-order integer systems, adaptive fractional control for a general class of uncertain integer-order nonlinear systems that without restriction on system order, subject to unknown external disturbances, and with prescribed performance, is conducted in this chapter. Also, different from [164] and [166, 167, 168], for preventing potential implementation issues, non-causal differential operation is avoided in the designed control scheme. Detailed explanations will be provided in the last step of controller design procedure and Remark 9.4 in Section 9.4.
3. It is theoretically proved based on Lyapunov analysis that under the developed fractional controller, the resulting closed-loop system is stable with signals remaining bounded. In addition, distinct from [166], this chapter guarantees that the output error converges to zero asymptotically without involving discontinuous sign function in control input.

The structure of this chapter is given as follows. Problem formulation for this chapter is outlined in Section 9.2. The transformed systems for prescribed performance are presented in Section 9.3. Section 9.4 provides the controller design procedure and stability analysis. In Section 9.5, the proposed scheme is verified by simulation studies, and the chapter is finally concluded in Section 9.6.

9.2 Problem Formulation

Take into account the following type of nonlinear nth-order systems that contain uncertain parameters and are subject to external time-changing disturbance:

$$\begin{cases} \dot{q}_1 = q_2 + \phi_1^T(q_1)\vartheta_1 \\ \dot{q}_2 = q_3 + \phi_2^T(q_1, q_2)\vartheta_2 \\ \quad\vdots \\ \dot{q}_{n-1} = q_n + \phi_{n-1}^T(q_1, q_2, \cdots, q_{n-1})\vartheta_{n-1} \\ \dot{q}_n = \mu g(q)u + \phi_n^T(q)\vartheta_n + d \\ y = q_1, \end{cases} \tag{9.1}$$

in which u and y represent the input and output of the system, respectively, and $q = [q_1, q_2, \cdots, q_n]^T \in \mathbb{R}^n$ is the state vector. The known smooth nonlinear functions are represented by $\phi_j \in \mathbb{R}^p$ $(j = 1, \cdots, n)$, and the unknown constants are denoted by $\vartheta_j \in \mathbb{R}^p$. The control coefficient μ is an unknown constant that is not equal to zero, and the function $g(q) \neq 0$ is a known continuous nonlinear function.

Assumption 9.1 *The reference signal y_r and its time derivatives $y_r^{(m)}$ $(m = 1, \cdots, n)$ are bounded, known, and smooth.*

Assumption 9.2 *The controller design can make use of the sign of μ.*

Assumption 9.3 *The unidentified external time-varying disturbance d is upper bounded by an unknown positive constant $\bar{d}$.*

Remark 9.1 *For the purpose of applying standard adaptive backstepping tracking control [86, 176, 217], Assumption 9.1 and Assumption 9.2 are essential. Additionally, Assumption 9.3 is needed when formulating the control input to counterbalance the external disturbance.*

The objective of control in this chapter is to construct a control scheme for the systems described in (9.1) that produces a stable closed-loop system, keeping all the closed-loop signals to remain bounded, and allowing the system output asymptotically reaches the desired reference signal y_r. Moreover, the output tracking error $e_h(t) = y(t) - y_r(t)$ can be maintained within the predetermined performance boundaries for $\forall t \geq 0$. The proposed approach is groundbreaking in the prescribed performance adaptive fractional backstepping control to resolve output tracking control issues of uncertain high-order nonlinear integer-order systems with improved system performance compared to the counterpart traditional integer-order control.

9.3 Transformed Systems for Prescribed Performance

In order to realize the aforementioned control objective, a smooth, decreasing function $h(t) = (h_0 - h_\infty)e^{-ct} + h_\infty$ with $c > 0$ and $h_0 > h_\infty > 0$ is first selected as the performance function which quantifies the required performance prescribed bound of the output tracking error [218, 219, 220]. The target of ensuring output transient performance can then be achieved under the satisfaction of the following condition with prescribed parameters $\underline{\lambda}$ and $\bar{\lambda}$, which are within $(0, 1]$,

$$-\underline{\lambda}h(t) \leq e_h(t) \leq \bar{\lambda}h(t). \tag{9.2}$$

Given the limited condition set out in (9.2), the problem can be reformulated to one that focuses solely on the boundedness of the signals. Besides, it is critical for output asymptotic tracking of the original systems to be realized that the reformulated system is stabilized asymptotically. With the smooth strictly increasing function $Y(z_1)$ designed as follows

$$Y(z_1) = \frac{\bar{\lambda}e^{(z_1+\tau)} - \underline{\lambda}e^{-(z_1+\tau)}}{e^{(z_1+\tau)} + e^{-(z_1+\tau)}}, \text{ where } \tau = \frac{\ln(\underline{\lambda}/\bar{\lambda})}{2}, \tag{9.3}$$

which possesses the following characterizations [221]

1. $-\underline{\lambda} < Y(z_1) < \bar{\lambda}$;
2. $\lim_{z_1 \to -\infty} Y(z_1) = -\underline{\lambda}$, $\lim_{z_1 \to +\infty} Y(z_1) = \bar{\lambda}$;
3. $Y(0) = 0$;

the performance requirement on $e_h(t)$ can be described as

$$e_h(t) = h(t)Y(z_1) \tag{9.4}$$

in accordance with the characterizations 1 and 2 of $Y(z_1)$, where z_1 denotes the transformed error. With initial condition that meets $-\underline{\lambda}h(0) < e_h(0) < \bar{\lambda}h(0)$, the boundedness of z_1 indicates (9.2) holds $\forall t \geq 0$, and $\lim_{t\to\infty} e_h(t) = 0$ if $\lim_{t\to\infty} z_1(t) = 0$ according to characterization 3 of $Y(z_1)$. Define $\xi = \frac{e_h}{h}$, the transformed error is expressed as

$$z_1 = Y^{-1}(\xi) = \frac{\ln(\bar{\lambda}\xi + \bar{\lambda}\underline{\lambda})}{2} - \frac{\ln(\underline{\lambda}\bar{\lambda} - \underline{\lambda}\xi)}{2}. \tag{9.5}$$

The derivative of z_1 is computed as

$$\dot{z}_1 = \frac{\partial Y^{-1}}{\partial \xi}\dot{\xi} = \eta\Big(\dot{e}_h - \frac{e_h\dot{h}}{h}\Big), \tag{9.6}$$

where $\eta = \frac{1}{2h}\left(\frac{1}{\xi+\underline{\lambda}} - \frac{1}{\xi-\bar{\lambda}}\right)$. Hence by substituting the equation of $\dot{q}_1$ in the original systems (9.1) with (9.6), the resulting transformed systems can be obtained as illustrated below

$$\begin{cases} \dot{z}_1 = \eta\Big(q_2 + \phi_1^T\vartheta_1 - \dot{y}_r - \frac{e_h\dot{h}}{h}\Big) \\ \dot{q}_k = q_{k+1} + \phi_k^T\vartheta_k,\ k = 2, \cdots, n-1 \\ \dot{q}_n = \mu g(q)u + \phi_n^T\vartheta_n + d. \end{cases} \tag{9.7}$$

9.4 Controller Design and Stability Analysis

9.4.1 Design of Controller

This subsection will outline the iterative backstepping procedure to be used in the design of a fractional-order adaptive controller.

Remark 9.2 *In this chapter, both fractional integral and fractional derivative relevant terms will be comprised in the designed PPB backstepping control signal. Accordingly, the transient and steady performance of the closed-loop system, compared with that under traditional integer backstepping prescribed performance control scheme, can be further improved via adjusting the additional design flexibility granted by fractional integration and differentiation.*

Introduce $\beta_i\,(i = 1, \cdots, n-1)$ as the virtual control signals that will be applied in the design procedure, and the virtual error variables $e_j\,(j = 1, \cdots, n)$ are defined as follows

$$e_1 = z_1, \tag{9.8}$$

$$e_l = q_l - \beta_{l-1}\,(l = 2 \cdots, n). \tag{9.9}$$

Step 1: Compute the derivative of a Lyapunov candidate function $\overline{W}_1 = \frac{1}{2}e_1^2$ as $\dot{\overline{W}}_1 = e_1\dot{e}_1$, where

$$\dot{e}_1 = \eta\Big(e_2 + \beta_1 + \phi_1^T\vartheta_1 - \dot{y}_r - \frac{e_h\dot{h}}{h}\Big), \tag{9.10}$$

thus

$$\dot{\overline{W}}_1 = \eta\Big(e_1e_2 + e_1\beta_1 + e_1\phi_1^T\vartheta_1 - e_1\dot{y}_r - e_1\frac{e_h\dot{h}}{h}\Big). \tag{9.11}$$

Construct function $W_1 = \overline{W}_1 + \frac{1}{2}\tilde{\vartheta}_1^T \Lambda_1^{-1} \tilde{\vartheta}_1$ with matrix Λ_1 being positive definite, $\tilde{\vartheta}_1 = \hat{\vartheta}_1 - \vartheta_1$, and $\hat{\vartheta}_1$ indicating the estimate of ϑ_1. By designing β_1 and the adaptive law at Step 1 respectively as

$$\beta_1 = -\frac{c_1 e_1}{\eta} + \dot{y}_r - \phi_1^T \hat{\vartheta}_1 + \frac{e_h \dot{h}}{h}, \tag{9.12}$$

$$\dot{\hat{\vartheta}}_1 = \eta e_1 \Lambda_1 \phi_1, \tag{9.13}$$

where $c_1 > 0$ is a design parameter, then the time derivative of W_1 is

$$\dot{W}_1 = \dot{\overline{W}}_1 + \tilde{\vartheta}_1^T \Lambda_1^{-1} \dot{\hat{\vartheta}}_1 = -c_1 e_1^2 + \eta e_1 e_2. \tag{9.14}$$

Step 2: From (9.12), it can be observed that the derivative of β_1 can be calculated as

$$\dot{\beta}_1 = \frac{\partial \beta_1}{\partial q_1}\dot{q}_1 + \frac{\partial \beta_1}{\partial \hat{\vartheta}_1}\dot{\hat{\vartheta}}_1 + \sum_{p=1}^{2}\left(\frac{\partial \beta_1}{\partial y_r^{(p-1)}} y_r^{(p)} + \frac{\partial \beta_1}{\partial h^{(p-1)}} h^{(p)}\right). \tag{9.15}$$

With (9.15), the computation of $e_2\dot{e}_2$ is

$$\begin{aligned} e_2\dot{e}_2 =& e_2 e_3 + e_2\beta_2 + e_2\phi_2^T\vartheta_2 - e_2\frac{\partial \beta_1}{\partial q_1}(q_2 + \phi_1^T\vartheta_1) \\ & - e_2\sum_{p=1}^{2}\left(\frac{\partial \beta_1}{\partial y_r^{(p-1)}} y_r^{(p)} + \frac{\partial \beta_1}{\partial h^{(p-1)}} h^{(p)}\right) - e_2\frac{\partial \beta_1}{\partial \hat{\vartheta}_1}\dot{\hat{\vartheta}}_1. \end{aligned} \tag{9.16}$$

Design β_2 as

$$\beta_2 = -\eta e_1 - c_2 e_2 - \phi_{v,2}^T \hat{\vartheta}_{v,2} - \psi_2 \tag{9.17}$$

with $c_2 > 0$,

$$\begin{aligned} \phi_{v,2} =& \left[\phi_2^T, \left(-\frac{\partial \beta_1}{\partial q_1}\right)\phi_1^T\right]^T, \ \vartheta_{v,2} = \left[\vartheta_2^T, \vartheta_1^T\right]^T, \\ \psi_2 =& -\frac{\partial \beta_1}{\partial q_1}q_2 - \frac{\partial \beta_1}{\partial \hat{\vartheta}_1}\dot{\hat{\vartheta}}_1 - \sum_{p=1}^{2}\left(\frac{\partial \beta_1}{\partial y_r^{(p-1)}} y_r^{(p)} + \frac{\partial \beta_1}{\partial h^{(p-1)}} h^{(p)}\right), \end{aligned}$$

and $\hat{\vartheta}_{v,2}$ indicates the estimate of $\vartheta_{v,2}$, the derivative of Lyapunov candidate function $\overline{W}_2 = W_1 + \frac{1}{2}e_2^2$ is then computed as

$$\dot{\overline{W}}_2 = -c_1 e_1^2 + e_2 e_3 - c_2 e_2^2 - e_2\phi_{v,2}^T \tilde{\vartheta}_{v,2}, \tag{9.18}$$

in which $\tilde{\vartheta}_{v,2} = \hat{\vartheta}_{v,2} - \vartheta_{v,2}$. Define Lyapunov candidate function $W_2 = \overline{W}_2 + \frac{1}{2}\tilde{\vartheta}_{v,2}^T \Lambda_2^{-1} \tilde{\vartheta}_{v,2}$, where Λ_2 is a positive definite constant matrix. With adaptive law constructed as

$$\dot{\hat{\vartheta}}_{v,2} = e_2 \Lambda_2 \phi_{v,2}, \tag{9.19}$$

the time derivative of W_2 is then calculated as

$$\begin{aligned} \dot{W}_2 =& -c_1 e_1^2 - c_2 e_2^2 + e_2 e_3 - e_2\phi_{v,2}^T \tilde{\vartheta}_{v,2} + \tilde{\vartheta}_{v,2}^T \Lambda_2^{-1} \dot{\hat{\vartheta}}_{v,2} \\ =& -c_1 e_1^2 - c_2 e_2^2 + e_2 e_3. \end{aligned} \tag{9.20}$$

Step i ($i \in \{3, \cdots, n-2\}$): The derivation of the design procedure from Step 3 to Step $n-2$ follows an analog pattern to Step 2, for brevity, the steps are not fully detailed here; the outcomes of the design are outlined below.

Virtual control signals:

$$\beta_i = -e_{i-1} - c_i e_i - \phi_{v,i}^T \hat{\vartheta}_{v,i} - \psi_i \tag{9.21}$$

with $\hat{\vartheta}_{v,i}$ denoting the estimate of $\vartheta_{v,i}$, $c_i > 0$, and

$$\begin{aligned}
\phi_{v,i} &= \left[\phi_i^T, \Big(-\frac{\partial \beta_{i-1}}{\partial q_{i-1}}\Big)\phi_{i-1}^T, \cdots, \Big(-\frac{\partial \beta_{i-1}}{\partial q_1}\Big)\phi_1^T\right]^T, \\
\vartheta_{v,i} &= \left[\vartheta_i^T, \vartheta_{i-1}^T, \cdots, \vartheta_1^T\right]^T, \ \psi_i = -\sum_{p=2}^{i} \frac{\partial \beta_{i-1}}{\partial q_{p-1}} q_p \\
&- \sum_{p=1}^{i} \left(\frac{\partial \beta_{i-1}}{\partial y_r^{(p-1)}} y_r^{(p)} + \frac{\partial \beta_{i-1}}{\partial h^{(p-1)}} h^{(p)} \right) - \sum_{p=1}^{i-1} \frac{\partial \beta_{i-1}}{\partial \hat{\vartheta}_{v,p}} \dot{\hat{\vartheta}}_{v,p}.
\end{aligned}$$

Adaptive laws:

$$\dot{\hat{\vartheta}}_{v,i} = e_i \Lambda_i \phi_{v,i}, \tag{9.22}$$

where Λ_i represents the positive definite constant matrix.

Constructing Lyapunov candidate function $W_{n-2} = \frac{1}{2}\sum_{p=1}^{n-2} \tilde{\vartheta}_{v,p}^T \Lambda_p^{-1} \tilde{\vartheta}_{v,p} + \frac{1}{2}\sum_{p=1}^{n-2} e_p^2$, in which $\vartheta_{v,1} \triangleq \vartheta_1$ and $\tilde{\vartheta}_{v,i} = \hat{\vartheta}_{v,i} - \vartheta_{v,i}$, the derivative of W_{n-2} is acquired as follows on the basis of the designed virtual control signals (9.21) and adaptive laws (9.22)

$$\dot{W}_{n-2} = -\sum_{p=1}^{n-2} c_p e_p^2 + e_{n-2} e_{n-1}. \tag{9.23}$$

Step $n-1$: Establishing Lyapunov candidate function $\overline{W}_{n-1} = W_{n-2} + \frac{e_{n-1}^2}{2}$ and computing the time derivative of it as

$$\dot{\overline{W}}_{n-1} = -\sum_{p=1}^{n-2} c_p e_p^2 + e_{n-2} e_{n-1} + e_{n-1}(\dot{q}_{n-1} - \dot{\beta}_{n-2}). \tag{9.24}$$

On the right-hand side of (9.24), we have

$$\begin{aligned}
e_{n-1}(\dot{q}_{n-1} - \dot{\beta}_{n-2}) =& e_{n-1}(q_n + \phi_{n-1}^T \vartheta_{n-1} - \dot{\beta}_{n-2}) \\
=& e_{n-1} e_n + e_{n-1}\beta_{n-1} + e_{n-1}\phi_{v,n-1}^T \vartheta_{v,n-1} + e_{n-1}\psi_{n-1},
\end{aligned} \tag{9.25}$$

where the definitions of $\phi_{v,n-1}$, $\vartheta_{v,n-1}$, and ψ_{n-1} can be found in the formulas given below (9.21) by replacing i with $n-1$ in them.

The virtual control signal at Step $n-1$ is designed as

$$\beta_{n-1} = -e_{n-2} - c_{n-1} e_{n-1} - \phi_{v,n-1}^T \hat{\vartheta}_{v,n-1} - \psi_{n-1} - a_0 I_t^{\alpha} e_{n-1}, \tag{9.26}$$

in which designed parameters $c_{n-1} > 0$, $a > 0$, and $\alpha \in [0,1]$, ${}_0I_t^{\alpha} e_{n-1}$ represents the α-order integral of e_{n-1} with the definition given in Definition 1.13, and $\hat{\vartheta}_{v,n-1}$ implies the estimate of $\vartheta_{v,n-1}$.

Remark 9.3 *The implementation of the fractional integral term ${}_0I_t^{\alpha}e_{n-1}$ in the virtual control signal (9.26) will result in a real control input u designed in the subsequent step that incorporates both fractional integral and derivative terms, i.e., ${}_0I_t^{\alpha}e_{n-1}$ and ${}_0\mathcal{D}_t^{1-\alpha}e_{n-1}$. Moreover, from Definition 1.11, it illustrates that for $\alpha \in (0,1)$, the fractional derivative can be seen as a weighted integral of the first-order derivative, showcasing the memory feature of fractional calculus and, therefore, allows for a more flexible and thorough consideration of the changing trend of system error when the proposed controller is applied to the system. As a result, with a recognized idea found in classic control theory, the system performance during transient and steady-state stages can be further enhanced by implementing fractional integration and differentiation operations with an adjustable fractional order α. On the other hand, due to the impracticality of realizing the non-causal differentiation, ${}_0\mathcal{D}_t^{1-\alpha}e_{n-1}$ will be processed to prevent the differential action in the control signal, as specified in Step n.*

Substituting (9.25) and (9.26) into (9.24) yields

$$\dot{\overline{W}}_{n-1} = -\sum_{p=1}^{n-1} c_p e_p^2 + e_{n-1}e_n - e_{n-1}\phi_{v,n-1}^T\tilde{\vartheta}_{v,n-1} - e_{n-1}a_0 {}_0I_t^{\alpha}e_{n-1} \tag{9.27}$$

with $\tilde{\vartheta}_{v,n-1} = \hat{\vartheta}_{v,n-1} - \vartheta_{v,n-1}$. Define Lyapunov candidate function $W_{n-1} = \overline{W}_{n-1} + \frac{1}{2}\tilde{\vartheta}_{v,n-1}^T\Lambda_{n-1}^{-1}\tilde{\vartheta}_{v,n-1}$, where Λ_{n-1} signifies a positive definite constant matrix, and design adaptive law as

$$\dot{\hat{\vartheta}}_{v,n-1} = e_{n-1}\Lambda_{n-1}\phi_{v,n-1}, \tag{9.28}$$

the derivative of W_{n-1} can be consequently computed as

$$\dot{W}_{n-1} = -\sum_{p=1}^{n-1} c_p e_p^2 + e_{n-1}e_n - e_{n-1}a_0 {}_0I_t^{\alpha}e_{n-1}. \tag{9.29}$$

Step n: Consider Lyapunov candidate function $\overline{W}_n = W_{n-1} + \frac{e_n^2}{2}$ and calculate its time derivative as

$$\dot{\overline{W}}_n = -\sum_{p=1}^{n-1} c_p e_p^2 + e_{n-1}e_n - e_{n-1}a_0 {}_0I_t^{\alpha}e_{n-1} + e_n(\mu g u + \phi_n^T\vartheta_n + d - \dot{\beta}_{n-1}), \tag{9.30}$$

where

$$\begin{aligned}\dot{\beta}_{n-1} =& \sum_{p=2}^{n} \frac{\partial\beta_{n-1}'}{\partial q_{p-1}} q_p + \sum_{p=1}^{n-1}\left(\frac{\partial\beta_{n-1}'}{\partial q_p}\phi_p^T\vartheta_p + \frac{\partial\beta_{n-1}'}{\partial\hat{\vartheta}_{v,p}}\dot{\hat{\vartheta}}_{v,p}\right) \\ &+ \sum_{p=1}^{n}\left(\frac{\partial\beta_{n-1}'}{\partial y_r^{(p-1)}}y_r^{(p)} + \frac{\partial\beta_{n-1}'}{\partial h^{(p-1)}}h^{(p)}\right) - a_0 {}_0\mathcal{D}_t^{1-\alpha}e_{n-1}\end{aligned} \tag{9.31}$$

with $\beta_{n-1}' = \beta_{n-1} + a_0 {}_0I_t^{\alpha}e_{n-1}$.

In order to keep the non-causal differential operation term ${}_0\mathcal{D}_t^{1-\alpha}e_{n-1}$ from being directly incorporated into the actual control signal u, further action as follows will be taken.

According to Property 1.2, ${}_0\mathcal{D}_t^{1-\alpha}e_{n-1}$ can be expressed as

$$
\begin{aligned}
{}_0\mathcal{D}_t^{1-\alpha}e_{n-1} =& \frac{1}{\Gamma(1-1+\alpha)}\Big(\int_0^t \frac{\dot{q}_{n-1}(s)}{(t-s)^{1-\alpha}}\mathrm{d}s - \int_0^t \frac{\dot{\beta}_{n-2}(s)}{(t-s)^{1-\alpha}}\mathrm{d}s\Big)\\
=& \frac{\int_0^t \dot{q}_{n-1}(t-s)^{\alpha-1}\mathrm{d}s - \int_0^t \dot{\beta}_{n-2}(t-s)^{\alpha-1}\mathrm{d}s}{\Gamma(1-1+\alpha)}\\
=& \frac{\int_0^t (q_n + \phi_{n-1}^T\vartheta_{n-1})(t-s)^{\alpha-1}\mathrm{d}s}{\Gamma(\alpha)} - \frac{1}{\Gamma(\alpha)}\int_0^t\Big(\sum_{p=1}^{n-2}\frac{\partial\beta_{n-2}}{\partial q_p}q_{p+1} + \sum_{p=1}^{n-2}\frac{\partial\beta_{n-2}}{\partial q_p}\\
& \phi_p^T\vartheta_p + \sum_{p=1}^{n-2}\frac{\partial\beta_{n-2}}{\partial\hat{\vartheta}_{v,p}}\dot{\hat{\vartheta}}_{v,p} + \sum_{p=1}^{n-1}\frac{\partial\beta_{n-2}}{\partial y_r^{(p-1)}}y_r^{(p)} + \sum_{p=1}^{n-1}\frac{\partial\beta_{n-2}}{\partial h^{(p-1)}}h^{(p)}\Big)(t-s)^{\alpha-1}\mathrm{d}s,
\end{aligned}
\tag{9.32}
$$

where from the second equation on the right-hand side, $\dot{q}_{n-1}(s)$ and $\dot{\beta}_{n-2}(s)$ are written as $\dot{q}_{n-1}$ and $\dot{\beta}_{n-2}$ for simplification. In light of (9.32) and Definition 1.13, (9.31) turns into

$$
\begin{aligned}
\dot{\beta}_{n-1} =& \sum_{p=2}^{n}\frac{\partial\beta_{n-1}'}{\partial q_{p-1}}q_p + \sum_{p=1}^{n-1}\Big(\frac{\partial\beta_{n-1}'}{\partial q_p}\phi_p^T\vartheta_p + \frac{\partial\beta_{n-1}'}{\partial\hat{\vartheta}_{v,p}}\dot{\hat{\vartheta}}_{v,p}\Big) + \sum_{p=1}^{n}\Big(\frac{\partial\beta_{n-1}'}{\partial y_r^{(p-1)}}y_r^{(p)}\\
& + \frac{\partial\beta_{n-1}'}{\partial h^{(p-1)}}h^{(p)}\Big) - a_0I_t^\alpha q_n - a_0I_t^\alpha\phi_{n-1}^T\vartheta_{n-1} + a_0I_t^\alpha\Big(\sum_{p=1}^{n-2}\frac{\partial\beta_{n-2}}{\partial q_p}q_{p+1}\\
& + \sum_{p=1}^{n-2}\frac{\partial\beta_{n-2}}{\partial q_p}\phi_p^T\vartheta_p + \sum_{p=1}^{n-2}\frac{\partial\beta_{n-2}}{\partial\hat{\vartheta}_{v,p}}\dot{\hat{\vartheta}}_{v,p} + \sum_{p=1}^{n-1}\frac{\partial\beta_{n-2}}{\partial y_r^{(p-1)}}y_r^{(p)} + \sum_{p=1}^{n-1}\frac{\partial\beta_{n-2}}{\partial h^{(p-1)}}h^{(p)}\Big),
\end{aligned}
\tag{9.33}
$$

which, with Lemma 2.1, results in

$$
\begin{aligned}
\dot{W}_n =& -\sum_{p=1}^{n-1}c_pe_p^2 + e_{n-1}e_n - e_{n-1}a_0I_t^\alpha e_{n-1} + e_n(\mu gu + d + \phi_{v,n}^T\vartheta_{v,n} + \psi_n)\\
\le& -\sum_{p=1}^{n-1}c_pe_p^2 + e_{n-1}e_n - e_{n-1}a_0I_t^\alpha e_{n-1} + e_n\mu gu + \delta\bar{d} + \frac{e_n^2}{\sqrt{e_n^2+\delta^2}}\bar{d}\\
& + e_n\phi_{v,n}^T\vartheta_{v,n} + e_n\psi_n
\end{aligned}
\tag{9.34}
$$

with $\delta(t) = e^{-rt}$ $(r > 0)$ and

$$
\begin{aligned}
\phi_{v,n} =& \Big[\phi_n^T, -\frac{\partial\beta_{n-1}'}{\partial q_{n-1}}\phi_{n-1}^T + a_0I_t^\alpha\phi_{n-1}^T, -\frac{\partial\beta_{n-1}'}{\partial q_{n-2}}\phi_{n-2}^T - a_0I_t^\alpha\frac{\partial\beta_{n-2}}{\partial q_{n-2}}\phi_{n-2}^T, \cdots,\\
& -\frac{\partial\beta_{n-1}'}{\partial q_1}\phi_1^T - a_0I_t^\alpha\frac{\partial\beta_{n-2}}{\partial q_1}\phi_1^T\Big]^T, \ \vartheta_{v,n} = \left[\vartheta_n^T, \cdots, \vartheta_1^T\right]^T,\\
\psi_n =& -\sum_{p=2}^{n}\frac{\partial\beta_{n-1}'}{\partial q_{p-1}}q_p - \sum_{p=1}^{n-1}\frac{\partial\beta_{n-1}'}{\partial\hat{\vartheta}_{v,p}}\dot{\hat{\vartheta}}_{v,p} - \sum_{p=1}^{n}\Big(\frac{\partial\beta_{n-1}'}{\partial y_r^{(p-1)}}y_r^{(p)} + \frac{\partial\beta_{n-1}'}{\partial h^{(p-1)}}h^{(p)}\Big)\\
& + a_0I_t^\alpha q_n - a_0I_t^\alpha\Big(\sum_{p=2}^{n-1}\frac{\partial\beta_{n-2}}{\partial q_{p-1}}q_p + \sum_{p=1}^{n-2}\frac{\partial\beta_{n-2}}{\partial\hat{\vartheta}_{v,p}}\dot{\hat{\vartheta}}_{v,p} + \sum_{p=1}^{n-1}\frac{\partial\beta_{n-2}}{\partial y_r^{(p-1)}}y_r^{(p)}\\
& + \sum_{p=1}^{n-1}\frac{\partial\beta_{n-2}}{\partial h^{(p-1)}}h^{(p)}\Big).
\end{aligned}
$$

The actual control input is designed as

$$u = \frac{\hat{\zeta}}{g}\beta_n, \tag{9.35}$$

in which $\hat{\zeta}$ stands for the estimate of $\zeta = \frac{1}{\mu}$, and

$$\beta_n = -e_{n-1} - c_n e_n - \frac{e_n}{\sqrt{e_n^2 + \delta^2}}\hat{d} - \psi_n + e_{n-1} a f_e(e_n, \nu)_0 I_t^\alpha e_{n-1} - \phi_{v,n}^T \hat{\vartheta}_{v,n} - u_c \tag{9.36}$$

with $\nu \in (0,1)$, $c_n > 0$, $\hat{\vartheta}_{v,n}$ and $\hat{d}$ respectively representing the estimates of $\vartheta_{v,n}$ and $\bar{d}$,

$$f_e(e_n, \nu) = \begin{cases} \frac{e_n}{e_n^2 - \nu^2 + \nu}, & |e_n| < \nu \\ \frac{1}{e_n}, & |e_n| \geq \nu, \end{cases} \tag{9.37}$$

and

$$u_c = \begin{cases} \omega(q)\hat{\chi}, & |e_n| < \nu \\ 0, & |e_n| \geq \nu, \end{cases} \tag{9.38}$$

where $\omega(q)$ is a known continuous function vector and $\hat{\chi}$ denotes the estimate of an unknown constant vector χ. Consider Lyapunov candidate function $\overline{W}_{nn} = \overline{W}_n + \frac{|\mu|}{2\gamma}\tilde{\zeta}^2$, where $\gamma > 0$ and $\tilde{\zeta} = \hat{\zeta} - \zeta$, and design adaptive law

$$\dot{\hat{\zeta}} = -\gamma \text{sign}(\mu) e_n \beta_n \tag{9.39}$$

based on Assumption 9.2 with sign(μ) indicating the sign of μ, the derivative of $\overline{W}_{nn}$ is obtained as follows

$$\begin{aligned} \dot{\overline{W}}_{nn} \leq & -\sum_{p=1}^{n-1} c_p e_p^2 + e_{n-1}e_n - e_{n-1} a_0 I_t^\alpha e_{n-1} + \delta\bar{d} + \frac{e_n^2}{\sqrt{e_n^2 + \delta^2}}\bar{d} \\ & + e_n \phi_{v,n}^T \vartheta_{v,n} + e_n \psi_n + e_n \mu(\zeta + \tilde{\zeta})\beta_n + \frac{|\mu|}{\gamma}\tilde{\zeta}\dot{\hat{\zeta}} \\ = & -\sum_{p=1}^{n} c_p e_p^2 - e_{n-1} a_0 I_t^\alpha e_{n-1} + \delta\bar{d} - \frac{e_n^2}{\sqrt{e_n^2 + \delta^2}}\tilde{d} \\ & + a e_{n-1} e_n f_e(e_n, \nu)_0 I_t^\alpha e_{n-1} - e_n \phi_{v,n}^T \tilde{\vartheta}_{v,n} - e_n u_c \end{aligned} \tag{9.40}$$

with $\tilde{\vartheta}_{v,n} = \hat{\vartheta}_{v,n} - \vartheta_{v,n}$ and $\tilde{d} = \hat{d} - \bar{d}$.

For $|e_n| \geq \nu$, the following inequality can be gained in accordance with (9.37) and (9.38)

$$\dot{\overline{W}}_{nn} \leq -\sum_{p=1}^{n} c_p e_p^2 + \delta\bar{d} - \frac{e_n^2}{\sqrt{e_n^2 + \delta^2}}\tilde{d} - e_n \phi_{v,n}^T \tilde{\vartheta}_{v,n}. \tag{9.41}$$

For $|e_n| < \nu$, from the definition of e_n and β_{n-1} yields

$$|e_n| = \left| q_n + e_{n-2} + c_{n-1}e_{n-1} + \phi_{v,n-1}^T \hat{\vartheta}_{v,n-1} + \psi_{n-1} + a_0 I_t^\alpha e_{n-1} \right| < \nu, \tag{9.42}$$

based on which indicates that the term $\left| a e_{n-1} {}_0 I_t^\alpha e_{n-1} \left(e_n \frac{e_n}{e_n^2 - \nu^2 + \nu} - 1 \right) \right|$ is upper bounded by a bound whose value is related to the value of e_n and the following condition satisfies in practice

$$\left| a e_{n-1} {}_0 I_t^\alpha e_{n-1} \left(e_n \frac{e_n}{e_n^2 - \nu^2 + \nu} - 1 \right) \right| \leq e_n \omega(q)\chi, \tag{9.43}$$

hence it together with (9.40) gives

$$\dot{\overline{W}}_{nn} \leq -\sum_{p=1}^{n} c_p e_p^2 + \delta\bar{d} - \frac{e_n^2}{\sqrt{e_n^2+\delta^2}}\tilde{d} - e_n\omega(q)\tilde{\chi} - e_n\phi_{v,n}^T\tilde{\vartheta}_{v,n} \tag{9.44}$$

where $\tilde{\chi} = \hat{\chi} - \chi$. Ultimately, the adaptive laws formulated below

$$\dot{\hat{\vartheta}}_{v,n} = e_n\Lambda_n\phi_{v,n}, \tag{9.45}$$

$$\dot{\hat{d}} = \kappa\frac{e_n^2}{\sqrt{e_n^2+\delta^2}}, \tag{9.46}$$

$$\dot{\hat{\chi}} = \iota e_n\omega(q), \tag{9.47}$$

with Λ_n being a positive definite constant matrix and $\kappa, \iota > 0$ result in the temporal derivative of the Lyapunov candidate function $W_n = \overline{W}_{nn} + \frac{1}{2}\tilde{\vartheta}_{v,n}^T\Lambda_n^{-1}\tilde{\vartheta}_{v,n} + \frac{1}{2\kappa}\tilde{d}^2 + \frac{1}{2\iota}\tilde{\chi}^2$ meets

$$\begin{aligned}\dot{W}_n \leq & -\sum_{p=1}^{n} c_p e_p^2 + \delta\bar{d} - \frac{e_n^2}{\sqrt{e_n^2+\delta^2}}\tilde{d} - e_n\omega(q)\tilde{\chi} - e_n\phi_{v,n}^T\tilde{\vartheta}_{v,n} \\ & + \tilde{\vartheta}_{v,n}^T\Lambda_n^{-1}e_n\Lambda_n\phi_{v,n} + \frac{1}{\kappa}\tilde{d}\kappa\frac{e_n^2}{\sqrt{e_n^2+\delta^2}} + \frac{1}{\iota}\tilde{\chi}e_n\omega(q)\iota \\ = & -\sum_{p=1}^{n} c_p e_p^2 + \delta\bar{d}. \end{aligned} \tag{9.48}$$

Remark 9.4 *In the studies [164, 165, 166, 167, 168] regarding the fractional prescribed performance control for continuous integer-order nonlinear systems, systems of order not higher than two are taken into account. In particular, the control law proposed in [166] utilizes the* sign *function to attain asymptotic stabilization of output error, yet it may lead to chattering. In contrast, such restrictions present in existing works are conquered in the proposed fractional-order PPB control scheme. Furthermore, the controllers developed in [166] and [164, 167, 168] respectively directly contain non-causal integer and fractional derivative operators, which could amplify system measurement noises or disturbances and potentially create difficulties in the practical implementation of control signals. For avoiding such implementation difficulty, the direct containment of differential operation in the control signal should be averted, however, finding a way to solve it is not straightforward. If following the design idea in [164, 167, 168] when developing our investigated fractional controller, the obtained control input will straightly include fractional derivative operation as shown below*

$$\begin{aligned}\underline{u} = & \frac{\hat{\varsigma}}{g}\Big[-e_{n-1} - c_n e_n - \frac{e_n}{\sqrt{e_n^2+\delta^2}}\hat{d} - \underline{\psi}_n - \underline{\phi}_{v,n}^T\hat{\vartheta}_{v,n} \\ & + e_{n-1}af_e(e_n,\nu)_0I_t^\alpha e_{n-1} - a_0D_t^{1-\alpha}e_{n-1} - u_c\Big]\end{aligned} \tag{9.49}$$

with $\underline{\psi}_n = -\sum_{p=2}^{n}\frac{\partial\beta'_{n-1}}{\partial q_{p-1}}q_p - \sum_{p=1}^{n-1}\frac{\partial\beta'_{n-1}}{\partial\hat{\vartheta}_{v,p}}\dot{\hat{\vartheta}}_{v,p} - \sum_{p=1}^{n}\left(\frac{\partial\beta'_{n-1}}{\partial y_r^{(p-1)}}y_r^{(p)} + \frac{\partial\beta'_{n-1}}{\partial h^{(p-1)}}h^{(p)}\right)$ *and* $\underline{\phi}_{v,n} = \left[\phi_n^T, -\frac{\partial\beta'_{n-1}}{\partial q_{n-1}}\phi_{n-1}^T, \cdots, -\frac{\partial\beta'_{n-1}}{\partial q_1}\phi_1^T\right]^T$*. Targeting circumventing the direct incorporation of the fractional derivative term in the control signal, this chapter offers an approach by introducing the action described in (1.26) and (9.32), resulting in the proposed control signal described in (9.35).*

9.4.2 Stability Analysis

The main results of this chapter are summarized in the following theorem.

Theorem 9.1 *Take into account the adaptive closed-loop system consisting of integer-order nonlinear system (9.1) and PPB adaptive fractional backstepping controller (9.35) with parameter update laws (9.13), (9.19), (9.22), (9.28), (9.39), (9.45), (9.46), and (9.47) under Assumption 9.1 to Assumption 9.3. All signals in the closed-loop system are ensured to be bounded with output tracking error* $e_h(t)$ *converging to zero asymptotically. Additionally, the system transient performance is assured with* $e_h(t)$ *always remaining within the given prescribed performance bound, meaning* $-\underline{\lambda}h(t) \leq e_h(t) \leq \bar{\lambda}h(t)$ *with* $t \geq 0$.

Proof: Taking integration of (9.48) gives

$$W_n(t) - W_n(0) \leq -\sum_{p=1}^{n} \int_0^t c_p e_p(s)^2 \mathrm{d}s + \frac{\bar{d}}{r} - \frac{\bar{d}}{r} e^{-rt} \leq \frac{\bar{d}}{r}, \tag{9.50}$$

which demonstrates the boundedness of W_n and thus e_p $(p = 1, \cdots, n)$, $\tilde{\vartheta}_{v,p}$, $\tilde{\zeta}$, $\tilde{d}$, and $\tilde{\chi}$ are assured to be bounded. By analyzing equations (9.12), (9.17), and (9.21) in conjunction with the boundedness stated in Assumption 9.1 on $y_r^{(p)}$, we can conclude that β_j $(j = 1, \cdots, n-2)$ are bounded. When $\alpha \in [0, 1)$, $t^{\alpha-1}$ will decrease to zero monotonically as t increases and it gives $t^{\alpha-1} = 1$ $(\forall t \geq 0)$ when $\alpha = 1$, thus the boundedness of ${}_0I_t^{\alpha} e_{n-1}$ can be maintained by the boundedness of e_{n-1} with ${}_0I_t^{\alpha} e_{n-1} = \frac{1}{\Gamma(\alpha)}(e_{n-1} * t^{\alpha-1})$ according to (1.29), in which $*$ signifies convolution operator. It follows from (9.26) that β_{n-1} keeps bounded and the boundedness of control signal u shown in (9.35) can be ensured analogously. Thus by the boundedness of h, e_p and β_j, q_p can be ensured to stay bounded.

From (9.8), we know $z_1 = e_1$ is bounded, thus according to the injective property of $Y(z_1)$, for arbitrary initial tracking error $e_h(0)$, (9.2) is met for $t \geq 0$ with $\underline{\lambda}$, $\bar{\lambda}$, and $h(0)$ being selected such that $-\underline{\lambda}h(0) \leq e_h(0) \leq \bar{\lambda}h(0)$ holds. Furthermore, from (9.50) it shows that $\lim_{t\to\infty} \sum_{p=1}^{n} \int_0^t c_p e_p(s)^2 \mathrm{d}s$ is bounded, hence by Barbalat's lemma, we obtain $\lim_{t\to\infty} e_p = 0, (p = 1, \cdots, n)$, which according to characterization 3 of function $Y(z_1)$, i.e., $Y(0) = 0$, $\lim_{t\to\infty} e_h = 0$ is achieved, indicating asymptotic output tracking. ■

Remark 9.5 *Significantly, fractional calculus is incorporated into the proposed control scheme (9.35) and (9.36), allowing the fractional operation-related parameters* a *and* α *to be adjusted for further improved control performance with respect to transient and steady-state. The proposed adaptive fractional PPB controller can be simplified back to the conventional integer-order one found in for example [222, 223, 224] by setting the control parameters* a *and* α *as zero.*

Remark 9.6 *In comparison with traditional integer-order control cases, the fractional-order calculus terms involved in the controller enable enhanced transient performance with a faster convergence rate and lower maximum overshoot of* $e_h(t)$, *as will be demonstrated in Section 9.5.*

Remark 9.7 *As clarified in the first contribution of this chapter, in Chapter 8, asymptotic output tracking to the given reference signal cannot be realized. Such difficulty of not being able to achieve asymptotic output tracking is mainly caused by the term related to the fractional integral of virtual error that appears in the last step of the backstepping-based controller design procedure. In our designed controller, a novel term* u_c *(9.38) is proposed targeting overcoming this mentioned challenge and thus resulting in asymptotic output tracking.*

9.5 Simulation Verification

The numerical tool named the Oustaloup filter presented in [225] is resorted to realize fractional integrals operation in this section. A presentation of the dynamic characteristics of the permanent magnet linear synchronous motor [20] model is given below

$$\begin{cases} \dot{q}_1 = q_2 + \phi_1^T(q_1)\vartheta_1 \\ \dot{q}_2 = \mu u + \phi_2^T(q)\vartheta_2 + d, \end{cases} \tag{9.51}$$

in which $\phi_1 = 0$, $\phi_2 = -115.025q_2$, $\mu = 180.33$, $\vartheta_2 = 1$, and d denotes the unknown changing disturbance. The two states q_1 and q_2 of the mover of the motor stand for the position and velocity of the winding respectively, with $y = q_1$ being the output and the winding current i_q serving as the control input u. The reference path is defined as $y_r = \sin(0.5t)$. The aim of the control is to guarantee that the closed-loop system is stable and the winding position $y = q_1$ can follow a given signal y_r asymptotically in the presence of system uncertainty and external disturbance with an ensured transient performance of $e_h(t) = y(t) - y_r(t)$.

9.5.1 Illustration of the Proposed Scheme and Comparisons with Integer-Order Control Scheme

To demonstrate the effectiveness of the proposed scheme, we first choose d as a combination of sine and cosine components with noise and a unit step function at $t = 2$ s, being equal to $0.01[\sin(t) + \cos(t) + U(t-2)] + Z(t)$, where $Z(t)$ denotes the white Gaussian noise with power 0.01.

The design of the control scheme, with prescribed performance bound, follows the procedure specified in Subsection 9.4.1. The initial values for the simulation are assigned to be $q(0) = [0.2, 0]^T$ and $\hat{\varsigma}(0) = 0.005$, $\hat{\vartheta}_2(0) = 0.8$, $\hat{d}(0) = 0.6$ respectively for system states and parameter estimates. A specific bound on output error performance is determined by setting $h(t) = 0.4e^{-0.5t} + 0.01$ with $\bar{\lambda} = 0.5$ and $\underline{\lambda} = 0.6$. The control parameters are chosen to be $c_1 = 0.5$, $c_2 = 0.05$, $a = 4$, $\delta(t) = e^{-rt}$ with $r = 0.05$, $\gamma = 10$, $\Lambda_2 = 10$, $\kappa = 1$, $\iota = 10$, and $\nu = 0.01$. For $\omega(q)$, it takes q_1 and q_2 as inputs and correspondingly defines six Gaussian membership functions uniformly distributed over $[-0.05, 0.05]$ for each input. The starting value for the estimation of χ is set to be $\hat{\chi}(0) = [1, \cdots, 1]^T \in \mathbb{R}^{36}$. As mentioned in Remark 9.5, setting parameters α and a to be zero results in the traditional integer-order PPB adaptive backstepping method [222, 223, 224]. For comparison, the tracking error performance using the traditional integer-order method is also considered with the same design parameters.

Fig. 9.1 and Fig. 9.2 both depict the tracking error $e_h(t) = y(t) - y_r(t)$. Fig. 9.1 demonstrates that when $\alpha = 1$, the designed controller is equivalent to one containing an integer integral term. Although the overall maximum overshoot can be reduced by the inclusion of the integer integral, it also causes considerable oscillation as a side effect. On the other hand, Fig. 9.2 illustrates that the proposed fractional adaptive PPB controller with fractional values of α set to be 0.2, 0.3 and 0.4 respectively results in faster convergence rate and lower overshoot of the output error, compared to the case with the traditional control method where $\alpha = a = 0$. In addition, no additional oscillation will be generated compared to the scenario where the controller has a value of $\alpha = 1$.

Fig. 9.3 shows the comparison between the typical performance index integral of squared error (ISE) when utilizing the proposed fractional adaptive PPB control scheme with various

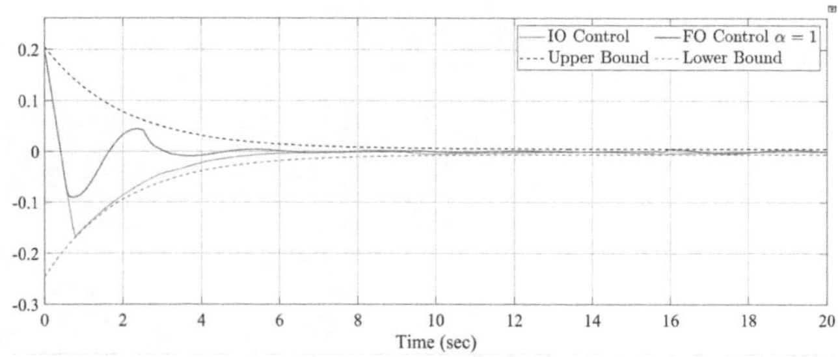

FIGURE 9.1: Tracking error of (9.51) under conventional integer controller ($\alpha = a = 0$) and proposed controller with integer integral $\alpha = 1$.

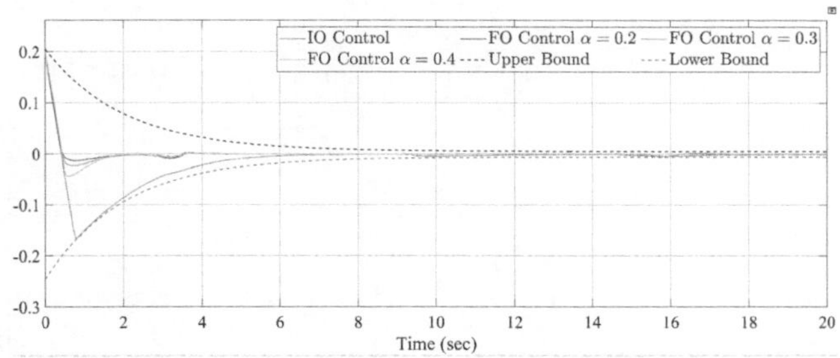

FIGURE 9.2: Tracking error of (9.51) under the proposed controller with various α and conventional integer controller ($\alpha = a = 0$).

α and the conventional integer design method, emphasizing the superiority of our designed controller. Furthermore, the magnitudes of fractional control inputs are even lower than the conventional integer ones, as demonstrated in Fig. 9.4, which, together with Fig. 9.1 to Fig. 9.3, suggests that the proposed control scheme can achieve better performances with less required control effort.

As observed from Fig. 9.1 to Fig. 9.4 that for the considered system (9.51) under our proposed fractional controller, the maximum overshoot of tracking error e_h, the value of ISE, and the maximum magnitude of the control signal will decrease with the reduction of fractional order α. Also, larger c_1, c_2, γ, Λ_2, κ, and ι and smaller ν can also reduce the value of ISE, and the largest overshoot of e_h, which, in contrast, will enlarge the value of control input. As a result, when selecting controller parameters for a specific system, the balance between achieving better control performance and the required amount of control effort should be considered.

Besides, to show the robustness of the proposed control method, another simulation study is carried out under the same controller parameters by handling an alternative disturbance to the system $d_2 = 8[\sin(20t) + \cos(20t) + U(t-8)] + Z(t)$, whose magnitude and frequency are respectively 800 and 20 times of d and the appearance of step function is later than that of d. As shown in Fig. 9.5, asymptotic output tracking is still ensured under the developed fractional control method with output error constantly remaining in the predefined bound even in the presence of external disturbance with dramatically higher amplitude and frequency. In contrast, slightly higher control effort is required to compensate for d_2, as displayed in Fig. 9.6, which are indeed practically reasonable.

To highlight the advantage of fractional control in improving control performance, an alternative fractional controller is designed under a further restricted prescribed bound on system output error $h_2(t) = 0.405e^{-1.3t} + 0.005$ with $\bar{\lambda} = 0.5$ and $\underline{\lambda} = 0.6$. Under the same design parameters as the afore-developed fractional controller, improved

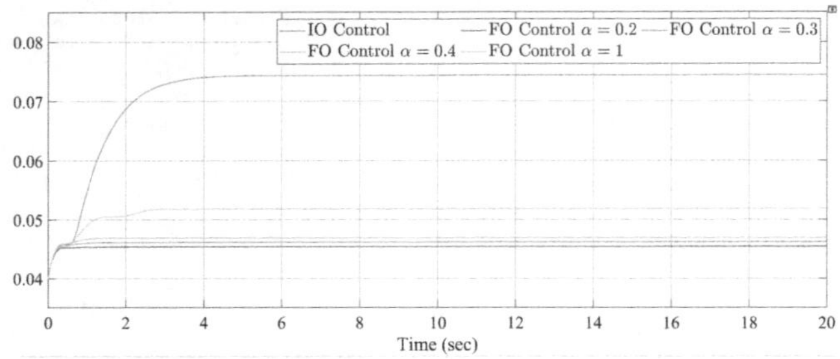

FIGURE 9.3: Performance index ISE with respect to tracking error of (9.51) under proposed fractional controller with various α and conventional integer controller ($\alpha = a = 0$).

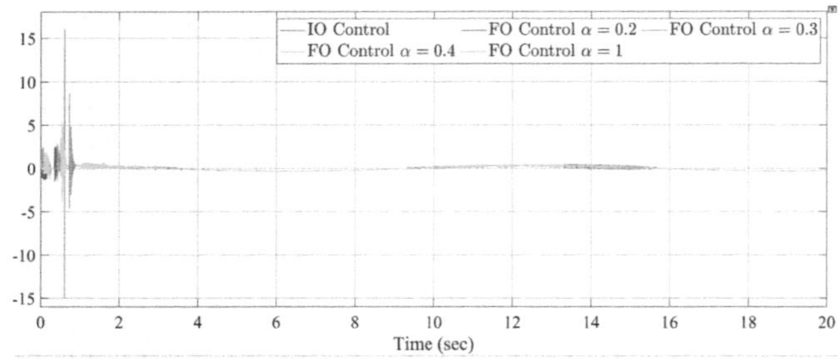

FIGURE 9.4: Control input of (9.51) under proposed fractional controller with various α and conventional integer controller ($\alpha = a = 0$).

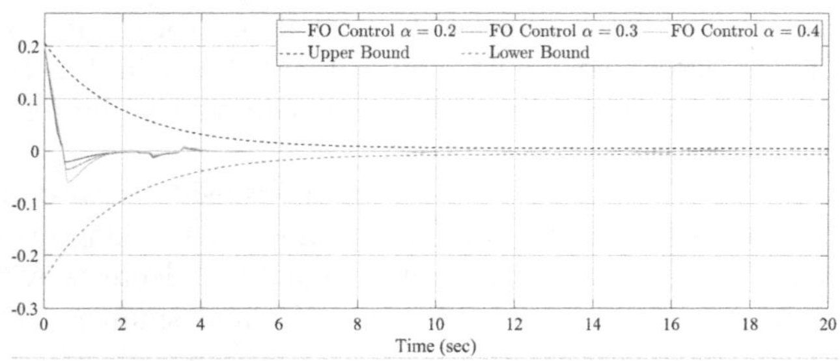

FIGURE 9.5: Tracking error of (9.51) with alternative disturbance d_2 under proposed fractional controller.

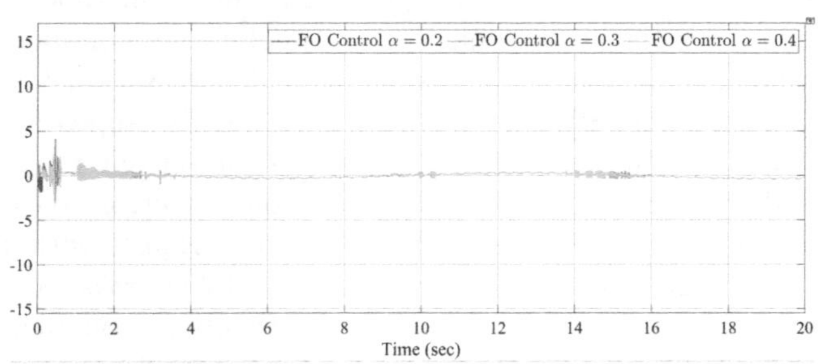

FIGURE 9.6: Control input of (9.51) with alternative disturbance d_2 under proposed fractional controller.

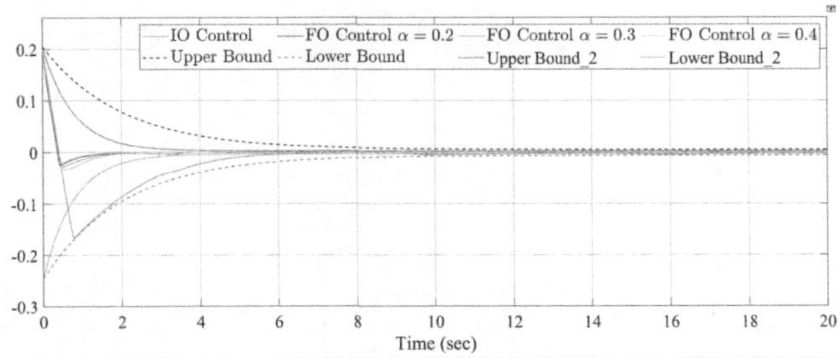

FIGURE 9.7: Tracking error of (9.51) under proposed fractional controller with alternative bound h_2 and conventional controller ($\alpha = a = 0$) with bound h.

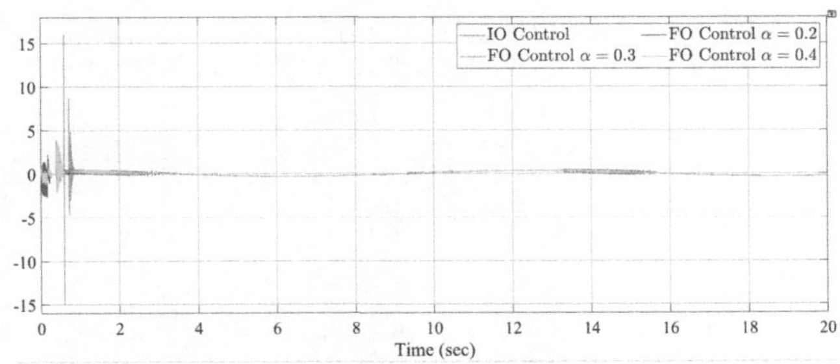

FIGURE 9.8: Control input of (9.51) under proposed fractional controller with alternative bound h_2 and conventional controller ($\alpha = a = 0$) with bound h.

transient performance of e_h in terms of decreased h_∞ and faster convergence rate (i.e., larger c of e^{-ct} in h_2) can be obtained, as shown in Fig. 9.7, where "Upper Bound_2" and "Lower Bound_2" represent the bounds determined by $h_2(t)$. For comparison, the system output error under the traditional integer control scheme that $\alpha = a = 0$ with predefined bound $h(t) = 0.4e^{-0.5t} + 0.01$ discussed previously is also included in it. By respectively comparing Fig. 9.7 with Fig. 9.2, Fig. 9.8 with Fig. 9.4, and Fig. 9.9 with Fig. 9.3, we can observe that the proposed fractional control strategy can provide a more advanced output error performance than the traditional controller with further constrained prescribed performance bounds (faster convergence rate and lower overshoot and h_∞), reduced control costs, and lower value of the performance index ISE.

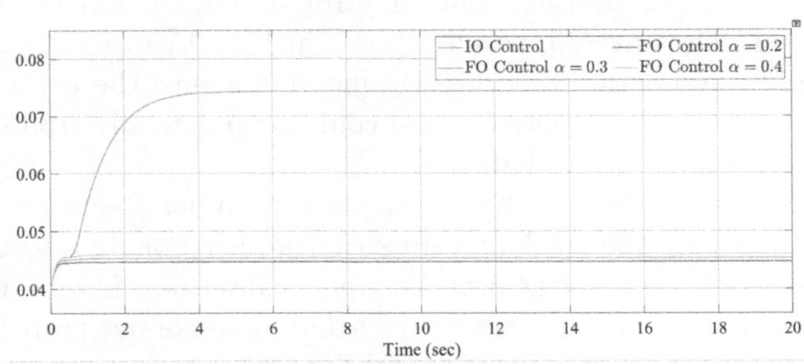

FIGURE 9.9: Performance index ISE with respect to tracking error of (9.51) under the proposed fractional controller with alternative bound h_2 and conventional controller ($\alpha = a = 0$) with bound h.

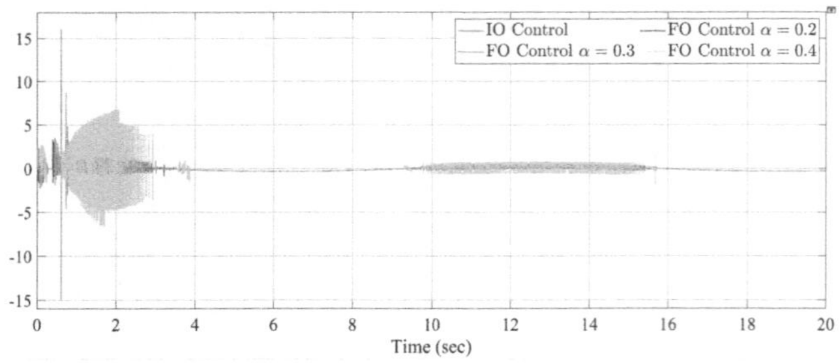

FIGURE 9.10: Control input $\underline{u}$ of (9.51) under the directly containing derivative operation fractional controller with various α and conventional integer controller ($\alpha = a = 0$).

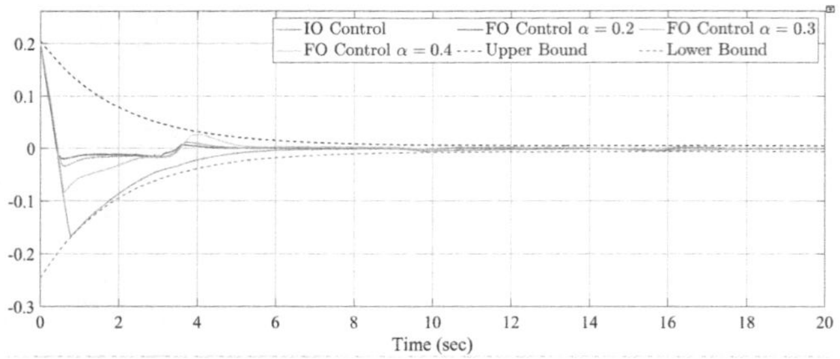

FIGURE 9.11: Tracking error of (9.51) under the directly containing derivative operation fractional controller with various α and conventional integer controller ($\alpha = a = 0$).

9.5.2 Other Comparative Studies

In addition to the comparison results with those under traditional integer control controller presented above, to further illustrate the effectiveness of the proposed control scheme, a comparative study with designed parameters remaining unchanged is conducted under control input $\underline{u}$ given in (9.49), where fractional derivative operation is straightly contained. By comparing Fig. 9.10 with Fig. 9.4, it can be found that control inputs $\underline{u}$ display considerably large amplitude chattering that persists for a long period of time, which demonstrates difficulties in practical implementation. Besides, Fig. 9.11 and Fig. 9.12 respectively show the system output error and the performance index ISE with control input $\underline{u}$. Compared with Fig. 9.2 and Fig. 9.3, we observe that under fractional controller directly including fractional derivative, larger transient output error overshoot and higher value of performance index appear. Thus overall, results shown in Fig. 9.10 to Fig. 9.12 indicate that if the fractional derivative operation is directly integrated into the controller as (9.49), not only the implementation of the control signal could be practically troublesome but weaken control performance might also be induced.

Additionally, Fig. 9.13 presents the tracking errors under the fractional controller proposed in Chapter 8, where the tracking errors can neither remain into the more restricted bound h_2 or the relaxed one h nor achieve asymptotically towards zero, which consequently further emphasizes and affirms the contributions and merits of our proposed control method in this chapter by comparing with Fig. 9.2 and Fig. 9.7.

The studies presented in this section corroborate the effectiveness of the developed fractional control method and demonstrate the established theoretical findings. As a result, even though the implementation of the fractional controller involving fractional-order integral based on the Oustaloup filter approximation method could be more intricate than that of integer-order ones, the additional complexity in numerically realizing fractional-order

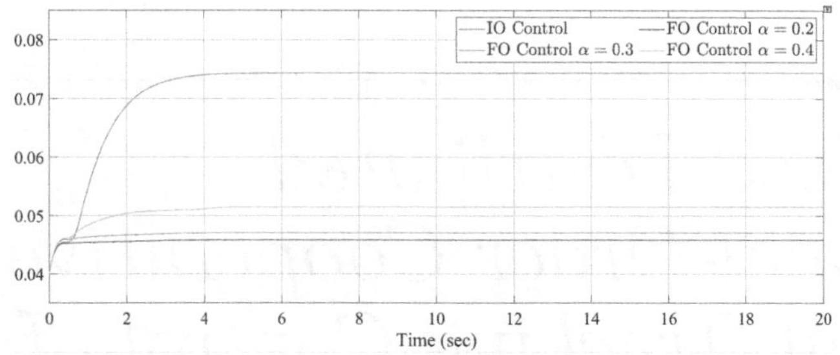

FIGURE 9.12: Performance index ISE with respect to tracking error of (9.51) under the directly containing derivative operation fractional controller with various α and conventional integer controller ($\alpha = a = 0$).

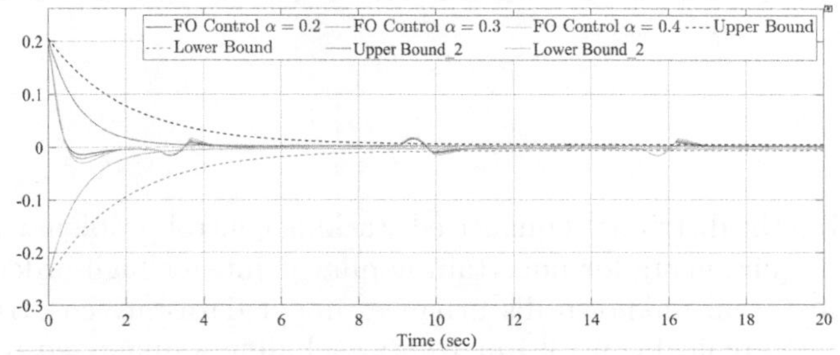

FIGURE 9.13: Tracking error of (9.51) under the fractional controller proposed in Chapter 8.

control on modern digital processors or computers that possess fast and mature computing capability can be insignificant, compared to the benefits brought by fractional control. Nevertheless, due to the complex nature of fractional calculus, the effort taken to reduce the computational burden associated with fractional control methods is worthwhile, making it a potential area of focus for future research.

9.6 Conclusions

A novel adaptive fractional backstepping control technique that ensures a prescribed performance for high-order nonlinear integer-order systems in the presence of systematic uncertainties and bounded unknown disturbances is developed in this chapter. The newly devised control approach eliminates the requirement for non-causal differential operations in the implementation of control signals, in comparison to current relevant studies that concentrate on second-order nonlinear integer-order systems. The proposed fractional-order control technology is capable of guaranteeing the stability of the closed-loop system and ensuring the system output tracks the desired trajectory asymptotically. Moreover, the prescribed transient performance on the output error is concurrently assured by the investigated fractional controller. Simulation findings affirm the validity of the theoretical outcomes and demonstrate the superiority of the novel developed fractional control method over the conventional integer approach.

10

Quantized Distributed Fractional-Order Cooperative Resilient Tracking Control of Nonlinear Uncertain Integer-Order MASs with Prescribed Performance

This chapter studies the distributed quantized tracking control problem with prescribed output performance requirements for uncertain nonlinear integer high-order MASs with both DoS attacks and external unknown disturbances under data-rate-constrained communication networks. A hierarchical control framework including a quantized resilient distributed observer layer and a decentralized quantized adaptive fractional-order controller layer is proposed for solving the considered control problem. Specifically, quantized resilient distributed observers are designed at the observer layer for each agent to guarantee the availability of observing the reference system under the influence of data-rate-constrained communication and DoS attacks. Based on the designed observers, smooth reference trajectories with the existence of high-order derivatives, which ensure the applicability of the backstepping technique, are then constructed at the decentralized controller layer, where a novel adaptive fractional-order PPB backstepping quantized control strategy is proposed for high-order integer uncertain systems, resulting in an enhanced control scheme with the integration of fractional calculus. It is theoretically proven that under data-rate constraints and DoS attacks, the output tracking error of each agent in relation to the reference signal will constantly strictly stay within the predefined performance limit under the proposed control scheme with all closed-loop signals remaining bounded. Simulation studies are provided as evidence to establish the efficacy of the proposed distributed quantized fractional adaptive PPB control strategy.

10.1 Introduction

In the realm of MASs, distributed tracking, whose goal is to enable all agents to successfully track a given reference to specific extents of interest, focuses on creating effective control strategies in a distributed manner and holds significant importance. One of its key applications lies in its applicability to address various engineering problems such as distributed secondary control for DC microgrids [226] and integrated energy system [227]. The effectiveness of tracking in MASs heavily hinges on the ability of agents to exchange information efficiently and reliably, as stated in the last paragraph of Subsection 1.3.3.

On the other hand, the open nature of the digital communication network also makes it susceptible to malicious cyber attacks, for instance, DoS attacks, as mentioned in the

DOI: 10.1201/9781003620372-10

fourth paragraph of Subsection 1.3.3. Motivated by addressing the controller design challenge caused by discontinuous signal transmission within a data-rate-constrained communication network under DoS attacks stated in the last paragraph of Subsection 1.3.3, developing a control scheme for handling tracking issues for general high-order uncertain nonlinear MASs to a given reference system is of significant practical importance. Besides, to the best of the authors' knowledge, no result on quantized adaptive fractional backstepping control for uncertain nonlinear MASs is available, let alone such control to the mentioned class of systems subject to data-rate-constrained communication and DoS attacks while simultaneously considering the guarantee of prescribed output performance. Therefore, given the restricted data transmission capacity in digital control systems, it becomes crucial in both theoretical and practical applications for enhancing control performance to design PPB quantized fractional-order control strategies for uncertain nonlinear MASs influenced by data-rate constraints and DoS attacks.

Inspired by the above discussions, the distributed quantized tracking control problem for uncertain nonlinear high-order MASs to a given reference system under adaptive quantized fractional-order backstepping technique within the DoS attacks-affected data-rate-constrained communication network is considered in this chapter. Compared to some existing works, the main contributions of this chapter are outlined below.

1. Different from [175], where a specific class of uncertain nonlinear MASs with local Lipschitz conditions is considered, on the basis of backstepping technique, this chapter presents the initial attempt to address the quantized adaptive fractional-order resilient distributed tracking problem for the general high-order uncertain nonlinear MASs without requirement on Lipschitz property, ensuring the agent output prescribed performance under data-rate-constrained communication networks with DoS attacks. A hierarchical control framework incorporating the quantized resilient distributed observer layer to observe the reference for each agent and the decentralized quantized adaptive fractional-order controller layer is put forward for solving the considered control problem.
2. By proposing a resilient observer for each agent, this chapter ensures that even not all agents have access to the reference information, the observer state converges to the reference signal generated by the reference system, showing distinguish from [175]. Due to the effect of non-continuous transmission of the MASs signals within the DoS attacks-influent and data-rate-constrained communication network, the time derivative of the observer state could be discontinuous, causing the implementation of a backstepping control strategy to be difficult. To tackle such a challenge not encountered in existing related work [175], inspired by the interpolation method, a novel smooth reference trajectory that owns high-order derivatives is constructed in this chapter at the local decentralized controller layer for each agent based on the designed observer.
3. A novel fractional adaptive quantized backstepping output PPB control solution for uncertain nonlinear high-order MASs is proposed at the local controller layer in this chapter to improve control performance, showing advancement from Chapter 8 which solely addresses limiting the system maximum output error. By such a method, remarkable enhancement in the control parameter adjustability and significant improvement in transient and steady-state output tracking performance can be achieved, outperforming the traditional integer backstepping control. Moreover, the designed fractional control scheme in this chapter deviates from those in previous works [167] and [168] by intentionally avoiding direct non-causal differential operations, thereby effectively addressing potential challenges in controller implementation.

The subsequent sections of this chapter are organized as follows. Section 10.2 offers the preliminaries and problem formulation for this chapter. The description of the data-rate constraints and DoS attacks is given in Section 10.3. Distributed observer and decentralized controller design procedure are presented in Section 10.4, provided with stability analysis. The verification of the proposed control scheme is conducted in Section 10.5 using simulation studies and Section 10.6 shows the conclusion of this chapter.

10.2 Problem Formulation

Communication Network Topology: The undirected communication graph $\mathcal{G} = \{\mathcal{V}, \mathcal{E}\}$ considered in this chapter consists of M agents. $\mathcal{V} = \{1, 2, \cdots, M\}$ and $\mathcal{E} = \{(i,j)|i,j \in \mathcal{V}\}$ respectively denote the set of agents and edges. $\mathcal{A}$ implies the adjacency matrix with $a_{ij} = 1$ if $(i,j) \in \mathcal{E}$ and $a_{ij} = 0$ otherwise, and a_{ii} in this chapter is defined to equal to 0, i.e., no self-loop in the communication graph. The undirected graph is connected if there is a sequence of connected edges between any two agents i and j. The in-degree matrix $\mathcal{D}$ is defined as $\mathcal{D} = \{deg_1, \cdots, deg_M\}$ with $deg_i = \sum_{j=1}^{M} a_{ij}$ $(i = 1, \cdots, M)$. The degree of $\mathcal{G}$ is $deg^* = \max_{1 \le i \le M} deg_i$, and the Laplacian matrix of $\mathcal{G}$ is $\mathcal{L} = \mathcal{D} - \mathcal{A}$.

10.2.1 Model Description

The dynamics of the ith uncertain nonlinear agent is

$$\begin{cases} \dot{x}_{i,q} = x_{i,q+1} + \varphi_{i,q}^T(\bar{x}_{i,q})\theta_{i,q},\ q = 1, \cdots, m-1 \\ \dot{x}_{i,m} = q(u_i) + \varphi_{i,m}^T(x_i)\theta_{i,m} + d_i,\ i = 1, \cdots, M, \end{cases} \tag{10.1}$$

with $y_i = x_{i,1}$, where $x_{i,q} \in \mathbb{R}$ denote the ith agent state with $\bar{x}_{i,q} = [x_{i,1}, \cdots, x_{i,q}]^T \in \mathbb{R}^q$ and $x_i = \bar{x}_{i,m}$, y_i and $q(u_i)$ are the output and quantized input of the agent, respectively. $\varphi_{i,j} \in \mathbb{R}^h$ $(j = 1, \cdots, m)$ represents the known smooth nonlinear function, the unknown constant vector is denoted by $\theta_{i,j} \in \mathbb{R}^h$, and d_i indicates the unknown time-varying external disturbance to the ith agent.

Assumption 10.1 *The maximum bound of the time-varying extrinsic disturbance d_i is specified by an unidentified positive constant $\bar{d}_i$.*

For the quantized control input $q(u_i)$, the hysteresis quantizer [117] is taken into account in this chapter. Define $u_{i,j} = \kappa_i^{1-j} u_{i,\min}$ $(j = 1, 2, \cdots)$, $\delta_i = \frac{1-\kappa_i}{1+\kappa_i}$, $0 < \kappa_i < 1$, $u_{i,\min} > 0$, and $q(u_i)$ is equal to

$$\begin{cases} u_{i,j}\mathrm{sign}(u_i), & \frac{u_{i,j}}{1+\delta_i} < |u_i| \le u_{i,j},\ \dot{u}_i < 0, \\ & \text{or } u_{i,j} < |u_i| \le \frac{u_{i,j}}{1-\delta_i},\ \dot{u}_i > 0, \\ u_{i,j}(1+\delta_i)\mathrm{sign}(u_i), & u_{i,j} < |u_i| \le \frac{u_{i,j}}{1-\delta_i},\ \dot{u}_i < 0, \\ & \text{or } \frac{u_{i,j}}{1-\delta_i} < |u_i| \le \frac{u_{i,j}(1+\delta_i)}{1-\delta_i},\ \dot{u}_i > 0, \\ 0, & 0 \le |u_i| < \frac{u_{i,\min}}{1+\delta_i},\ \dot{u}_i < 0, \\ & \text{or } \frac{u_{i,\min}}{1+\delta_i} \le u_i \le u_{i,\min},\ \dot{u}_i > 0, \\ q(u_i(t^-)), & \dot{u}_i = 0, \end{cases}$$

where $q(u_i)$ belongs to the set $Q = \{0, \pm u_{i,j}, \pm u_{i,j}(1+\delta_i)\}$.

Lemma 10.1 ([117]) *The quantization error satisfies*

$$|q(u_i) - u_i| \leq \delta_i |u_i| + u_{i,\min}. \tag{10.2}$$

The reference system for each agent to follow is

$$\dot{r} = Fr, \; y_r = Br, \tag{10.3}$$

with known matrix $F \in \mathbb{R}^{n\times n}$ and row vector $B \in \mathbb{R}^{1\times n}$.

Assumption 10.2 *The eigenvalues of matrix F lie on the imaginary axis.*

Assumption 10.3 *The undirected graph $\mathcal{G}$ is connected and there exists at least one agent that can obtain information from the reference.*

Remark 10.1 *The control method for uncertain nonlinear MASs developed in [175] mandates the nonlinearities to be locally Lipschitz continuous and the disturbances as well as their derivatives remain bounded. This chapter covers a wider array of MASs without the necessity of imposing Lipschitz conditions on nonlinearities or bounding the derivative of disturbance, in contrast to reference [175]. Besides, it is typical to incorporate Assumption 10.2 when dealing with the distributed tracking issue, which is often employed in control systems to even out irregular oscillations, decrease periodic instabilities stemming from issues like friction and saturation, strengthen system stability, and make the system more robust [228]. If Assumption 10.3 holds, then the eigenvalues of $\mathcal{P} = \mathcal{L} + \mathcal{H}$ satisfy $0 < \lambda_1 \leq \lambda_2 \leq \cdots \leq \lambda_M$, where $\mathcal{H} = diag\{a_{10}, \cdots, a_{M0}\}$ with $a_{i0} = 1\,(i = 1, \cdots, M)$ indicating the ith agent can receive information from the reference, and $a_{i0} = 0$ otherwise [229]. In addition, Lemma 10.1 and Lemma 2.1 will be applied for designing the decentralized quantized adaptive fractional-order controller.*

The control problem of this chapter is to design a hierarchical control method for the uncertain nonlinear high-order MASs (10.1), which includes a resilient distributed observation layer and local decentralized quantized adaptive fractional tracking layer, such that all agent outputs $y_i\,(i = 1, \cdots, M)$ can track the reference signal y_r in (10.3) with prescribed performance, even under the influence of data-rate constraints and DoS attacks, which will be further described in Section 10.3, and also not all agents have access to the reference information. In addition, improving the transient and steady-state system output performance compared to the traditional integer backstepping ones under the newly proposed PPB fractional backstepping control strategy is the other control objective of this chapter.

10.2.2 Transformed Agent System for Prescribed Performance in Local Controller

To address the above control problem, define $z_{i,1} = y_i - y_{vi}\,(i = 1, \cdots, M)$, where y_{vi} is the smooth trajectory of the observer of the reference signal to be designed later, the target of ensuring the prescribed performance [221] of each agent is realized by satisfying the following condition

$$-\underline{\lambda}_i h_i(t) \leq z_{i,1}(t) \leq \bar{\lambda}_i h_i(t), \tag{10.4}$$

where prescribed parameters $\underline{\lambda}_i < \bar{\lambda}_i$ are within $(0, 1]$, and $h_i(t) = (h_{i,0} - h_{i,\infty})e^{-c_i t} + h_{i,\infty}$ with $c_i > 0$ and $h_{i,0} > h_{i,\infty} > 0$ is a decreasing smooth function.

Provided the condition in (10.4), the prescribed performance problem can be reframed to the one that considers the boundedness of signals. By achieving asymptotic stabilization of the reformulated system, the output asymptotic tracking of the original systems can then be effectively accomplished. Define $\zeta_{i,1} = \frac{z_{i,1}}{h_i}$, $\tau_i = \frac{1}{2}\ln\left(\frac{\underline{\lambda}_i}{\bar{\lambda}_i}\right)$, and a strictly increasing smooth function $Y(a_{i,1}) = \frac{\bar{\lambda}_i e^{\tau_i + a_{i,1}} - \underline{\lambda}_i e^{-(\tau_i + a_{i,1})}}{e^{\tau_i + a_{i,1}} + e^{-(\tau_i + a_{i,1})}}$ with characterizations 1) $-\underline{\lambda}_i < Y(a_{i,1}) < \bar{\lambda}_i$; 2) $\lim_{a_{i,1} \to -\infty} Y(a_{i,1}) = -\underline{\lambda}_i$, $\lim_{a_{i,1} \to +\infty} Y(a_{i,1}) = \bar{\lambda}_i$; 3) $Y(0) = 0$, then the transformed error and its derivative are respectively represented as $a_{i,1} = Y^{-1}(\zeta_{i,1}) = \frac{\ln(\bar{\lambda}_i \zeta_{i,1} + \bar{\lambda}_i \underline{\lambda}_i)}{2} - \frac{\ln(\underline{\lambda}_i \bar{\lambda}_i - \underline{\lambda}_i \zeta_{i,1})}{2}$ and $\dot{a}_{i,1} = \frac{\partial Y^{-1}}{\partial \zeta_{i,1}} \dot{\zeta}_{i,1} = \eta_i \left(\dot{z}_{i,1} - \frac{z_{i,1} \dot{h}_i}{h_i} \right)$, where $\eta_i = \frac{1}{2h_i} \left(\frac{1}{\zeta_{i,1} + \underline{\lambda}_i} - \frac{1}{\zeta_{i,1} - \bar{\lambda}_i} \right)$. Therefore the resulting transformed agent i can be obtained as

$$\begin{cases} \dot{a}_{i,1} = \eta_i \left(x_{i,2} + \varphi_{i,1}^T (x_{i,1}) \theta_{i,1} - \dot{y}_{vi} - \dfrac{z_{i,1} \dot{h}_i}{h_i} \right) \\ \dot{x}_{i,q} = x_{i,q+1} + \varphi_{i,q}^T (\bar{x}_{i,q}) \theta_{i,q}, \; q = 2, \cdots, m-1 \\ \dot{x}_{i,m} = q(u_i) + \varphi_{i,m}^T (x_i) \theta_{i,m} + d_i, \; i = 1, \cdots, M. \end{cases} \tag{10.5}$$

10.3 Data-Rate Constraints and DoS Attacks

The MASs (10.1) and reference system (10.3) are connected by a digital data-rate-constrained network. In order to effectively communicate among agents and reference, the information is required to undergo a process of quantization and encoding at the transmitting end, and subsequently be decoded at the receiving end.

At the resilient distributed observation layer, the finite-level uniform symmetric quantizer [169] will be applied

$$\bar{q}(a) = \begin{cases} 0 & -\frac{1}{2} < a < \frac{1}{2} \\ z & \frac{2z-1}{2} \leq a < \frac{2z+1}{2}, \; z = 1, \cdots, L-1 \\ L & \frac{2L-1}{2} \leq a \\ -\bar{q}(-a) & a \leq -\frac{1}{2}, \end{cases} \tag{10.6}$$

where $L \in \{1, 2, \cdots\}$. In this chapter, the transmission attempts for each agent and reference occur periodically at $t = k\Delta \, (k \in \mathbb{N}_0)$, where $\Delta > 0$ denotes the sampling interval. DoS attack, which affects all agents and the reference, refers to the event that all the encoded signals cannot be received by the decoder, that is, the DoS attack may occur in any communication channel between any two agents or between the reference and agents in the graph, as shown in Fig. 10.1. The main constraint on DoS attacks is stated in Assumption 10.4. Due to DoS attacks, not all the transmission attempts at $k\Delta$ succeed, and $\{s_0\Delta, s_1\Delta, \cdots\} = \{s_q\Delta\} \subseteq \{k\Delta\} \, (q = 0, 1, \cdots)$ is defined as the sequence of successfully transmitted time instants and $H_s \triangleq \{s_q\}$.

Assumption 10.4 *The duration of the time interval* $[s_j\Delta, s_{j+1}\Delta) \, (j = 0, 1, \cdots)$ *is bounded.*

Remark 10.2 *As stated above,* $s_j\Delta$ *is defined as the* j*th successful transmit time instant, and the time interval between any two adjacent successful transmission instants implies a transmission-failed interval caused by DoS attacks. Thus by Assumption 10.4, it indicates that the maximal number of DoS attacks-caused consecutive transmission fails is restricted, which shows correlations with DoS attack models in related works such as [151] and [175].*

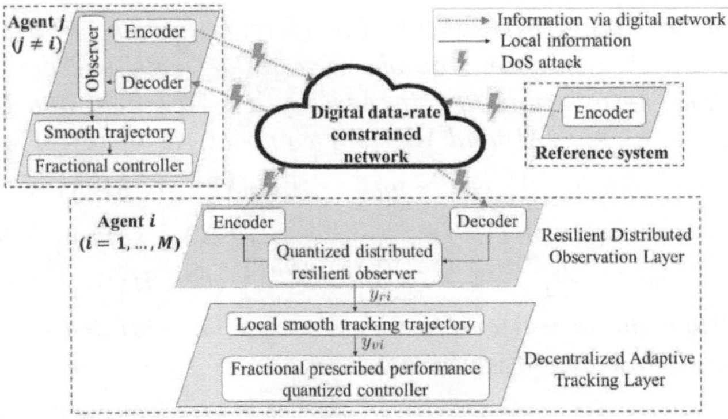

FIGURE 10.1: The structure of the designed control framework.

10.4 Controller Design and Stability Analysis

10.4.1 Quantized Distributed Resilient Observer Design

Due to the no accessibility for all agents to the reference system, data-rate-constrained communication, and DoS attacks, the resilient observer for each agent to observe the reference information is designed as

$$\dot{r}_i(t) = Fr_i(t) + \sum_{j\in\bar{\mathcal{N}}_i(k\Delta)} \rho a_{ij}\big(\hat{r}_{ji}(k\Delta) - \mu_i(k\Delta)\big), \tag{10.7}$$

where $t \in [k\Delta, (k+1)\Delta)$, $i = 1, \cdots, M$, $\rho \in \mathbb{R}^+$, and $\bar{\mathcal{N}}_i(k\Delta) = \mathcal{N}_i$ if $k \in H_s$ and $\bar{\mathcal{N}}_i(k\Delta) = \emptyset$ if $k \notin H_s$, in which $\mathcal{N}_i$ denotes the neighbor of agent i and the reference if $a_{i0} = 1$. The definitions of $\hat{r}_{ji}(k\Delta)$ and $\mu_i(k\Delta)$ will be provided later.

The data to be transmitted via the digital network at $t = k\Delta$ is $r_j(k\Delta)$, and the encoder for agent j is designed as follows. For convenience, here the reference is defined as agent 0,

$$\begin{cases} \mu_j((k+1)\Delta) = \mu_j(k\Delta) + \bar{\gamma}(k\Delta)\Omega_j((k+1)\Delta) \text{ if } k+1 \in H_s, \\ \mu_j((k+1)\Delta) = \mu_j(k\Delta) \text{ if } k+1 \notin H_s, \end{cases} \tag{10.8}$$

where $\mu_j(0) = 0$, $j = 0, 1, \cdots, M$, $r_0 = r$, μ_j stands for the encoder internal state, $\Omega_j((k+1)\Delta) = \bar{q}\Big(\frac{r_j((k+1)\Delta)-\mu_j(k\Delta)}{\bar{\gamma}(k\Delta)}\Big)$ is the output of the encoder, and $\bar{\gamma}(k\Delta)$ is the scaling function initialized as $\bar{\gamma}(0) = \bar{\gamma}_0 > 0$ and updated by

$$\bar{\gamma}((k+1)\Delta) = \beta\bar{\gamma}(k\Delta),\ \beta = \begin{cases} 0 < \bar{\beta} < 1 & \text{if } k+1 \in H_s, \\ \quad 1 & \text{if } k+1 \notin H_s. \end{cases} \tag{10.9}$$

The output of agent j, i.e., $\Omega_j\,(j = 0, \cdots, M)$ will be transmitted via the DoS attacks-affected digital network if agent i is a neighbor of agent j. The decoder for agent $i\,(i = 1, \cdots, M)$ for reconstructing r_j is designed as

$$\begin{cases} \hat{r}_{ji}((k+1)\Delta) = \hat{r}_{ji}(k\Delta) + \bar{\gamma}(k\Delta)\Omega_j((k+1)\Delta) \text{ if } k+1 \in H_s, \\ \hat{r}_{ji}((k+1)\Delta) = \hat{r}_{ji}(k\Delta) \text{ if } k+1 \notin H_s, \text{ with } \hat{r}_{ji}(0) = 0, \end{cases} \tag{10.10}$$

where $\hat{r}_{ji}$ implies the estimate of r_j obtained by agent i.

Lemma 10.2 *Consider the MASs (10.1) with reference system (10.3) under a digital network subject to DoS attacks. Given the observer (10.7) with encoder (10.8) and decoder (10.10), suppose that Assumption 10.2 to Assumption 10.4 hold and the initial condition* $r_i(0) \in W_r$, *where* $i = 1, \cdots, M$ *and* W_r *is a compact set of* $\mathbb{R}^n$. *For any given* $L \geq 1$, *let* $b \in \left(0, \min\left\{\frac{2}{\Delta(\lambda_1+\lambda_M)}, b_m\right\}\right)$, *where* $b_m = 2Lm_0\lambda_1^{-1}\Delta[\sqrt{M}\lambda_M^2 + 2m_0\lambda_1 deg^* + (2L+1)(1-m_0)m_0\lambda_1^2]^{-1}$ *and* $m_0 \in (0,1)$, $\bar{\beta} = 1-(1-m_0)b\Delta\lambda_1$, *scaling function being updated by (10.9), and* $\bar{\gamma}_0 > \max\left\{\frac{H_b}{L+\frac{1}{2}}, \frac{2H_b(\bar{\beta}-\sigma_h)(b\Delta\lambda_M+2\bar{\beta})}{b\Delta\lambda_M}\right\}$ *with* $H_b \geq \max_{1\leq i\leq M}\|r_i(0)\|_\infty$ *and* $\sigma_h = 1-b\Delta\lambda_1$. *Then the quantizer (10.6) will never be saturated and the observer error* $r_i - r$ *will converge to* 0 *asymptotically.*

Proof: The proof follows a similar technique to the proof of Theorem 3 in [175], with details omitted due to space limits. ■

10.4.2 Smooth Tracking Trajectory Design

It follows from (10.7) that the derivatives of r_i and further $y_{ri} = Br_i$ might be discontinuous due to the DoS attack-affected data-rate-constrained network transmission. By using the backstepping technique [86], the up to mth-order derivative of the tracking trajectory should be applied to designing controller. To handle this challenge, inspired by the Hermite interpolation method [230], a novel smooth trajectories $y_{vi}(t)\,(i = 1, \cdots, M)$ with up to mth-order derivatives are designed in this chapter as follows for agents to follow

$$y_{vi}(t) = \begin{cases} y_{ri}(h_{l-1}) & t = h_l \\ \sum_{p=1}^{2m+2} \psi_{i,p,l} t^{2m+2-p} & t \in (h_l, h_{l+1}) \end{cases} \tag{10.11}$$

where $h_l = lh$, $l \in \mathbb{N}$, and h is a small positive constant. $y_{vi}(t) = y_{ri}(0)$ for $t \in [-h, 0)$, $h_0 = 0$, $\psi_{i,p,l} = \mathcal{Z}_{i,l}[p]$ where $\mathcal{Z}_{i,l}[p]$ is the pth row of the vector $\mathcal{Z}_{i,l}$ with $\mathcal{Z}_{i,l} = \mathcal{H}_{il}^{-1}\mathcal{R}_{il}$. $\mathcal{H}_{il} = \mathrm{col}(\mathcal{H}_{il,1}, \ldots, \mathcal{H}_{il,2m+2})$ and $\mathcal{R}_{il} = \mathrm{col}(\mathcal{R}_{il,1}, \ldots, \mathcal{R}_{il,2m+2})$, where $\mathcal{H}_{il,1} = [h_l^{2m+1}, h_l^{2m}, \ldots, h_l, 1]$, $\mathcal{H}_{il,2} = [h_{l+1}^{2m+1}, h_{l+1}^{2m}, \ldots, h_{l+1}, 1]$, $\mathcal{H}_{il,2k-1} = [\frac{(2m+1)!}{(2m+2-k)!}h_l^{2m+2-k}, \frac{(2m)!}{(2m+1-k)!}h_l^{2m+1-k}, \ldots, h_l, 1, 0, 0, \ldots]$ and $\mathcal{H}_{il,2k} = [\frac{(2m+1)!}{(2m+2-k)!}h_{l+1}^{2m+2-k}, \ldots, h_{l+1}, 1, 0, 0, \ldots]$ for $k = 2, \ldots, m+1$. $\mathcal{R}_{il,1} = y_{ri}(h_{l-1})$, $\mathcal{R}_{il,2} = y_{ri}(h_l)$, and $\mathcal{R}_{il,2k-1} = \mathcal{R}_{il,2k} = 0$ for $k = 2, \ldots, m+1$.

Lemma 10.3 *The up to* m*th-order derivatives of* $y_{vi}(t)$ *designed in (10.11) exist and* $|y_{ri}(t) - y_{vi}(t)| \leq \bar{v}$ *for* $\forall t \geq 0$, *where* $\bar{v}$ *is a small positive constant that can be tuned small enough by properly designing parameter* h.

Proof: Now we divide the proof into two parts. Specifically, we first prove the existence of $y_{vi}(t)$ and then prove that $|y_{ri}(t) - y_{vi}(t)| \leq \bar{v}$ for $\forall t \geq 0$.

Part (1): From the definition of $y_{vi}(t)$ in (10.11), it is easy to show that the first mth-order derivatives of $y_{vi}(t)$ exist exclude for the zero-measure set $\cup_{l\in\mathbb{N}}\{h_l\}$. Thus, only the existence of the jth-order derivatives for $j = 1, \ldots, m$ of $y_{vi}(t)$ on $\cup_{l\in\mathbb{N}}\{h_l\}$ needs to be considered.

For $t \in (h_l, h_{l+1})$, we first consider the continuity of $y_{vi}(t)$ in (10.11). Specifically, from (10.11), we get that $y_{vi}(h_l^+) = \sum_{p=1}^{2m+2} h_l^{2m+2-p}\psi_{i,p,l}$ and $y_{vi}(h_{l+1}^-) = \sum_{p=1}^{2m+2} h_{l+1}^{2m+2-p}\psi_{i,p,l}$. Together with $\begin{bmatrix} \mathcal{H}_{i,l,1} \\ \mathcal{H}_{i,l,2} \end{bmatrix} \mathcal{Z}_{i,l} = \begin{bmatrix} \mathcal{R}_{i,l,1} \\ \mathcal{R}_{i,l,2} \end{bmatrix}$, we obtain that $y_{vi}(h_l^+) = y_{vi}(h_l)$ and $y_{vi}(h_{l+1}^-) = y_{vi}(h_{l+1})$, which ensures the continuity of $y_{vi}(t)$ at the interval $t \in [h_l, h_{l+1})$.

Then, we consider the continuity of $y_{vi}^{(k)}(t)$ $(k = 1, \ldots, m)$. Specifically, the kth-order derivative of $y_{vi}(t)$ is computed as

$$y_{vi}^{(k)}(t) = \sum_{p=1}^{2m+2-k} \frac{(2m+2-p)!}{(2m+2-p-k)!} \psi_{i,p,l} t^{2m+2-k-p}.$$

We then have

$$y_{vi}^{(k)}(h_l^+) = \sum_{p=1}^{2m+2-k} \frac{(2m+2-p)!}{(2m+2-p-k)!} h_l^{2m+2-k-p} \psi_{i,p,l}$$

and

$$y_{vi}^{(k)}(h_{l+1}^-) = \sum_{p=1}^{2m+2-k} \frac{(2m+2-p)!}{(2m+2-p-k)!} h_{l+1}^{2m+2-k-p} \psi_{i,p,l}.$$

Together with $\begin{bmatrix} \mathcal{H}_{i,l,2k-1} \\ \mathcal{H}_{i,l,2k-2} \end{bmatrix} \mathcal{Z}_{i,l} = \begin{bmatrix} \mathcal{R}_{i,l,2k-1} \\ \mathcal{R}_{i,l,2k-2} \end{bmatrix}$, we obtain that $y_{vi}^{(k)}(h_l^+) = y_{vi}^{(k)}(h_l)$ and $y_{vi}^{(k)}(h_{l+1}^-) = y_{vi}^{(k)}(h_{l+1})$, which ensures the continuity of $y_{vi}^{(k)}(t)$ at the interval $t \in [h_l, h_{l+1})$. Thus, we have shown that the up to mth-order derivatives of $y_{vi}(t)$ in (10.11) exist.

Part (2): According to (10.3) and Assumption 10.2, the solution of $r(t)$ can be written as $r(t) = e^{iF_I t} r(0)$, where F_I indicates the imaginary part of F, i.e., $F = iF_I$. Therefore, the difference of $r(t)$ between two neighboring time instants h_l and h_{l+1} is obtained as follows

$$\begin{aligned} r(h_{l+1}) - r(h_l) =& e^{iF_I lh} r(0)(e^{iF_I h} - 1) \\ =& r(0)\big[\cos(F_I lh) + i\sin(F_I lh)\big]\big[\cos(F_I h) - 1 + i\sin(F_I h)\big] \\ =& r(0)\Big\{\big[\cos((l+1)F_I h) - \cos(lF_I h)\big] + i\big[\sin((l+1)F_I h) - \sin(lF_I h)\big]\Big\}, \end{aligned} \tag{10.12}$$

which induces that

$$\begin{aligned} |r((l+1)h) - r(lh)| =& |r(0)|\Big\{\big[\cos((l+1)F_I h) - \cos(lF_I h)\big]^2 \\ &+ \big[\sin((l+1)F_I h) - \sin(lF_I h)\big]^2\Big\}^{\frac{1}{2}} \\ =& \sqrt{2}|r(0)|\sqrt{1 - \cos(F_I h)}. \end{aligned} \tag{10.13}$$

By choosing h such that $\frac{1}{2} < \cos(F_I h) < 1$, it has $0 \leq |r((l+1)h) - r(lh)| < |r(0)|$. Besides, it can be observed from (10.13) that $|r((l+1)h) - r(lh)| \propto h$. As $r_i - r$ will asymptotically converge to 0 in accordance with Lemma 10.2, thus the interpolation error $|y_{ri}(t) - y_{vi}(t)|$ $(\forall t \geq 0)$ is upper bounded by $\bar{v} \triangleq |y_{ri}((l+1)h) - y_{ri}(lh)| \propto h$, which can tend to 0 if h is designed small enough. ■

Remark 10.3 *Note that the influence of the DoS attacks and data-rate constraints might lead to the non-existence of high-order derivatives of the reference signal $y_{ri}(t)$ at the intermittent communication instants. However, the presence of these high-order derivatives is of importance for the design process of the recursive backstepping control. In order to tackle such a challenge, a smooth reference trajectory $y_{vi}(t)$ possesses high-order derivatives is put forth in this chapter in (10.11), which, as proved in the proposed Lemma 10.3, can accurately track the observer of reference systems with a small adjustable error.*

10.4.3 Local Fractional Prescribed Performance Quantized Adaptive Control

This subsection will present a novel PPB quantized fractional-order backstepping decentralized controller for the agent that incorporates fractional derivative and integral terms.

Remark 10.4 *Given the extra design flexibility introduced through fractional calculus, the transient and steady-state performance of the closed-loop system, relative to that of the conventional integer backstepping control method, can be further enhanced in this chapter.*

Step 1: For the ith agent ($i = 1, \cdots, M$), define error variables as $e_{i,1} = a_{i,1}$ and $e_{i,j} = x_{i,j} - \gamma_{i,j-1}$ ($j = 2, \cdots, m$), where $\gamma_{i,j-1}$ is the virtual control signal to be designed later. Compute the time derivative of the Lyapunov candidate function $\overline{U}_{i,1} = \frac{1}{2}e_{i,1}^2$ as $\dot{\overline{U}}_{i,1} = \eta_i\big(e_{i,1}e_{i,2} + e_{i,1}\gamma_{i,1} + e_{i,1}\varphi_{i,1}^T\theta_{i,1} - e_{i,1}\dot{y}_{vi} - e_{i,1}\frac{z_{i,1}\dot{h}_i}{h_i}\big)$. Define Lyapunov candidate function $U_{i,1} = \overline{U}_{i,1} + \frac{1}{2}\tilde{\theta}_{i,1}^T\Lambda_{i,1}^{-1}\tilde{\theta}_{i,1}$ with $\hat{\theta}_{i,1}$ being the estimate of $\theta_{i,1}$, $\tilde{\theta}_{i,1} = \hat{\theta}_{i,1} - \theta_{i,1}$, and matrix $\Lambda_{i,1}$ being positive definite. By respectively designing virtual control signal $\gamma_{i,1}$ and adaptive law as

$$\gamma_{i,1} = -\frac{c_{i,1}e_{i,1}}{\eta_i} - \varphi_{i,1}^T\hat{\theta}_{i,1} + \frac{z_{i,1}\dot{h}_i}{h_i} + \dot{y}_{vi}, \tag{10.14}$$

$$\dot{\hat{\theta}}_{i,1} = \eta_i e_{i,1}\Lambda_{i,1}\varphi_{i,1}, \tag{10.15}$$

where design parameter $c_{i,1} > 0$, thus the time derivative of $U_{i,1}$ is computed as

$$\dot{U}_{i,1} = \eta_i e_{i,1}e_{i,2} - c_{i,1}e_{i,1}^2. \tag{10.16}$$

Step k ($k \in \{2, \cdots, m-2\}$)*:* The design process for Steps 2 to $m-2$ are condensed here for the sake of brevity, and the overview of the outcomes is provided below.

Virtual control signals:

$$\gamma_{i,k} = -\eta_k e_{i,k-1} - c_{i,k}e_{i,k} - \varphi_{iv,k}^T\hat{\theta}_{iv,k} - \psi_{i,k}, \tag{10.17}$$

where $c_{i,k} > 0$, $\eta_k = \eta_i$ ($k = 2$), $\eta_k = 1$ ($k = 3, \cdots, m-2$), $\hat{\theta}_{iv,k}$ stands for the estimate of $\theta_{iv,k}$, $\varphi_{iv,k} = \left[\varphi_{i,k}^T, -\frac{\partial\gamma_{i,k-1}}{\partial x_{i,k-1}}\varphi_{i,k-1}^T, \cdots, -\frac{\partial\gamma_{i,k-1}}{\partial x_{i,1}}\varphi_{i,1}^T\right]^T$, $\theta_{iv,k} = \left[\theta_{i,k}^T, \theta_{i,k-1}^T, \cdots, \theta_{i,1}^T\right]^T$, and $\psi_{i,k} = -\sum_{p=2}^{k}\frac{\partial\gamma_{i,k-1}}{\partial x_{i,p-1}}x_{i,p} - \sum_{p=1}^{k}\left(\frac{\partial\gamma_{i,k-1}}{\partial y_{vi}^{(p-1)}}y_{vi}^{(p)} + \frac{\partial\gamma_{i,k-1}}{\partial h_i^{(p-1)}}h_i^{(p)}\right) - \sum_{p=1}^{k-1}\frac{\partial\gamma_{i,k-1}}{\partial\hat{\theta}_{iv,p}}\dot{\hat{\theta}}_{iv,p}$.

Adaptive laws:

$$\dot{\hat{\theta}}_{iv,k} = e_{i,k}\Lambda_{i,k}\varphi_{iv,k}, \tag{10.18}$$

where $\Lambda_{i,k}$ denotes the positive definite constant matrix. Defining $\theta_{iv,1} = \theta_{i,1}$ and Lyapunov candidate function $U_{i,m-2} = \frac{1}{2}\sum_{p=1}^{m-2}\tilde{\theta}_{iv,p}^T\Lambda_{i,p}^{-1}\tilde{\theta}_{iv,p} + \frac{1}{2}\sum_{p=1}^{m-2}e_{i,p}^2$, where $\tilde{\theta}_{iv,p} = \hat{\theta}_{iv,p} - \theta_{iv,p}$, according to the investigated virtual control signals (10.17) and adaptive laws (10.18), the derivative of $U_{i,m-2}$ can be obtained as

$$\dot{U}_{m-2} = -\sum_{p=1}^{m-2}c_{i,p}e_{i,p}^2 + e_{i,m-2}e_{i,m-1}. \tag{10.19}$$

Step m − 1: The virtual control signal at this step is designed as

$$\gamma_{i,m-1} = -e_{i,m-2} - c_{i,m-1}e_{i,m-1} - \varphi_{iv,m-1}^T\hat{\theta}_{iv,m-1} - \psi_{i,m-1} - k_{i0}I_t^{\alpha_i}e_{i,m-1}, \tag{10.20}$$

where design parameters $k_i > 0$, $c_{i,m-1} > 0$, $\hat{\theta}_{iv,m-1}$ represents the estimate of $\theta_{iv,m-1}$, $\alpha_i \in [0,1]$. The definitions of $\theta_{iv,m-1}$, $\varphi_{iv,m-1}$, and $\psi_{i,m-1}$ can be found in Step k by replacing k with $m-1$.

Design adaptive law as

$$\dot{\hat{\theta}}_{iv,m-1} = e_{i,m-1}\Lambda_{i,m-1}\varphi_{iv,m-1}, \tag{10.21}$$

with $\Lambda_{i,m-1}$ standing for a positive definite constant matrix, then the derivative of the Lyapunov candidate function $U_{i,m-1} = U_{i,m-2} + \frac{e_{i,m-1}^2}{2} + \frac{1}{2}\tilde{\theta}_{iv,m-1}^T\Lambda_{i,m-1}^{-1}\tilde{\theta}_{iv,m-1}$ is

$$\dot{U}_{i,m-1} = -\sum_{p=1}^{m-1} c_{i,p}e_{i,p}^2 + e_{i,m-1}e_{i,m} - k_i e_{i,m-1}\,{}_0I_t^{\alpha_i} e_{i,m-1}.$$

Step m: Establish Lyapunov candidate function $\overline{U}_{i,m} = U_{i,m-1} + \frac{e_{i,m}^2}{2}$ and calculate its time derivative as

$$\begin{aligned}\dot{\overline{U}}_{i,m} =& -\sum_{p=1}^{m-1} c_{i,p}e_{i,p}^2 - k_i e_{i,m-1}\,{}_0I_t^{\alpha_i} e_{i,m-1} + e_{i,m-1}e_{i,m}\\ &+ e_{i,m}\big(q(u_i) + \varphi_{i,m}^T\theta_{i,m} + d_i - \dot{\gamma}_{i,m-1}\big),\end{aligned} \tag{10.22}$$

in which the term $\dot{\gamma}_{i,m-1}$ can be expressed as follows by defining $\gamma_{i,m-1} = \gamma'_{i,m-1} - k_{i0}\,{}I_t^{\alpha_i} e_{i,m-1}$

$$\begin{aligned}\dot{\gamma}_{i,m-1} =& \sum_{p=2}^{m}\frac{\partial\gamma'_{i,m-1}}{\partial x_{i,p-1}}x_{i,p} + \sum_{p=1}^{m-1}\left(\frac{\partial\gamma'_{i,m-1}}{\partial x_{i,p}}\varphi_{i,p}^T\theta_{i,p} + \frac{\partial\gamma'_{i,m-1}}{\partial\hat{\theta}_{iv,p}}\dot{\hat{\theta}}_{iv,p}\right)\\ &+ \sum_{p=1}^{m}\left(\frac{\partial\gamma'_{i,m-1}}{\partial y_{vi}^{(p-1)}}y_{vi}^{(p)} + \frac{\partial\gamma'_{i,m-1}}{\partial h_i^{(p-1)}}h_i^{(p)}\right) - k_{i0}I_t^{\alpha_i}\varphi_{i,m-1}^T\theta_{i,m-1}\\ &- k_{i0}I_t^{\alpha_i}x_{i,m} + k_{i0}I_t^{\alpha_i}\left(\sum_{p=1}^{m-2}\frac{\partial\gamma_{i,m-2}}{\partial x_{i,p}}x_{i,p+1} + \sum_{p=1}^{m-2}\frac{\partial\gamma_{i,m-2}}{\partial x_{i,p}}\varphi_{i,p}^T\theta_{i,p}\right.\\ &\left.+ \sum_{p=1}^{m-2}\frac{\partial\gamma_{i,m-2}}{\partial\hat{\theta}_{iv,p}}\dot{\hat{\theta}}_{iv,p} + \sum_{p=1}^{m-1}\frac{\partial\gamma_{i,m-2}}{\partial y_{vi}^{(p-1)}}y_{vi}^{(p)} + \sum_{p=1}^{m-1}\frac{\partial\gamma_{i,m-2}}{\partial h_i^{(p-1)}}h_i^{(p)}\right).\end{aligned} \tag{10.23}$$

Remark 10.5 *Different from existing works [167] and [168], to avoid direct differential action ${}_0\mathcal{D}_t^{1-\alpha_i}e_{i,m-1}$ in the control signal, the following action is taken in this chapter, combined with which results in the terms in (10.23) involving fractional integral.*

$$\begin{aligned}{}_0\mathcal{D}_t^{1-\alpha_i}e_{i,m-1} =& {}_0\mathcal{D}_t^{1-\alpha_i}x_{i,m-1} - {}_0\mathcal{D}_t^{1-\alpha_i}\gamma_{i,m-2}\\ =& \frac{\int_0^t(x_{i,m} + \varphi_{i,m-1}^T\theta_{i,m-1})(t-a)^{\alpha_i-1}\mathrm{d}a}{\Gamma(\alpha_i)} - \frac{1}{\Gamma(\alpha_i)}\int_0^t\left(\sum_{p=1}^{m-2}\right.\\ &\frac{\partial\gamma_{i,m-2}}{\partial x_{i,p}}x_{i,p+1} + \sum_{p=1}^{m-2}\frac{\partial\gamma_{i,m-2}}{\partial x_{i,p}}\varphi_{i,p}^T\theta_{i,p} + \sum_{p=1}^{m-2}\frac{\partial\gamma_{i,m-2}}{\partial\hat{\theta}_{iv,p}}\dot{\hat{\theta}}_{iv,p}\\ &\left.+ \sum_{p=1}^{m-1}\frac{\partial\gamma_{i,m-2}}{\partial y_{vi}^{(p-1)}}y_{vi}^{(p)} + \sum_{p=1}^{m-1}\frac{\partial\gamma_{i,m-2}}{\partial h_i^{(p-1)}}h_i^{(p)}\right)(t-a)^{\alpha_i-1}\mathrm{d}a.\end{aligned}$$

In accordance with Lemma 2.1 and (10.23), (10.22) can be expressed as

$$\dot{U}_{i,m} \leq -\sum_{p=1}^{m-1} c_{i,p}e_{i,p}^2 + e_{i,m-1}e_{i,m} - k_i e_{i,m-1} {}_0I_t^{\alpha_i} e_{i,m-1} + e_{i,m}q(u_i) + n_i\bar{d}_i \\ + \frac{e_{i,m}^2}{\sqrt{n_i^2 + e_{i,m}^2}}\bar{d}_i + e_{i,m}\psi_{i,m} + e_{i,m}\varphi_{iv,m}^T\theta_{iv,m}, \tag{10.24}$$

where $n_i = e^{-b_i t}\,(b_i > 0)$, $\varphi_{iv,m} = \Big[\varphi_{i,m}^T, -\frac{\partial\gamma'_{i,m-1}}{\partial x_{i,m-1}}\varphi_{i,m-1}^T + k_{i0}I_t^{\alpha_i}\varphi_{n-1}^T, \cdots, -\frac{\partial\gamma'_{i,m-1}}{\partial x_{i,1}}\varphi_{i,1}^T - k_{i0}I_t^{\alpha_i}\frac{\partial\gamma_{i,m-2}}{\partial x_{i,1}}\varphi_{i,1}^T\Big]^T$, $\theta_{iv,m} = \big[\theta_{i,m}^T, \cdots, \theta_{i,1}^T\big]^T$, and $\psi_{i,m} = -\sum_{p=2}^m \frac{\partial\gamma'_{i,m-1}}{\partial x_{i,p-1}}x_{i,p} - \sum_{p=1}^{m-1}\frac{\partial\gamma'_{i,m-1}}{\partial\hat{\theta}_{iv,p}}\dot{\hat{\theta}}_{iv,p} - \sum_{p=1}^m\Big(\frac{\partial\gamma'_{i,m-1}}{\partial y_{vi}^{(p-1)}}y_{vi}^{(p)} + \frac{\partial\gamma'_{i,m-1}}{\partial h_i^{(p-1)}}h_i^{(p)}\Big) + k_{i0}I_t^{\alpha_i}x_{i,m} - k_{i0}I_t^{\alpha_i}\Big(\sum_{p=2}^{m-1}\frac{\partial\gamma_{i,m-2}}{\partial x_{i,p-1}}x_{i,p} + \sum_{p=1}^{m-2}\frac{\partial\gamma_{i,m-2}}{\partial\hat{\theta}_{iv,p}}\dot{\hat{\theta}}_{iv,p} + \sum_{p=1}^{m-1}\frac{\partial\gamma_{i,m-2}}{\partial y_{vi}^{(p-1)}}y_{vi}^{(p)} + \sum_{p=1}^{m-1}\frac{\partial\gamma_{i,m-2}}{\partial h_i^{(p-1)}}h_i^{(p)}\Big)$.

For term $e_{i,m}q(u_i)$ in (10.24), based on Lemma 10.1, it has

$$e_{i,m}q(u_i) \leq e_{i,m}u_i + \delta_i|e_{i,m}u_i| + u_{i,\min}|e_{i,m}|, \tag{10.25}$$

and the control signal u_i is designed as

$$u_i = \frac{-e_{i,m}u_{ni}}{\sqrt{n_i^2 + e_{i,m}^2}},\ u_{ni} = \frac{-\gamma_{i,m}\sqrt{n_i^2 + e_{i,m}^2} + \chi_i e_{i,m}}{(1-\delta_i)e_{i,m}} \tag{10.26}$$

$$\gamma_{i,m} = -e_{i,m} - c_{i,m}e_{i,m} - \frac{e_{i,m}}{\sqrt{n_i^2 + e_{i,m}^2}}\hat{d}_i - \psi_{i,m} - u_{ei} \\ + k_i e_{i,m-1} g_e(e_{i,m}, \nu_i) {}_0I_t^{\alpha_i} e_{i,m-1} - \varphi_{iv,m}^T\hat{\theta}_{iv,m}, \tag{10.27}$$

where $\chi_i \geq u_{i,\min}$, $c_{i,m} > 0$, $\nu_i \in (0,1)$, $\hat{d}_i$ and $\hat{\theta}_{iv,m}$ are the estimates of $\bar{d}_i$ and $\theta_{iv,m}$, respectively,

$$u_{ei} = \begin{cases} 0, & |e_{i,m}| \geq \nu_i, \\ \omega_i(x_i)\hat{\eta}_{ei}, & |e_{i,m}| < \nu_i \end{cases} \tag{10.28}$$

with $\omega_i(x_i)$ implying the known continuous function vector and $\hat{\eta}_{ei}$ the estimate of the unknown constant vector η_{ei}, and

$$g_e(e_{i,m}, \nu_i) = \begin{cases} \frac{1}{e_{i,m}}, & |e_{i,m}| \geq \nu_i, \\ \frac{e_{i,m}}{\nu_i - \nu_i^2 + e_{i,m}^2}, & |e_{i,m}| < \nu_i. \end{cases} \tag{10.29}$$

Then on the basis of (10.25), (10.26), and Lemma 2.1, it yields

$$e_{i,m}q(u_i) \leq e_{i,m}\gamma_{i,m} + \chi_i n_i. \tag{10.30}$$

Define $\tilde{d}_i = \hat{d}_i - \bar{d}_i$, $\tilde{\eta}_{ei} = \hat{\eta}_{ei} - \eta_{ei}$, and $\tilde{\theta}_{iv,m} = \hat{\theta}_{iv,m} - \theta_{iv,m}$, then in light of the designed controller (10.26) and (10.27), (10.24) can be calculated as

$$\dot{U}_{i,m} \leq -\sum_{p=1}^{m} c_{i,p}e_{i,p}^2 + n_i(\bar{d}_i + \chi_i) - e_{i,m}\varphi_{iv,m}^T\tilde{\theta}_{iv,m} - \frac{e_{i,m}^2}{\sqrt{n_i^2 + e_{i,m}^2}}\tilde{d}_i - e_{i,m}\omega_i(x_i)\tilde{\eta}_{ei}. \tag{10.31}$$

Construct the final Lyapunov function $U_{i,m} = \overline{U}_{i,m} + \frac{1}{2}\tilde{\theta}_{iv,m}^T\Lambda_{i,m}^{-1}\tilde{\theta}_{iv,m} + \frac{1}{2\iota_i}\tilde{d}_i^2 + \frac{1}{2q_i}\tilde{\eta}_{ei}^2$, where $\Lambda_{i,m}$ is positive definite and $\iota_i > 0$, $q_i > 0$, then by designing adaptive laws

$$\dot{\hat{\theta}}_{iv,m} = e_{i,m}\Lambda_{i,m}\varphi_{iv,m}, \tag{10.32}$$

$$\dot{\hat{d}}_i = \frac{\iota_i e_{i,m}^2}{\sqrt{n_i^2 + e_{i,m}^2}}, \tag{10.33}$$

$$\dot{\hat{\eta}}_{ei} = q_i e_{i,m}\omega_i(x_i), \tag{10.34}$$

the derivative of $U_{i,m}$ satisfies

$$\dot{U}_{i,m} \leq -\sum_{p=1}^{m} c_{i,p}e_{i,p}^2 + n_i(\bar{d}_i + \chi_i). \tag{10.35}$$

Fig. 10.1 illustrates the framework diagram of the proposed control scheme.

10.4.4 Stability Analysis

Theorem 10.1 *Under Assumption 10.1 to Assumption 10.4, the prescribed performance output tracking problem of the uncertain high-order nonlinear MASs (10.1) to the reference system (10.3) under data-rate constraints and DoS attacks can be solved with all closed-loop signals remain bounded if the quantized decentralized fractional controllers (10.26) with adaptive laws (10.15), (10.18), (10.21), (10.32) - (10.34) are designed based on the quantized distributed observer (10.7) and the smooth tracking trajectory (10.11) with conditions in Lemma 10.2 and Lemma 10.3 being satisfied.*

Proof: Taking integration of (10.35) yields

$$U_{i,m}(t) - U_{i,m}(0) \leq -\sum_{p=1}^{m}\int_0^t c_{i,p}e_{i,p}(a)^2\mathrm{d}a + \int_0^t e^{-b_i a}(\bar{d}_i + \chi_i)\mathrm{d}a \leq \frac{\bar{d}_i + \chi_i}{b_i}, \tag{10.36}$$

which demonstrates the boundedness of $U_{i,m}$ and thus $e_{i,q}$ $(i = 1, \cdots, M, q = 1, \cdots, m)$, $\tilde{\theta}_{iv,q}$, $\tilde{d}_i$, and $\tilde{\eta}_{ei}$ are bounded. By examining (10.14) and (10.17) while considering the boundedness of $y_{vi}^{(q)}$, it can be inferred that the values of $\gamma_{i,j}$ $(j = 1, \cdots, m-2)$ stay bounded. For $\alpha_i \in [0,1)$, the value of $t^{\alpha_i - 1}$ will monotonically diminish to zero as t progressively grows, and $t^{\alpha_i-1} = 1$ for $\alpha_i = 1$, thus the boundedness of ${}_0I_t^{\alpha_i}e_{i,m-1}$ can be maintained by the boundedness of $e_{i,m-1}$ with ${}_0I_t^{\alpha_i}e_{i,m-1} = \frac{1}{\Gamma(\alpha_i)}(e_{i,m-1} * t^{\alpha_i-1})$ according to (1.29), in which $*$ signifies the convolution operator. It can be inferred by (10.20) that $\gamma_{i,m-1}$ remains bounded, and similarly, the boundedness of the control signal u_i as demonstrated in (10.26) can be guaranteed. As h_i, $e_{i,q}$, and $\gamma_{i,j}$ are bounded, it ensures that $x_{i,q}$ is bounded.

Given that $a_{i,1} = e_{i,1}$ is bounded, it follows from the injective characteristic of $Y(a_{i,1})$ that for any initial error $z_{i,1}(0)$, (10.4) will hold for $t \geq 0$ by suitably selecting values for $\underline{\lambda}_i$, $\bar{\lambda}_i$, and $h_i(0)$, ensuring $-\underline{\lambda}_i h_i(0) \leq z_{i,1}(0) \leq \bar{\lambda}_i h_i(0)$. Besides, based on (10.36), it is evident that $\lim_{t\to\infty}\sum_{p=1}^{m}\int_0^t c_{i,p}e_{i,p}(a)^2\mathrm{d}a$ remains bounded, thus by applying Barbalat's lemma, we can conclude that $\lim_{t\to\infty} e_{i,p} = 0\,(p = 1, \cdots, m)$, which further on the basis of the third characterization of function $Y(a_{i,1})$, specifically, $Y(0) = 0$, implying that $\lim_{t\to\infty} z_{i,1} = 0$ is realized, that is $\lim_{t\to\infty}(y_i - y_{vi}) = 0$. Additionally, based on Lemma 10.3 that $|y_{ri} - y_{vi}| \leq \bar{v}$ holds for $\forall t \geq 0$ and Lemma 10.2 that $\lim_{t\to\infty}(y_{ri} - y_r) = 0$, we get $\limsup_{t\to\infty}|y_i - y_r| \leq \bar{v}$, indicating that the output y_i of each agent i can track the reference y_r with tunable small error $\bar{v}$ that can be adjusted small enough by properly selecting parameter h. Thus, based on our proposed hierarchical observer-based adaptive fractional quantized controller, all agent outputs y_i can track the reference signal y_r with prescribed performance under the DoS attacks affected data-rate-constrained communication network. ■

10.5 Simulation Verification

In this section, a system with one reference and four agents, whose communication topology is given in Fig. 10.2 with 0 denoting the reference system, is considered as a simulation example for demonstrating the effectiveness of the proposed control scheme in this chapter.

10.5.1 Illustration of Proposed Control Scheme

The system matrices and initial condition for the reference system are $F = \begin{bmatrix} 0 & -1 \\ 1 & 0 \end{bmatrix}$, $B = [1\ \ 0]$, $r(0) = [1\ \ 0]^T$, from where it can be observed that Assumption 10.2 is satisfied. Each agent $i\,(i = 1, 2, 3, 4)$ is a permanent magnet linear synchronous motor [20] model with dynamic characteristics

$$\begin{cases} \dot{y}_1 = \dot{x}_{i,1} = x_{i,2} \\ \dot{x}_{i,2} = Kq(u_i) + Qx_{i,2}\theta_{i,2} + d_i, \end{cases} \tag{10.37}$$

where $K = 180.33$, $Q = -115.025$, and $d_i = d_{i1}\sin(t) + d_{i2}\cos(t) + d_{i3}U(t - d_{i4}) + G(t)$ with $U(t-d_{i4})$ and $G(t)$ respectively standing for the unit step function at $t = d_{i4}\,\text{s}$ and the Gaussian noise with power equals to 0.01. The values of d_{ij} $(j = 1, 2, 3, 4)$ in this simulation verification are as follows: $d_{11} = 0.02$, $d_{12} = 0.03, d_{13} = 0.05$, $d_{14} = 4$, $d_{21} = 0.03$, $d_{22} = 0.01$, $d_{23} = 0.02$, $d_{24} = 7$, $d_{31} = 0.01$, $d_{32} = 0.02$, $d_{33} = 0.03$, $d_{34} = 1$, $d_{41} = 0.02$, $d_{42} = 0.01$, $d_{43} = 0.01$, and $d_{44} = 10$.

For simulation purposes, the DoS attack time intervals for the data-rate-constrained communication network among agents and the reference system are $(0.3\,\text{s}, 1.4\,\text{s})$, $(2.1\,\text{s}, 3.2\,\text{s})$, $(4.9\,\text{s}, 6.1\,\text{s})$, $(10.7\,\text{s}, 11.6\,\text{s})$, and $(15.5\,\text{s}, 16.7\,\text{s})$, ensuring that Assumption 10.4 is met. For the proposed quantized distributed resilient observer (10.7) with sampling interval $\triangle = 0.01\,\text{s}$, initial values $r_1(0) = [0.55, 0.55]^T$, $r_2(0) = [0.5, 0.5]^T$, $r_3(0) = [0.35, 0.35]^T$, $r_4(0) = [0.4, 0.4]^T$, encoder (10.8), and decoder (10.10), the corresponding parameters $\rho = 0.05$, $L = 21$, $\bar{\beta} = 0.99$, and $\bar{\gamma}_0 = 1$ are designed in accordance with the satisfaction of Lemma 10.2. As shown in Fig. 10.3, under the investigated distributed observer, the observer error to the reference system for each agent $r_i - r$ can asymptotically converge to 0 even with data-rate-constrained communication and DoS attacks. Besides, by the proposed smooth tracking reference y_{vi} in (10.11) with $h = 0.001\,\text{s}$, it can be verified by Fig. 10.4 that the smooth trajectory y_{vi} for agent i can perfectly track the reference y_r.

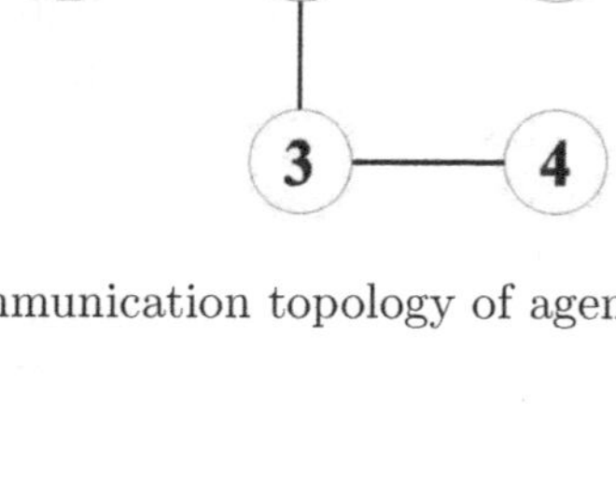

FIGURE 10.2: Communication topology of agents and the reference.

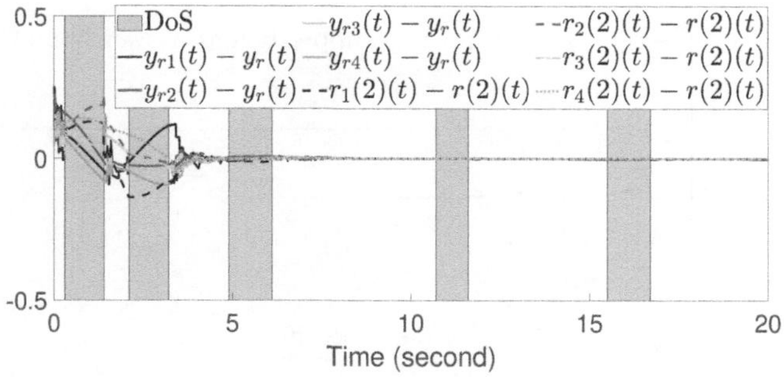

FIGURE 10.3: Trajectories of $y_{ri} - y_r$ and $r_i(2) - r(2)$ $(i = 1, 2, 3, 4)$.

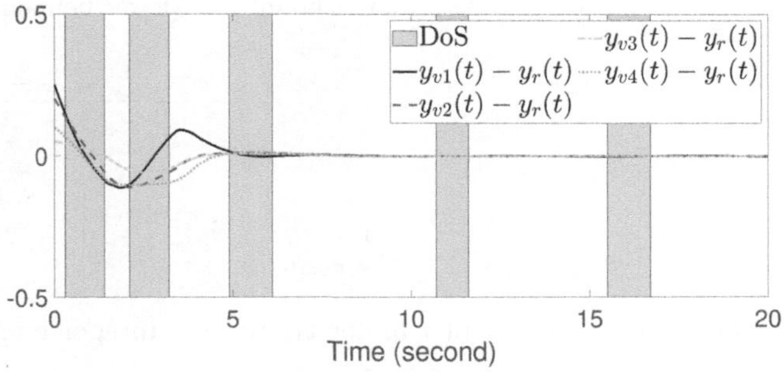

FIGURE 10.4: Trajectories of $y_{vi} - y_r$.

The local adaptive fractional-order backstepping-based quantizer controller for agent i that ensures its output winding position $x_{i,1} = y_i$ can track the reference signal y_r with prescribed performance is designed based on our proposed method provided in Subsection 10.4.3. Under the specified performance on output $h_i(t) = (0.79 - 0.01)e^{-1.5t} + 0.01$ along with prescribed parameters $\underline{\lambda}_i = \bar{\lambda}_i = 0.5$ and system initial values $x_1 = [0.6, 0]^T$, $x_2 = [0.4, 0]^T$, $x_3 = [0.5, 0]^T$, $x_4 = [0.3, 0]^T$, the designed parameters are chosen as $\chi_i = 0.1$, $\delta_i = 0.2$, $c_{i,1} = 1$, $c_{i,2} = 10$, $k_i = 16$, $\Lambda_{i,2} = 10$, $\iota_i = 1$, $q_i = 10$, $\alpha_i = 0.2$, $\nu_i = 0.05$, $n_i(t) = e^{-b_i t}$ with $b_i = 0.06$, and $\omega_i(x_i)$ defined with six Gaussian membership functions uniformly distributed over $[-0.05, 0.05]$. The initial values for the estimates of $\theta_{i,2}$, d_i, and η_{ei} are selected as $\hat{\theta}_{1,2}(0) = 0.1$, $\hat{\theta}_{2,2}(0) = 0.15$, $\hat{\theta}_{3,2}(0) = 0.2$, $\hat{\theta}_{4,2}(0) = -0.4$, $\hat{d}_1(0) = 0.3$, $\hat{d}_2(0) = 0.6$, $\hat{d}_3(0)$, $\hat{d}_4(0) = 0.1$, and $\hat{\eta}_{ei}(0) = [1, \cdots, 1]^T \in \mathbb{R}^{36}$, respectively.

Fig. 10.5 presents the output error $z_{i,1} = y_i - y_{vi}$ of agent i, it can be seen from Fig. 10.4 and Fig. 10.5 that even under data-rate-constrained communication subject to DoS attacks, by our proposed quantized control scheme, all agents that system uncertainties and external disturbances exist can track the reference signal asymptotically without violating the predefined output performance bound.

10.5.2 Comparison Studies Regarding System Transient and Steady-State Performance

For comparison, the agent output tracking performance to reference signal utilizing conventional integer-order backstepping method [222] in the local decentralized tracking layer, i.e., setting the fractional calculus-related control parameter $k_i = 0$ while other design parameters remaining the same, is studied in this simulation to highlight the advantages of the

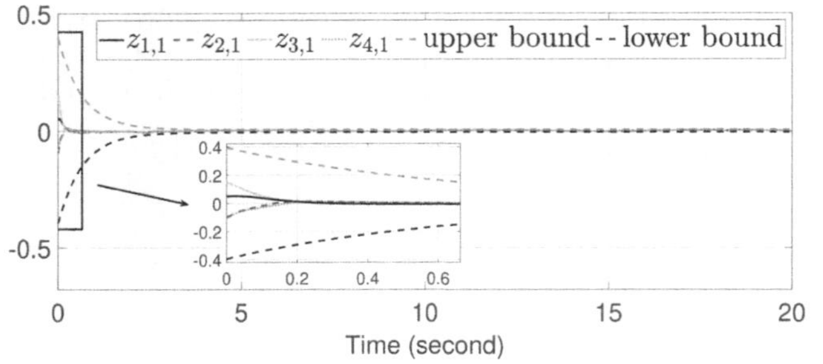

FIGURE 10.5: Output error z_{i1} of agent i.

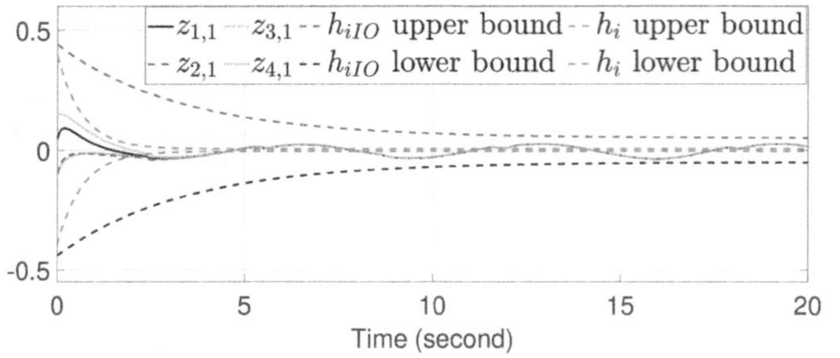

FIGURE 10.6: Output error z_{i1} of agent i under traditional integer backstepping method.

fractional control scheme as described in Remark 10.4 in improving transient and steady-state performance compared to the traditional integer counterpart. Fig. 10.6 accordingly displays the agent output error $z_{i,1}$, from where it can be noticed that under the traditional backstepping control method, merely a more relaxed prescribed output performance $h_{iIO}(t) = (0.88 - 0.1)e^{-0.3t} + 0.1$ with slower convergence rate and larger steady-state error bound in comparison to $h_i(t)$ can be met, and the asymptotic tracking to reference signal of agent outputs is failed to achieve. In order to more intuitively demonstrate the benefits of fractional-order control regarding enhancing system transient and steady-state performance, the upper and lower bounds with $h_i(t)$, which is more restricted with a faster prescribed converge rate and smaller steady error, are also displayed in Fig. 10.6.

In addition, another study that involves comparing our proposed controller with the one in existing works like [167] and [168] that control signal directly contains fractional derivative operation while maintaining unchanged designed parameters is undertaken to further showcase the effectiveness of our proposed control strategy. As shown in Fig. 10.7, under the controller directly contains derivative operation, the output error can only be maintained in an extremely relaxed prescribed bound $h_{iDD}(t) = (3 - 0.1)e^{-0.05t} + 0.1$ compared to the bound $h_i(t)$ under our designed controller. Besides, it can be found in Fig. 10.8 that the quantized control signals under the fractional controller that directly involves fractional derivative operation, i.e., $q(u_i)_{DD}$, display significant large amplitude chattering that persists for the entire process, causing difficulties in practical implementation and potential damage to instruments. Moreover, by comparing Fig. 10.9 with Fig. 10.5, the advantage of our proposed control method can be further highlighted as the output error under the fractional controller in Chapter 8 cannot be contained in the strict predefined bound $h_i(t)$. All studies presented demonstrate the established theoretical results and confirm the effectiveness of the developed control method.

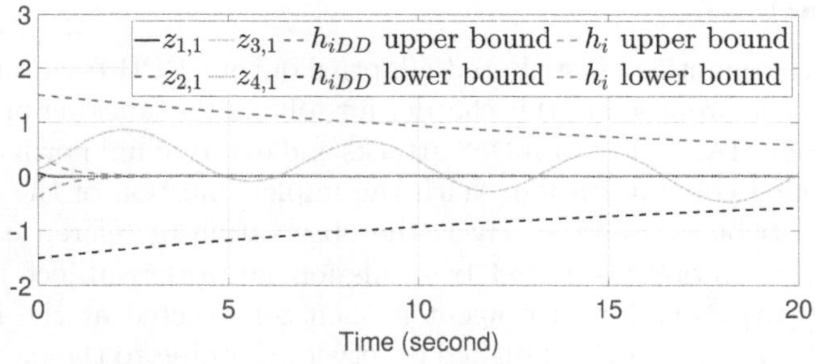

FIGURE 10.7: Output error z_{i1} of agent i under direct derivative contain fractional backstepping method.

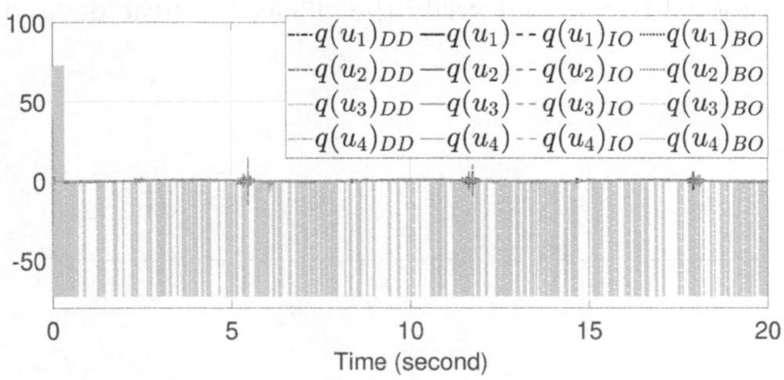

FIGURE 10.8: Quantized control signals of agent i under proposed controller ($q(u_i)$), integer controller ($q(u_i)_{IO}$), direct derivative contain fractional controller ($q(u_i)_{DD}$), and fractional controller in Chapter 8 ($q(u_i)_{BO}$).

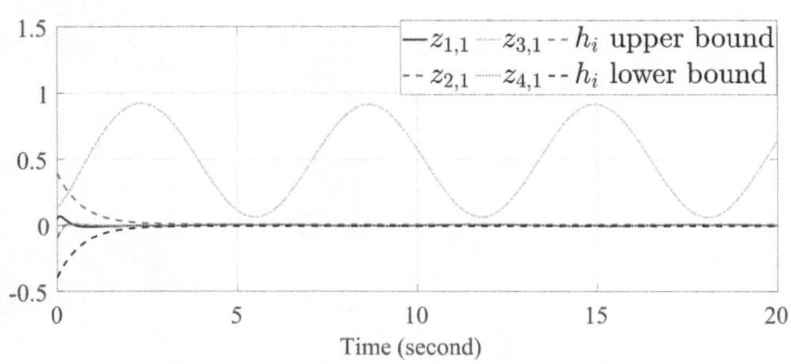

FIGURE 10.9: Output error z_{i1} of agent i under fractional backstepping method in Chapter 8.

10.6 Conclusions

A novel hierarchical quantized distributed adaptive output PPB fractional-order backstepping control scheme is proposed in this chapter for solving the tracking problem of uncertain nonlinear high-order MASs with both DoS attacks and external unknown disturbances under data-rate-constrained communication. With the implementation of the quantized resilient observer at the distributed observer layer, the observation of reference states is available for all agents within data-rate-limited transmission networks subject to DoS attacks. A smooth reference trajectory for each agent is then constructed at the local decentralized controller layer on the basis of the designed observer according to the interpolation method. In addition, an adaptive fractional-order backstepping-based quantized control scheme that guarantees prescribed output performance for high-order uncertain nonlinear integer-order systems with unknown time-varying external disturbances is investigated at the local controller layer in this chapter. The newly proposed fractional-order control technology can not only ensure the stability of the closed-loop systems but also guarantee that each agent output perfectly tracks the reference signal with the prescribed transient and steady-state output performance being met. The outcomes obtained from the simulation confirm the validity of the theoretical results and verify the efficiency of the devised control method.

11

Conclusions and Future Works

The book will be summarized in this chapter with some interesting future topics regarding the control for fractional-order systems and fractional-order control provided.

11.1 Conclusions

This book has developed control schemes for fractional-order nonlinear uncertain systems and linear fractional-order systems and MASs. The fractional-order control schemes for integer-order nonlinear uncertain systems and MASs have also been investigated in this book. The following results have been obtained in the book.

In Chapter 2, the adaptive backstepping control problem for a class of SISO commensurate fractional-order nonlinear systems in strict feedback form with system uncertainties and unknown time-varying external disturbance is studied. Unlike existing works, an auxiliary function instead of noncontinuous functions such as the sign function is employed to gain a smooth control input for compensating arbitrary bounded external disturbances and realizing asymptotic output tracking without causing chattering in the control signals. In addition, due to the absence of a concise form of the time derivative of the composite function obtained by the chain rule as in the integer-order case, we proposed an approach to approximate the fractional-order time derivatives of virtual control signals during the backstepping-based controller design procedure, and adaptive laws are designed accordingly to estimate the bounds of the approximation errors. The global asymptotic stability of the resulted closed-loop system can be ensured with our designed backstepping controller under an assumption that the approximation errors between the actual values of fractional derivatives of virtual control signals and their approximated values stay bounded. Another adaptive smooth backstepping control scheme is afterward designed to relax such an assumption. In the sense that all the closed-loop signals are uniformly ultimately bounded and the output tracking error can be driven to a small neighborhood around the origin, it is proved that the resulting closed-loop system is stable under such a proposed controller.

In Chapter 3, an analog chain rule computation method to calculate the fractional derivatives of virtual control signals that exist in the adaptive backstepping controller design procedure for uncertain fractional nonlinear systems is developed. It is rigorously demonstrated that when applying the quantized adaptive control input proposed in Chapter 3, the output of the associated closed-loop uncertain nonlinear fractional system in the presence of unknown external disturbances is able to accurately track the reference signal while also keeping all closed-loop signals remain bounded, ensuring overall system stability.

In Chapter 4, an investigation of the incommensurate fractional-order nonlinear system, which is the general form and of practically useful yet seldom been studied, is conducted. An adaptive backstepping controller for a class of incommensurate fractional-order nonlinear systems with uncertainties and time-varying disturbance is proposed in this chapter to fill

DOI: 10.1201/9781003620372-11

up certain research gaps. Similar to Chapter 2, a smooth controller is designed to get rid of chattering and prolong the lifespan of actuators potentially. The investigated control strategy can guarantee the boundedness of all the closed-loop signals and the asymptotic convergence of the system output to the reference signal.

Chapter 5 and Chapter 6 focus on the sampled-data control for linear fractional-order systems. A consensus scheme based on sampled-data control is firstly proposed in Chapter 5 for linear fractional-order MASs in single-integrator form. Theoretical stability analysis is presented by considering the hereditary and infinite memory properties of fractional-order calculus, which improves the existing works related to sampled-data consensus control of fractional-order MASs where such properties are not taken into account when designing controllers. It is shown that with the proposed distributed control scheme, all the closed-loop signals are guaranteed to be globally bounded with system asymptotic consensus being achieved. Subsequently, we came up with a design procedure in Chapter 5 for a finite-dimensional sampled-data controller for a class of linear single-integrator fractional-order systems, which guarantees the associated closed-loop system is globally ISS. A quantitative condition for determining the storage space for historical control inputs required to compute the current control signal in accordance with given control accuracy is introduced, which further reveals the trade-off between control performance and control effort. Furthermore, a novel sampled-data control scheme for asymptotically stabilizing a more general class of linear fractional-order systems is presented in Chapter 6.

The security control issue for discrete-time fractional-order linear systems under DoS attacks is addressed in Chapter 7. A finite-dimensional controller that requires the information of a finite number of the previous state, which could be practically useful, is proposed in this chapter. With strict stability analysis being presented, the global input-to-state stability of the associated closed-loop system is ensured.

The fractional-order adaptive backstepping control problem for a class of integer-order nonlinear systems subject to systematic uncertainties and unknown external disturbance is addressed in Chapter 8. It is indicated that better control performance can be obtained by combining fractional-order calculus with the traditional backstepping controller design procedure. Different from existing results only for second-order nonlinear integer systems, our designed fractional-order controller is applicable to systems without any restriction on the orders. Under such a controller, all the corresponding closed-loop signals are ensured to be globally uniformly ultimately bounded and the system output tracking error can be driven to a tunable arbitrarily small bound around the origin.

In Chapter 9, a PPB adaptive fractional backstepping control technique is developed for high-order nonlinear integer-order systems subject to systematic uncertainties and bounded unknown disturbances. The proposed fractional-order control technology is able to guarantee the stability of the closed-loop system and ensure the system output tracks the desired trajectory asymptotically, and the prescribed transient performance on the output error is concurrently assured by the investigated fractional controller.

For addressing the tracking problem of uncertain nonlinear high-order MASs with both DoS attacks and external unknown disturbances under data-rate-constrained communication, a hierarchical quantized distributed adaptive output PPB fractional-order backstepping control scheme is proposed in Chapter 10. The newly proposed hierarchical fractional-order control technology can not only ensure the stability of the closed-loop systems but also guarantee that each agent output perfectly tracks the reference signal with the prescribed transient and steady-state output performance being met.

11.2 Recommendations for Future Research

Based on the research works presented in this book, some future research aspects regarding control for fractional-order systems and fractional-order control worth exploring are suggested in this section as follows.

11.2.1 Adaptive Backstepping Control for Nonlinear Fractional-Order Systems

- Whether the control protocols that came up in Chapter 2 to Chapter 4 are effective for fractional-order systems with order greater than 1 is yet to be determined, providing an interesting research avenue for future work.

- In Chapter 2 to Chapter 4, it can be noticed from the system models that each state $x_i\,(i = 1, \cdots, n)$ corresponds to one unknown constant vector θ_i, which should be estimated accordingly at each step i. Nevertheless, the issue of over parameterization arises in designing adaptive controller based on backstepping technique since not only $\theta_j\,(j = 2, \cdots, i)$ but also θ_1 to θ_{j-1} will be estimated at step j. Solving such a problem will certainly simplify the controller. For either integer or fractional-order systems, when all the unknown θ_i are equal, i.e., $\theta_i = \theta$, the tuning function method can be applied to avoid the over parameterization issue. Addressing such an issue exists in estimating different system parametric uncertainties θ_i will be meaningful for whether fractional or integer-order systems when designing backstepping-based adaptive controllers.

- Extending the obtained results in Chapter 2 to Chapter 4 to the fixed-time convergence counterpart could be another practically meaningful future direction.

- Due to the incommensurate structure, it is practically meaningful yet still challenging to extend the developed control scheme in Chapter 4 to a general class of incommensurate fractional systems.

11.2.2 Sampled-Data Control for Linear Fractional-Order Systems

- In Chapter 5, a sufficient condition (5.12) for linear fractional MASs to guarantee the boundedness of control signals while strictly considering the unique properties of fractional calculus is provided. It can be observed from (5.13) that if the sampling period h tends to 0, the value of β could tend to 1, which might cause the satisfaction of condition (5.12) to be difficult. To overcome such a challenge, a trial between the choice of h and control parameter γ could be made, yet as noticed from (5.11) to (5.13), the selection of h and γ is also related to the system parameter α and the communication topology information, condition (5.12) possesses certain constraints on the applicable class of systems of the proposed control method. Therefore, relieving the conservativeness of condition (5.12) is of high potential practical worth.

- As mentioned in Remark 6.5, there is conservatism in the designed control scheme in Chapter 6, for practical requirements, reducing the conservativeness of the bound of the proposed sampled-data control signals as well as developing finite-dimensional sampled-data control law for a general class of linear fractional-order systems are worthy to be studied in the future.

- In both Chapter 5 and Chapter 6, the periodic sampling scheme is applied. In practice, the use of a non-periodic sampling method will be more suitable for further reducing the communication cost of transmitting control signals. Consequently, constructing an aperiodic sampled-data-based control scheme for fractional-order systems while rigorously taking into account the unique properties of fractional calculus by treating the non-fixed sampling periods as integer multiples of one base sampling period is possible yet requires further investigation.

- In Chapter 5 to Chapter 7, it can be noticed that in practice, it is possible that the system matrices or parameters could not be known precisely, which consequently implies an interesting future topic for considering the study of adaptive sampled-data or finite-dimensional control for fractional linear systems by introducing adaptive techniques in the obtained work in Chapter 5 to Chapter 7 to estimate system uncertainties.

11.2.3 Fractional-Order Control for Nonlinear Integer-Order Systems

- When investigating the control problem for integer systems using fractional-order controllers in Chapter 8, it is found that because of the complexity and properties of fractional-order calculus, designing fractional-order controllers and comparing them with the conventional integer-order controllers systematically and fairly to demonstrate the advantages of fractional-order calculus are complicated and not straightforward, as also stated in [14]. Therefore, formulating quantitative indicators for evaluating the control performance of integer-order nonlinear systems under fractional-order controllers regarding illustrating the potentials and merits of fractional control will be challenging yet significant.

- As the event-triggered control scheme works as an effective one for reducing the computation and communication load for control, developing an event-triggered-based fractional control strategy could be a significant future work to overcome the challenge mentioned at the end of Section 9.5 in Chapter 9 that the computation of implementing fractional control is complicated than the integer counterpart.

- The results in Chapter 10 are established based on the assumption of undirected graphs. Efforts in future research should be directed toward expanding the established methods to handle more challenging situations with directed communication networks.

Bibliography

[1] H. Liu, Y. Pan, S. Li, and Y. Chen, "Adaptive fuzzy backstepping control of fractional-order nonlinear systems," *IEEE Transactions on Systems, Man, and Cybernetics: Systems*, vol. 47, pp. 2209–2217, Aug. 2017.

[2] S. Song, J. H. Park, B. Zhang, and X. Song, "Observer-based adaptive hybrid fuzzy resilient control for fractional-order nonlinear systems with time-varying delays and actuator failures," *IEEE Transactions on Fuzzy Systems*, vol. 29, pp. 471–485, Mar. 2021.

[3] M. C. Tripathy, D. Mondal, K. Biswas, and S. Sen, "Design and performance study of phase-locked loop using fractional-order loop filter," *International Journal of Circuit Theory and Applications*, vol. 43, pp. 776–792, Jun. 2015.

[4] Z. Yu, H. Jiang, C. Hu, and J. Yu, "Necessary and sufficient conditions for consensus of fractional-order multiagent systems via sampled-data control," *IEEE Transactions on Cybernetics*, vol. 47, pp. 1892–1901, Aug. 2017.

[5] A. A. A. Kilbas, H. M. Srivastava, and J. J. Trujillo, *Theory and Applications of Fractional Differential Equations*, vol. 204. Amsterdam; Boston: Elsevier, 2006.

[6] K. S. Miller and B. Ross, *An Introduction to the Fractional Calculus and Fractional Differential Equations.* New York, USA: Wiley, 1993.

[7] I. Petráš, *Fractional-Order Nonlinear Systems: Modeling, Analysis and Simulation.* Berlin, Heidelberg: Springer-Verlag, 2011.

[8] I. Podlubny, *Fractional Differential Equations: An Introduction to Fractional Derivatives, Fractional Differential Equations, to Methods of Their Solution and Some of Their Applications*, vol. 198. San Diego, CA, USA: Academic, 1999.

[9] K. Oldham and J. Spanier, *The Fractional Calculus Theory and Applications of Differentiation and Integration to Arbitrary Order*, vol. 111. New York, USA: Academic Press, 1974.

[10] S. Manabe, "The non-integer integral and its application to control systems," *The Journal of the Institute of Electrical Engineers of Japan*, vol. 80, pp. 589–597, May 1960.

[11] M. Caputo and F. Mainardi, "A new dissipation model based on memory mechanism," *Pure and Applied Geophysics*, vol. 91, pp. 134–147, Dec. 1971.

[12] I. Podlubny, "Fractional-order systems and $PI^{\lambda}D^{\mu}$-controllers," *IEEE Transactions on Automatic Control*, vol. 44, pp. 208–214, Jan. 1999.

[13] A. Oustaloup, *La Commande CRONE: Commande robuste d'ordre non entier.* Paris, France: Hermès, 1991.

[14] Y. Luo and Y. Chen, *Fractional-Order Motion Controls.* London, UK: John Wiley & Sons, 2012.

[15] B. J. West, *Fractional Calculus and the Future of Science*, vol. 23. MDPI, Apr. 2022.

[16] A. Azimi, F. Bakhtiari-Nejad, and W. Zhu, "Fractional-order control with second-order sliding mode algorithm and disturbance estimation for vibration suppression of marine riser," *Journal of the Franklin Institute*, vol. 358, pp. 6545–6565, Sep. 2021.

[17] M. P. Aghababa, "A fractional sliding mode for finite-time control scheme with application to stabilization of electrostatic and electromechanical transducers," *Applied Mathematical Modelling*, vol. 39, pp. 6103–6113, Oct. 2015.

[18] Y. Wang, L. Gu, Y. Xu, and X. Cao, "Practical tracking control of robot manipulators with continuous fractional-order nonsingular terminal sliding mode," *IEEE Transactions on Industrial Electronics*, vol. 63, pp. 6194–6204, May 2016.

[19] G. Sun, L. Wu, Z. Kuang, Z. Ma, and J. Liu, "Practical tracking control of linear motor via fractional-order sliding mode," *Automatica*, vol. 94, pp. 221–235, Aug. 2018.

[20] S.-Y. Chen, T.-H. Li, and C.-H. Chang, "Intelligent fractional-order backstepping control for an ironless linear synchronous motor with uncertain nonlinear dynamics," *ISA Transactions*, vol. 89, pp. 218–232, Jun. 2019.

[21] N. Nikdel, M. Badamchizadeh, V. Azimirad, and M. A. Nazari, "Fractional-order adaptive backstepping control of robotic manipulators in the presence of model uncertainties and external disturbances," *IEEE Transactions on Industrial Electronics*, vol. 63, pp. 6249–6256, Oct. 2016.

[22] N. Nikdel and M. A. Badamchizadeh, "Fractional-order adaptive backstepping control of a class of uncertain systems with external disturbances," *International Journal of Control*, vol. 92, pp. 1344–1353, Jun. 2019.

[23] Z. Yu, Y. Zhang, B. Jiang, C.-Y. Su, J. Fu, Y. Jin, and T. Chai, "Decentralized fractional-order backstepping fault-tolerant control of multi-UAVs against actuator faults and wind effects," *Aerospace Science and Technology*, vol. 104, p. 105939, Sep. 2020.

[24] Z. Yu, Y. Zhang, B. Jiang, C.-Y. Su, J. Fu, Y. Jin, and T. Chai, "Nussbaum-based finite-time fractional-order backstepping fault-tolerant flight control of fixed-wing UAV against input saturation with hardware-in-the-loop validation," *Mechanical Systems and Signal Processing*, vol. 153, p. 107406, May 2021.

[25] R. L. Magin, *Fractional Calculus in Bioengineering*, vol. 2. Redding, USA: Begell House Publishers, 2006.

[26] R. Hilfer *et al.*, *Applications of Fractional Calculus in Physics*, vol. 35. Singapore: World Scientific, 2000.

[27] C. A. Monje, Y. Chen, B. M. Vinagre, D. Xue, and V. Feliu-Batlle, *Fractional-Order Systems and Controls: Fundamentals and Applications.* London, UK: Springer, 2010.

[28] A. Oustaloup and B. Dubuisson, *Diversity and Non-Integer Differentiation for System Dynamics.* London, UK: Wiley, 2014.

[29] D. Xue, *Fractional-Order Control Systems: Fundamentals and Numerical Implementations*, vol. 1. Berlin, Germany: De Gruyter, 2017.

[30] I. Petráš, *Handbook of Fractional Calculus with Applications*, vol. 6. Berlin, Boston: De Gruyter, 2019.

[31] I. Podlubny, "Fractional-order systems and fractional-order controllers," *Institute of Experimental Physics, Slovak Academy of Sciences, Kosice*, vol. 12, pp. 1–18, Nov. 1994.

[32] J. Wang, "Realizations of generalized Warburg impedance with RC ladder networks and transmission lines," *Journal of Electrochemical Society*, vol. 134, no. 8, pp. 1915–1920, 1987.

[33] S. Das, *Functional Fractional Calculus*, vol. 1. Berlin, Heidelberg: Springer, 2011.

[34] J.-L. Battaglia, O. Cois, L. Puigsegur, and A. Oustaloup, "Solving an inverse heat conduction problem using a non-integer identified model," *International Journal of Heat and Mass Transfer*, vol. 44, pp. 2671–2680, Jul. 2001.

[35] H. Cao, Z. Deng, X. Li, J. Yang, and Y. Qin, "Dynamic modeling of electrical characteristics of solid oxide fuel cells using fractional derivatives," *International Journal of Hydrogen Energy*, vol. 35, pp. 1749–1758, Feb. 2010.

[36] Y. Wang, M. Li, and Z. Chen, "Experimental study of fractional-order models for lithium-ion battery and ultra-capacitor: Modeling, system identification, and validation," *Applied Energy*, vol. 278, p. 115736, Nov. 2020.

[37] C. Zou, L. Zhang, X. Hu, Z. Wang, T. Wik, and M. Pecht, "A review of fractional-order techniques applied to lithium-ion batteries, lead-acid batteries, and supercapacitors," *Journal of Power Sources*, vol. 390, pp. 286–296, Jun. 2018.

[38] D. Guo, G. Yang, X. Feng, X. Han, L. Lu, and M. Ouyang, "Physics-based fractional-order model with simplified solid phase diffusion of lithium-ion battery," *Journal of Energy Storage*, vol. 30, p. 101404, Aug. 2020.

[39] Y. Jiang and B. Zhang, "High-power fractional-order capacitor with $1 < \alpha < 2$ based on power converter," *IEEE Transactions on Industrial Electronics*, vol. 65, pp. 3157–3164, Apr. 2018.

[40] R. L. Magin, "Fractional calculus in bioengineering: A tool to model complex dynamics," in *Proceedings of the 13th International Carpathian Control Conference (ICCC)*, pp. 464–469, IEEE, 2012.

[41] G. Gupta, S. Pequito, and P. Bogdan, "Re-thinking EEG-based non-invasive brain interfaces: Modeling and analysis," in *2018 ACM/IEEE 9th International Conference on Cyber-Physical Systems (ICCPS)*, pp. 275–286, IEEE, 2018.

[42] G. Gupta, S. Pequito, and P. Bogdan, "Dealing with unknown unknowns: Identification and selection of minimal sensing for fractional dynamics with unknown inputs," in *2018 Annual American Control Conference (ACC)*, pp. 2814–2820, IEEE, 2018.

[43] J. Kytariolos, A. Dokoumetzidis, and P. Macheras, "Power law IVIVC: An application of fractional kinetics for drug release and absorption," *European Journal of Pharmaceutical Sciences*, vol. 41, pp. 299–304, Oct. 2010.

[44] R. Magin and M. Ovadia, "Modeling the cardiac tissue electrode interface using fractional calculus," *Journal of Vibration and Control*, vol. 14, pp. 1431–1442, Sep. 2008.

[45] T. J. Anastasio, "The fractional-order dynamics of brainstem vestibulo-oculomotor neurons," *Biological Cybernetics*, vol. 72, pp. 69–79, Nov. 1994.

[46] C. Friedrich, "Relaxation and retardation functions of the Maxwell model with fractional derivatives," *Rheologica Acta*, vol. 30, pp. 151–158, Mar. 1991.

[47] R. Koeller, "Applications of fractional calculus to the theory of viscoelasticity," *Journal of Applied Mechanics*, vol. 51, pp. 299–307, Jun. 1984.

[48] X.-J. Wu and S.-L. Shen, "Chaos in the fractional-order Lorenz system," *International Journal of Computer Mathematics*, vol. 86, pp. 1274–1282, Jun. 2009.

[49] Y. Yu, H.-X. Li, S. Wang, and J. Yu, "Dynamic analysis of a fractional-order Lorenz chaotic system," *Chaos, Solitons & Fractals*, vol. 42, pp. 1181–1189, Oct. 2009.

[50] I. Petráš, "Control of fractional-order Chua's system," *Journal of ELECTRICAL ENGINEERING*, vol. 53, no. 7-8, pp. 219–222, 2002.

[51] H. Zhu, S. Zhou, and J. Zhang, "Chaos and synchronization of the fractional-order Chua's system," *Chaos, Solitons & Fractals*, vol. 39, pp. 1595–1603, Feb. 2009.

[52] V. Daftardar-Gejji and S. Bhalekar, "Chaos in fractional ordered Liu system," *Computers & Mathematics with Applications*, vol. 59, pp. 1117–1127, Feb. 2010.

[53] X.-Y. Wang and M.-J. Wang, "Dynamic analysis of the fractional-order Liu system and its synchronization," *Chaos: An Interdisciplinary Journal of Nonlinear Science*, vol. 17, p. 033106, Aug. 2007.

[54] J. G. Lu and G. Chen, "A note on the fractional-order Chen system," *Chaos, Solitons & Fractals*, vol. 27, pp. 685–688, Feb. 2006.

[55] M. M. Asheghan, M. T. H. Beheshti, and M. S. Tavazoei, "Robust synchronization of perturbed Chen's fractional-order chaotic systems," *Communications in Nonlinear Science and Numerical Simulation*, vol. 16, pp. 1044–1051, Feb. 2011.

[56] R. El-Khazali and W. Ahmad, "Fractional-order phase-locked loop," in *2007 9th International Symposium on Signal Processing and Its Applications*, pp. 1–4, IEEE, 2007.

[57] K. Biswas, G. Bohannan, R. Caponetto, A. M. Lopes, and J. A. T. Machado, *Fractional-Order Devices*. Cham, Switzerland: Springer, 2017.

[58] K. Cao and Y. Chen, *Fractional Order Crowd Dynamics: Cyber-Human System Modeling and Control*, vol. 4. Boston; Berlin: De Gruyter, 2018.

[59] H. Sun, Y. Zhang, D. Baleanu, W. Chen, and Y. Chen, "A new collection of real world applications of fractional calculus in science and engineering," *Communications in Nonlinear Science and Numerical Simulation*, vol. 64, pp. 213–231, Nov. 2018.

[60] Y. Cao, Y. Li, W. Ren, and Y. Chen, "Distributed coordination of networked fractional-order systems," *IEEE Transactions on Systems, Man, and Cybernetics, Part B (Cybernetics)*, vol. 40, pp. 362–370, Apr. 2010.

[61] W. Sun, Y. Li, C. Li, and Y. Chen, "Convergence speed of a fractional order consensus algorithm over undirected scale-free networks," *Asian Journal of Control*, vol. 13, pp. 936–946, Nov. 2011.

[62] J. Shen and J. Cao, "Necessary and sufficient conditions for consensus of delayed fractional-order systems," *Asian Journal of Control*, vol. 14, pp. 1690–1697, Feb. 2012.

[63] W. Yu, Y. Li, G. Wen, X. Yu, and J. Cao, "Observer design for tracking consensus in second-order multi-agent systems: Fractional order less than two," *IEEE Transactions on Automatic Control*, vol. 62, pp. 894–900, Feb. 2017.

[64] L. Chen, Y.-W. Wang, W. Yang, and J.-W. Xiao, "Robust consensus of fractional-order multi-agent systems with input saturation and external disturbances," *Neurocomputing*, vol. 303, pp. 11–19, Aug. 2018.

[65] M. Shahvali, M.-B. Naghibi-Sistani, and H. Modares, "Distributed consensus control for a network of incommensurate fractional-order systems," *IEEE Control Systems Letters*, vol. 3, pp. 481–486, Apr. 2019.

[66] P. Gong, Q.-L. Han, and W. Lan, "Finite-time consensus tracking for incommensurate fractional-order nonlinear multiagent systems with directed switching topologies," *IEEE Transactions on Cybernetics*, vol. 52, pp. 65–76, Jan. 2022.

[67] D. Matignon, "Generalized fractional differential and difference equations: Stability properties and modelling issues," in *Mathematical Theory of Networks and Systems Symposium*, pp. 503–506, Citeseer, 1998.

[68] Y. Li, Y. Chen, and I. Podlubny, "Mittag–Leffler stability of fractional order nonlinear dynamic systems," *Automatica*, vol. 45, pp. 1965–1969, Aug. 2009.

[69] Y. Li, Y. Chen, and I. Podlubny, "Stability of fractional-order nonlinear dynamic systems: Lyapunov direct method and generalized Mittag–Leffler stability," *Computers & Mathematics with Applications*, vol. 59, pp. 1810–1821, Mar. 2010.

[70] N. Aguila-Camacho, M. A. Duarte-Mermoud, and J. A. Gallegos, "Lyapunov functions for fractional order systems," *Communications in Nonlinear Science and Numerical Simulation*, vol. 19, pp. 2951–2957, Sep. 2014.

[71] M. Ö. Efe, "Fractional order systems in industrial automation - A survey," *IEEE Transactions on Industrial Informatics*, vol. 7, pp. 582–591, Nov. 2011.

[72] J. Fei and Z. Feng, "Fractional-order finite-time super-twisting sliding mode control of micro gyroscope based on double-loop fuzzy neural network," *IEEE Transactions on Systems, Man, and Cybernetics: Systems*, vol. 51, pp. 7692–7706, Dec. 2021.

[73] Y. Liu and H. Su, "Necessary and sufficient conditions for containment in fractional-order multiagent systems via sampled data," *IEEE Transactions on Systems, Man, and Cybernetics: Systems*, vol. 52, pp. 238–246, Jan. 2022.

[74] A. Saif, R. Fareh, S. Sinan, and M. Bettayeb, "Fractional synergetic tracking control for robot manipulator," *Journal of Control and Decision*, pp. 1–14, Nov. 2022.

[75] Y. Chen, I. Petras, and D. Xue, "Fractional order control - A tutorial," in *2009 American Control Conference*, pp. 1397–1411, 2009.

[76] B. M. Vinagre, C. A. Monje, A. J. Calderón, and J. I. Suárez, "Fractional PID controllers for industry application. A brief introduction," *Journal of Vibration and Control*, vol. 13, pp. 1419–1429, Sep. 2007.

[77] A. Dumlu and K. Erenturk, "Trajectory tracking control for a 3-DOF parallel manipulator using fractional-order $PI^{\lambda}D^{\mu}$ control," *IEEE Transactions on Industrial Electronics*, vol. 61, pp. 3417–3426, Jul. 2014.

[78] G. Gurumurthy and D. Kumar Das, "A novel fractional order controller design algorithm for a class of linear systems," *Journal of Control and Decision*, vol. 9, pp. 218–225, Sep. 2022.

[79] J. Fei, H. Wang, and Y. Fang, "Novel neural network fractional-order sliding-mode control with application to active power filter," *IEEE Transactions on Systems, Man, and Cybernetics: Systems*, vol. 52, pp. 3508–3518, Jun. 2022.

[80] Y. Chu, S. Hou, C. Wang, and J. Fei, "Recurrent-neural-network-based fractional order sliding mode control for harmonic suppression of power grid," *IEEE Transactions on Industrial Informatics*, vol. 19, pp. 9979–9990, Oct. 2023.

[81] J. Sabatier, C. Farges, and J.-C. Trigeassou, "Fractional systems state space description: Some wrong ideas and proposed solutions," *Journal of Vibration and Control*, vol. 20, pp. 1076–1084, May 2014.

[82] C. Li and W. Deng, "Remarks on fractional derivatives," *Applied Mathematics and Computation*, vol. 187, pp. 777–784, Apr. 2007.

[83] V. Lakshmikantham, S. Leela, and J. V. Devi, *Theory of Fractional Dynamic Systems.* Cambridge, U.K.: CSP, 2009.

[84] J. A. Gallegos, M. A. Duarte-Mermoud, N. Aguila-Camacho, and R. Castro-Linares, "On fractional extensions of Barbalat lemma," *Systems & Control Letters*, vol. 84, pp. 7–12, Oct. 2015.

[85] R. Zhang and Y. Liu, "A new Barbalat's lemma and Lyapunov stability theorem for fractional order systems," in *2017 29th Chinese Control and Decision Conference (CCDC)*, pp. 3676–3681, 2017.

[86] M. Krstić, I. Kanellakopoulos, and P. V. Kokotović, *Nonlinear and Adaptive Control Design.* New York, USA: Wiley, 1995.

[87] W. Wang, C. Wen, and J. Zhou, *Adaptive Backstepping Control of Uncertain Systems with Actuator Failures, Subsystem Interactions, and Nonsmooth Nonlinearities.* Boca Raton, FL, USA: CRC Press, 2017.

[88] J. Zhou, C. Wen, W. Wang, and F. Yang, "Adaptive backstepping control of nonlinear uncertain systems with quantized states," *IEEE Transactions on Automatic Control*, vol. 64, pp. 4756–4763, Nov. 2019.

[89] Y. Zhang, C. Wen, and Y. C. Soh, "Adaptive backstepping control design for systems with unknown high-frequency gain," *IEEE Transactions on Automatic Control*, vol. 45, pp. 2350–2354, Dec. 2000.

[90] Y. Zhang, C. Wen, and Y. C. Soh, "Discrete-time robust backstepping adaptive control for nonlinear time-varying systems," *IEEE Transactions on Automatic Control*, vol. 45, pp. 1749–1755, Sep. 2000.

[91] J. Zhou, C. Wen, and Y. Zhang, "Adaptive backstepping control of a class of uncertain nonlinear systems with unknown backlash-like hysteresis," *IEEE Transactions on Automatic Control*, vol. 49, pp. 1751–1759, Oct. 2004.

[92] D. Baleanu, J. A. T. Machado, and A. C. Luo, *Fractional Dynamics and Control.* New York, NY, USA: Springer, 2011.

[93] Y. Wei, Y. Chen, S. Liang, and Y. Wang, "A novel algorithm on adaptive backstepping control of fractional order systems," *Neurocomputing*, vol. 165, pp. 395–402, Oct. 2015.

[94] D. Sheng, Y. Wei, S. Cheng, and Y. Wang, "Adaptive backstepping state feedback control for fractional order systems with input saturation," *IFAC PapersOnLine*, vol. 50, pp. 6996–7001, Jul. 2017.

[95] Y. Wei, W. T. Peter, Z. Yao, and Y. Wang, "Adaptive backstepping output feedback control for a class of nonlinear fractional order systems," *Nonlinear Dynamics*, vol. 86, pp. 1047–1056, Jul. 2016.

[96] J.-C. Trigeassou, N. Maamri, J. Sabatier, and A. Oustaloup, "A Lyapunov approach to the stability of fractional differential equations," *Signal Processing*, vol. 91, pp. 437–445, Mar. 2011.

[97] M. K. Shukla and B. Sharma, "Backstepping based stabilization and synchronization of a class of fractional order chaotic systems," *Chaos, Solitons & Fractals*, vol. 102, pp. 274–284, Sep. 2017.

[98] M. K. Shukla and B. Sharma, "Control and synchronization of a class of uncertain fractional order chaotic systems via adaptive backstepping control," *Asian Journal of Control*, vol. 20, pp. 707–720, Mar. 2018.

[99] M. K. Shukla and B. Sharma, "Stabilization of a class of fractional order chaotic systems via backstepping approach," *Chaos, Solitons & Fractals*, vol. 98, pp. 56–62, May 2017.

[100] Q. Wang, J. Zhang, D. Ding, and D. Qi, "Adaptive Mittag-Leffler stabilization of a class of fractional order uncertain nonlinear systems," *Asian Journal of Control*, vol. 18, pp. 2343–2351, Nov. 2016.

[101] D. Ding, D. Qi, and Q. Wang, "Non-linear Mittag–Leffler stabilisation of commensurate fractional-order non-linear systems," *IET Control Theory & Applications*, vol. 9, pp. 681–690, Mar. 2015.

[102] D. Ding, D. Qi, Y. Meng, and L. Xu, "Adaptive Mittag-Leffler stabilization of commensurate fractional-order nonlinear systems," in *53rd IEEE Conference on Decision and Control*, pp. 6920–6926, 2014.

[103] H. Liu, Y. Pan, and J. Cao, "Composite learning adaptive dynamic surface control of fractional-order nonlinear systems," *IEEE Transactions on Cybernetics*, vol. 50, pp. 2557–2567, Jun. 2020.

[104] Z. Ma and H. Ma, "Adaptive fuzzy backstepping dynamic surface control of strict-feedback fractional order uncertain nonlinear systems," *IEEE Transactions on Fuzzy Systems*, vol. 28, pp. 122–133, Jan. 2020.

[105] F. Zouari, "Neural network based adaptive backstepping dynamic surface control of drug dosage regimens in cancer treatment," *Neurocomputing*, vol. 366, pp. 248–263, Nov. 2019.

[106] S. Song, B. Zhang, X. Song, and Z. Zhang, "Neuro-fuzzy-based adaptive dynamic surface control for fractional-order nonlinear strict-feedback systems with input constraint," *IEEE Transactions on Systems, Man, and Cybernetics: Systems*, vol. 51, pp. 3575–3586, Jun. 2021.

[107] Y.-X. Li, M. Wei, and S. Tong, "Event-triggered adaptive neural control for fractional-order nonlinear systems based on finite-time scheme," *IEEE Transactions on Cybernetics*, vol. 52, pp. 9481–9489, Sep. 2022.

[108] W. Yang, W. Yu, Y. Lv, L. Zhu, and T. Hayat, "Adaptive fuzzy tracking control design for a class of uncertain nonstrict-feedback fractional-order nonlinear SISO systems," *IEEE Transactions on Cybernetics*, vol. 51, pp. 3039–3053, Jun. 2021.

[109] Y.-X. Li, Q.-Y. Wang, and S. Tong, "Fuzzy adaptive fault-tolerant control of fractional-order nonlinear systems," *IEEE Transactions on Systems, Man, and Cybernetics: Systems*, vol. 51, pp. 1372–1379, Mar. 2021.

[110] S. Pashaei and M. Badamchizadeh, "A new fractional-order sliding mode controller via a nonlinear disturbance observer for a class of dynamical systems with mismatched disturbances," *ISA Transactions*, vol. 63, pp. 39–48, Jul. 2016.

[111] M. Chen, S.-Y. Shao, P. Shi, and Y. Shi, "Disturbance-observer-based robust synchronization control for a class of fractional-order chaotic systems," *IEEE Transactions on Circuits and Systems II: Express Briefs*, vol. 64, pp. 417–421, Apr. 2017.

[112] S. Song, B. Zhang, J. Xia, and Z. Zhang, "Adaptive backstepping hybrid fuzzy sliding mode control for uncertain fractional-order nonlinear systems based on finite-time scheme," *IEEE Transactions on Systems, Man, and Cybernetics: Systems*, vol. 50, pp. 1559–1569, Apr. 2020.

[113] Z. Wang, "Synchronization of an uncertain fractional-order chaotic system via backstepping sliding mode control," *Discrete Dynamics in Nature and Society*, vol. 2013, pp. 1–6, Jun. 2013.

[114] P. Gong and W. Lan, "Adaptive robust tracking control for multiple unknown fractional-order nonlinear systems," *IEEE Transactions on Cybernetics*, vol. 49, pp. 1365–1376, Apr. 2019.

[115] T. Hayakawa, H. Ishii, and K. Tsumura, "Adaptive quantized control for linear uncertain discrete-time systems," *Automatica*, vol. 45, pp. 692–700, Mar. 2009.

[116] T. Hayakawa, H. Ishii, and K. Tsumura, "Adaptive quantized control for nonlinear uncertain systems," *Systems & Control Letters*, vol. 58, pp. 625–632, Sep. 2009.

[117] J. Zhou, C. Wen, and G. Yang, "Adaptive backstepping stabilization of nonlinear uncertain systems with quantized input signal," *IEEE Transactions on Automatic Control*, vol. 59, pp. 460–464, Feb. 2014.

[118] W.-H. Wang, Y.-W. Wang, X.-K. Liu, and Z.-W. Liu, "Quantized control for networked switched systems under denial-of-service attacks via a barrier event-triggered mechanism," *Nonlinear Analysis: Hybrid Systems*, vol. 49, p. 101343, Aug. 2023.

[119] C. Hua, J. Ning, G. Zhao, and Y. Li, "Output feedback NN tracking control for fractional-order nonlinear systems with time-delay and input quantization," *Neurocomputing*, vol. 290, pp. 229–237, May 2018.

[120] S. Song, J. H. Park, B. Zhang, X. Song, and Z. Zhang, "Adaptive command filtered neuro-fuzzy control design for fractional-order nonlinear systems with unknown control directions and input quantization," *IEEE Transactions on Systems, Man, and Cybernetics: Systems*, vol. 51, pp. 7238–7249, Nov. 2021.

[121] Y. Fan, X. Huang, Z. Wang, J. Xia, and H. Shen, "Quantized control for synchronization of delayed fractional-order memristive neural networks," *Neural Processing Letters*, vol. 52, pp. 403–419, May 2020.

[122] D. Sheng, Y. Wei, S. Cheng, and Y. Wang, "Observer-based adaptive backstepping control for fractional order systems with input saturation," *ISA Transactions*, vol. 82, pp. 18–29, Nov. 2018.

[123] D. Sheng, Y. Wei, S. Cheng, and J. Shuai, "Adaptive backstepping control for fractional order systems with input saturation," *Journal of the Franklin Institute*, vol. 354, pp. 2245–2268, Mar. 2017.

[124] Y. Wei, D. Sheng, Y. Chen, and Y. Wang, "Fractional order chattering-free robust adaptive backstepping control technique," *Nonlinear Dynamics*, vol. 95, pp. 2383–2394, Feb. 2019.

[125] Z. Ruo-Xun and Y. Shi-Ping, "Adaptive stabilization of an incommensurate fractional order chaotic system via a single state controller," *Chinese Physics B*, vol. 20, p. 110506, Jun. 2011.

[126] R. Luo and H. Su, "The stability of impulsive incommensurate fractional order chaotic systems with Caputo derivative," *Chinese Journal of Physics*, vol. 56, pp. 1599–1608, Aug. 2018.

[127] C. Yin, S. Dadras, S.-M. Zhong, and Y. Chen, "Control of a novel class of fractional-order chaotic systems via adaptive sliding mode control approach," *Applied Mathematical Modelling*, vol. 37, pp. 2469–2483, Feb. 2013.

[128] H. T. Toivonen, "Sampled-data control of continuous-time systems with an H_∞ optimality criterion," *Automatica*, vol. 28, pp. 45–54, Jan. 1992.

[129] L. Hou, A. N. Michel, and H. Ye, "Some qualitative properties of sampled-data control systems," *IEEE Transactions on Automatic Control*, vol. 42, pp. 1721–1725, Dec. 1997.

[130] L.-S. Hu, T. Bai, P. Shi, and Z. Wu, "Sampled-data control of networked linear control systems," *Automatica*, vol. 43, pp. 903–911, May 2007.

[131] W. Zhang, Q.-L. Han, Y. Tang, and Y. Liu, "Sampled-data control for a class of linear time-varying systems," *Automatica*, vol. 103, pp. 126–134, May 2019.

[132] W. Yu, W. X. Zheng, G. Chen, W. Ren, and J. Cao, "Second-order consensus in multi-agent dynamical systems with sampled position data," *Automatica*, vol. 47, pp. 1496–1503, Jul. 2011.

[133] H. Su, Y. Ye, X. Chen, and H. He, "Necessary and sufficient conditions for consensus in fractional-order multiagent systems via sampled data over directed graph," *IEEE Transactions on Systems, Man, and Cybernetics: Systems*, vol. 51, pp. 2501–2511, Apr. 2021.

[134] H. Liu, G. Xie, and Y. Gao, "Consensus of fractional-order double-integrator multi-agent systems," *Neurocomputing*, vol. 340, pp. 110–124, May 2019.

[135] Y. Ye, H. Su, J. Chen, and Y. Peng, "Consensus in fractional-order multi-agent systems with intermittence sampled data over directed networks," *IEEE Transactions on Circuits and Systems II: Express Briefs*, vol. 67, pp. 365–369, Feb. 2020.

[136] Y. Ye and H. Su, "Consensus of delayed fractional-order multiagent systems with intermittent sampled data," *IEEE Transactions on Industrial Informatics*, vol. 16, pp. 3828–3837, Jun. 2020.

[137] J. Chen, B. Chen, and Z. Zeng, "Synchronization and consensus in networks of linear fractional-order multi-agent systems via sampled-data control," *IEEE Transactions on Neural Networks and Learning Systems*, vol. 31, pp. 2955–2964, Aug. 2020.

[138] Y. Chen, G. Wen, Z. Peng, and A. Rahmani, "Consensus of fractional-order multiagent system via sampled-data event-triggered control," *Journal of the Franklin Institute*, vol. 356, pp. 10241–10259, Nov. 2019.

[139] F. Wang and Y. Yang, "Leader-following consensus of nonlinear fractional-order multi-agent systems via event-triggered control," *International Journal of Systems Science*, vol. 48, no. 3, pp. 571–577, 2017.

[140] Y. Ye, H. Wei, R. Lu, H. Su, and Y. Wu, "Containment control for networked fractional-order systems with sampled position data," *IEEE Transactions on Circuits and Systems I: Regular Papers*, vol. 68, pp. 3881–3889, Sep. 2021.

[141] D. Wu, Q. An, Y. Sun, Y. Liu, and H. Su, "Containment control in fractional-order multi-agent systems with intermittent sampled data over directed networks," *Neurocomputing*, vol. 442, pp. 209–220, Jun. 2021.

[142] T. Kaczorek, "Practical stability of positive fractional discrete-time linear systems," *Bulletin of the Polish Academy of Sciences: Technical Sciences*, vol. 56, no. 4, pp. 313–317, 2008.

[143] S. Guermah, S. Djennoune, and M. Bettayeb, "Discrete-time fractional-order systems: Modeling and stability issues," in *Advances in Discrete Time Systems*, pp. 183–212, InTech, 2012.

[144] M. BusłOwicz and T. Kaczorek, "Simple conditions for practical stability of positive fractional discrete-time linear systems," *International Journal of Applied Mathematics and Computer Science*, vol. 19, pp. 263–269, Jul. 2009.

[145] S. Guermah, S. Djennoune, and M. Bettayeb, "A new approach for stability analysis of linear discrete-time fractional-order systems," in *New Trends in Nanotechnology and Fractional Calculus Applications*, pp. 151–162, Dordrecht, The Netherlands: Springer, 2010.

[146] P. Sopasakis, S. Ntouskas, and H. Sarimveis, "Robust model predictive control for discrete-time fractional-order systems," in *2015 23rd Mediterranean Conference on Control and Automation (MED)*, pp. 384–389, IEEE, 2015.

[147] A. Alessandretti, S. Pequito, G. J. Pappas, and A. P. Aguiar, "Finite-dimensional control of linear discrete-time fractional-order systems," *Automatica*, vol. 115, p. 108512, May 2020.

[148] P. Sopasakis and H. Sarimveis, "Stabilising model predictive control for discrete-time fractional-order systems," *Automatica*, vol. 75, pp. 24–31, Jan. 2017.

[149] C. De Persis and P. Tesi, "Input-to-state stabilizing control under denial-of-service," *IEEE Transactions on Automatic Control*, vol. 60, pp. 2930–2944, Nov. 2015.

[150] Z. Lian, P. Shi, and C.-C. Lim, "Adaptive resilient control for cyber-physical systems under cyberattack and input saturation," *IEEE Transactions on Industrial Informatics*, vol. 19, pp. 6513–6524, May 2023.

[151] X. Li, C. Wen, J. Wang, L. Xing, and X. Li, "Jamming-resilient synchronization of networked Lagrangian systems with quantized sampling data," *IEEE Transactions on Industrial Informatics*, vol. 18, pp. 8724–8734, Dec. 2022.

[152] L. Zhao and G.-H. Yang, "Adaptive fault-tolerant control for nonlinear multi-agent systems with DoS attacks," *Information Sciences*, vol. 526, pp. 39–53, Jul. 2020.

[153] D. Zhang, L. Liu, and G. Feng, "Consensus of heterogeneous linear multiagent systems subject to aperiodic sampled-data and DoS attack," *IEEE Transactions on Cybernetics*, vol. 49, pp. 1501–1511, Apr. 2019.

[154] Z. Feng and G. Hu, "Secure cooperative event-triggered control of linear multiagent systems under DoS attacks," *IEEE Transactions on Control Systems Technology*, vol. 28, no. 3, pp. 741–752, 2019.

[155] B. Chang, X. Mu, Z. Yang, and J. Fang, "Event-based secure consensus of muti-agent systems under asynchronous DoS attacks," *Applied Mathematics and Computation*, vol. 401, p. 126120, Jul. 2021.

[156] G. Narayanan, M. S. Ali, H. Alsulami, G. Stamov, I. Stamova, and B. Ahmad, "Impulsive security control for fractional-order delayed multi-agent systems with uncertain parameters and switching topology under DoS attack," *Information Sciences*, vol. 618, pp. 169–190, Dec. 2022.

[157] J. Bai, H. Wu, and J. Cao, "Secure synchronization and identification for fractional complex networks with multiple weight couplings under DoS attacks," *Computational and Applied Mathematics*, vol. 41, pp. 1–18, May 2022.

[158] G. Narayanan, M. S. Ali, and S. Ahamad, "Cyber secure consensus of discrete-time fractional-order multi-agent systems with distributed delayed control against attacks," in *2021 IEEE International Conference on Systems, Man, and Cybernetics (SMC)*, pp. 2191–2196, 2021.

[159] W. Zheng, Y. Luo, X. Wang, Y. Pi, and Y. Chen, "Fractional order $PI^{\lambda}D^{\mu}$ controller design for satisfying time and frequency domain specifications simultaneously," *ISA Transactions*, vol. 68, pp. 212–222, May 2017.

[160] J. Ma, H. Wang, and J. Qiao, "Adaptive neural fixed-time tracking control for high-order nonlinear systems," *IEEE Transactions on Neural Networks and Learning Systems*, vol. 35, pp. 708 – 717, Jan. 2024.

[161] H. Wang, K. Xu, and H. Zhang, "Adaptive finite-time tracking control of nonlinear systems with dynamics uncertainties," *IEEE Transactions on Automatic Control*, vol. 68, pp. 5737–5744, Sep. 2023.

[162] Q. Wang, J. Cao, and H. Liu, "Adaptive fuzzy control of nonlinear systems with predefined time and accuracy," *IEEE Transactions on Fuzzy Systems*, vol. 30, pp. 5152–5165, Dec. 2022.

[163] S. Shao and M. Chen, "Adaptive neural discrete-time fractional-order control for a UAV system with prescribed performance using disturbance observer," *IEEE Transactions on Systems, Man, and Cybernetics: Systems*, vol. 51, pp. 742–754, Feb. 2021.

[164] Z. Yu, H. Badihi, Y. Zhang, Y. Ma, B. Jiang, and C.-Y. Su, "Fractional-order sliding-mode fault-tolerant neural adaptive control of fixed-wing UAV with prescribed tracking performance," in *2020 2nd International Conference on Industrial Artificial Intelligence (IAI)*, pp. 1–6, 2020.

[165] J. Sun, J. Wang, P. Yang, and S. Guo, "Fractional-order prescribed performance sliding-mode control with time-delay estimation for wearable exoskeletons," *IEEE Transactions on Industrial Informatics*, vol. 19, pp. 8274–8284, Jul. 2023.

[166] J. Sun, J. Wang, P. Yang, and Y. Geng, "Model-free fractional-order adaptive backstepping prescribed performance control for wearable exoskeletons," *International Journal of Intelligent Robotics and Applications*, vol. 5, pp. 590–605, Jun. 2021.

[167] Z. Yu, Y. Zhang, B. Jiang, C.-Y. Su, J. Fu, Y. Jin, and T. Chai, "Enhanced recurrent fuzzy neural fault-tolerant synchronization tracking control of multiple unmanned airships via fractional calculus and fixed-time prescribed performance function," *IEEE Transactions on Fuzzy Systems*, vol. 30, pp. 4515–4529, Oct. 2022.

[168] S. Huang, J. Wang, L. Xiong, J. Liu, P. Li, and Z. Wang, "Distributed predefined-time fractional-order sliding mode control for power system with prescribed tracking performance," *IEEE Transactions on Power Systems*, vol. 37, pp. 2233–2246, May 2022.

[169] T. Li, M. Fu, L. Xie, and J.-F. Zhang, "Distributed consensus with limited communication data rate," *IEEE Transactions on Automatic Control*, vol. 56, pp. 279–292, Feb. 2011.

[170] C. Gao, Z. Wang, X. He, and H. Dong, "Fault-tolerant consensus control for multi-agent systems: An encryption-decryption scheme," *IEEE Transactions on Automatic Control*, vol. 67, pp. 2560–2567, May 2022.

[171] C. Gao, Z. Wang, X. He, Y. Liu, and D. Yue, "Differentially private consensus control for discrete-time multiagent systems: Encoding–decoding schemes," *IEEE Transactions on Automatic Control*, vol. 69, pp. 5554–5561, Feb. 2024.

[172] W. Liu, J. Sun, G. Wang, F. Bullo, and J. Chen, "Resilient control under quantization and denial-of-service: Codesigning a deadbeat controller and transmission protocol," *IEEE Transactions on Automatic Control*, vol. 67, pp. 3879–3891, Aug. 2022.

[173] S. Feng and H. Ishii, "Dynamic quantized consensus of general linear multiagent systems under Denial-of-Service attacks," *IEEE Transactions on Control of Network Systems*, vol. 9, pp. 562–574, Jun. 2022.

[174] S. Feng, M. Ran, B. Zhang, L. Xie, and S. Xu, "The bottleneck and ceiling effects in quantized tracking control of heterogeneous multi-agent systems under DoS attacks," *Automatica*, vol. 160, p. 111424, Feb. 2024.

[175] M. Ran, S. Feng, J. Li, and L. Xie, "Quantized consensus under data-rate constraints and DoS attacks: A zooming-in and holding approach," *IEEE Transactions on Automatic Control*, vol. 68, pp. 5397–5412, Sep. 2023.

[176] W. Wang, J. Huang, C. Wen, and H. Fan, "Distributed adaptive control for consensus tracking with application to formation control of nonholonomic mobile robots," *Automatica*, vol. 50, pp. 1254–1263, Apr. 2014.

[177] J. Li, Y. Liu, and J. Yu, "A new result on semi-synchronous event-triggered backstepping robust control for a class of non-Lipschitzian networked systems," *Applied Mathematics and Computation*, vol. 424, p. 127027, Jul. 2022.

[178] C. Deng, C. Wen, W. Wang, X. Li, and D. Yue, "Distributed adaptive tracking control for high-order nonlinear multiagent systems over event-triggered communication," *IEEE Transactions on Automatic Control*, vol. 68, pp. 1176–1183, Feb. 2023.

[179] J. Huang, W. Wang, C. Wen, J. Zhou, and G. Li, "Distributed adaptive leader–follower and leaderless consensus control of a class of strict-feedback nonlinear systems: A unified approach," *Automatica*, vol. 118, p. 109021, Aug. 2020.

[180] S.-S. Sun, Y.-X. Li, and Z. Hou, "Prescribed performance-based resilient model-free adaptive control for CPSs against aperiodic DoS attacks," *International Journal of Robust & Nonlinear Control*, vol. 34, pp. 3335–3350, Mar. 2024.

[181] S. Kamali, S. M. Tabatabaei, M. M. Arefi, and S. Yin, "Prescribed performance quantized tracking control for a class of delayed switched nonlinear systems with actuator hysteresis using a filter-connected switched hysteretic quantizer," *IEEE Transactions on Neural Networks and Learning Systems*, vol. 33, pp. 61–74, Jan. 2022.

[182] D. Yang, G. Zong, and H. R. Karimi, "H_∞ refined anti-disturbance control of switched LPV systems with application to aero-engine," *IEEE Transactions on Industrial Electronics*, vol. 67, pp. 3180–3190, Apr. 2020.

[183] G. Zong, Y. Li, and H. Sun, "Composite anti-disturbance resilient control for Markovian jump nonlinear systems with general uncertain transition rate," *Science China Information Sciences*, vol. 62, p. 22205, Jan. 2019.

[184] H. Sun, Y. Li, G. Zong, and L. Hou, "Disturbance attenuation and rejection for stochastic Markovian jump system with partially known transition probabilities," *Automatica*, vol. 89, pp. 349–357, Mar. 2018.

[185] H. Zhang, J. Han, C. Luo, and Y. Wang, "Fault-tolerant control of a nonlinear system based on generalized fuzzy hyperbolic model and adaptive disturbance observer," *IEEE Transactions on Systems, Man, and Cybernetics: Systems*, vol. 47, pp. 2289–2300, Aug. 2017.

[186] Z. Zuo and C. Wang, "Adaptive trajectory tracking control of output constrained multi-rotors systems," *IET Control Theory & Applications*, vol. 8, pp. 1163–1174, Sep. 2014.

[187] G. Jumarie, "Modified Riemann-Liouville derivative and fractional Taylor series of nondifferentiable functions further results," *Computers & Mathematics with Applications*, vol. 51, pp. 1367–1376, May 2006.

[188] J. Wang, Y. Wen, Y. Gou, Z. Ye, and H. Chen, "Fractional-order gradient descent learning of BP neural networks with Caputo derivative," *Neural Networks*, vol. 89, pp. 19–30, May 2017.

[189] I. Petráš, "A note on the fractional-order Chua's system," *Chaos, Solitons & Fractals*, vol. 38, pp. 140–147, Oct. 2008.

[190] D. Cvijović, "New identities for the partial Bell polynomials," *Applied Mathematics Letters*, vol. 24, pp. 1544–1547, Sep. 2011.

[191] M. M. Polycarpou, "Stable adaptive neural control scheme for nonlinear systems," *IEEE Transactions on Automatic Control*, vol. 41, pp. 447–451, Mar. 1996.

[192] A. Yousefpour, H. Jahanshahi, J. M. Munoz-Pacheco, S. Bekiros, and Z. Wei, "A fractional-order hyper-chaotic economic system with transient chaos," *Chaos, Solitons & Fractals*, vol. 130, p. 109400, Jan. 2020.

[193] V. Silva-García, R. Flores-Carapia, C. Rentería-Márquez, B. Luna-Benoso, and M. Aldape-Pérez, "Substitution box generation using chaos: An image encryption application," *Applied Mathematics and Computation*, vol. 332, pp. 123–135, Sep. 2018.

[194] M. Zhang, H. Zang, and L. Bai, "A new predefined-time sliding mode control scheme for synchronizing chaotic systems," *Chaos, Solitons & Fractals*, vol. 164, p. 112745, Nov. 2022.

[195] J. C. Sprott, "Some simple chaotic flows," *Physical Review E*, vol. 50, p. R647, Aug. 1994.

[196] S. Nimmo and A. K. Evans, "The effects of continuously varying the fractional differential order of chaotic nonlinear systems," *Chaos, Solitons & Fractals*, vol. 10, pp. 1111–1118, Jul. 1999.

[197] M. Shukla and B. Sharma, "Stabilization of a class of uncertain fractional order chaotic systems via adaptive backstepping control," in *2017 Indian Control Conference (ICC)*, pp. 462–467, IEEE, 2017.

[198] Z. Yu, Y. Zhang, B. Jiang, J. Fu, Y. Jin, and T. Chai, "Composite adaptive disturbance observer-based decentralized fractional-order fault-tolerant control of networked UAVs," *IEEE Transactions on Systems, Man, and Cybernetics: Systems*, vol. 52, pp. 799–813, Feb. 2022.

[199] S. Lu, X. Wang, and Y. Li, "Adaptive neural network finite-time command filtered tracking control of fractional-order permanent magnet synchronous motor with input saturation," *Journal of the Franklin Institute*, vol. 357, pp. 13707–13733, Dec. 2020.

[200] J. Ni, L. Liu, C. Liu, and X. Hu, "Fractional order fixed-time nonsingular terminal sliding mode synchronization and control of fractional order chaotic systems," *Nonlinear Dynamics*, vol. 89, pp. 2065–2083, Aug. 2017.

[201] J. Sabatier, O. P. Agrawal, and J. T. Machado, *Advances in Fractional Calculus: Theoretical Developments and Applications in Physics and Engineering*, vol. 4. Dordrecht, The Netherlands: Springer, 2007.

[202] K. Biswas, S. Sen, and P. K. Dutta, "Realization of a constant phase element and its performance study in a differentiator circuit," *IEEE Transactions on Circuits and Systems II: Express Briefs*, vol. 53, pp. 802–806, Sep. 2006.

[203] T. J. Freeborn, B. Maundy, and A. S. Elwakil, "Measurement of supercapacitor fractional-order model parameters from voltage-excited step response," *IEEE Journal on Emerging and Selected Topics in Circuits and Systems*, vol. 3, pp. 367–376, Sep. 2013.

[204] R. Olfati-Saber, J. A. Fax, and R. M. Murray, "Consensus and cooperation in networked multi-agent systems," *Proceedings of the IEEE*, vol. 95, pp. 215–233, Jan. 2007.

[205] A. G. Radwan, "Resonance and quality factor of the $RL_\alpha C_\alpha$ fractional circuit," *IEEE Journal on Emerging and Selected Topics in Circuits and Systems*, vol. 3, pp. 377–385, Sep. 2013.

[206] A. Pyrkin and A. Isidori, "Adaptive output regulation of right-invertible MIMO LTI systems, with application to vessel motion control," *European Journal of Control*, vol. 46, pp. 63–79, Mar. 2019.

[207] W. P. Hunek, "Perfect control for right-invertible Grünwald-Letnikov plants–an innovative approach to practical implementation," *Fractional Calculus and Applied Analysis*, vol. 22, pp. 424–443, May 2019.

[208] W. P. Hunek and T. Feliks, "A new extension of inverse model control design to non-full rank state-space plants," in *2020 European Control Conference (ECC)*, pp. 1783–1788, IEEE, 2020.

[209] S. Bańka and K. J. Latawiec, "On steady–state error–free regulation of right–invertible LTI MIMO plants," *IFAC Proceedings Volumes*, vol. 42, no. 13, pp. 378–383, 2009.

[210] R. Schmid and L. Ntogramatzidis, "On the design of non-overshooting linear tracking controllers for right-invertible systems," in *Proceedings of the 48h IEEE Conference on Decision and Control (CDC) held jointly with 2009 28th Chinese Control Conference*, pp. 5323–5327, IEEE, 2009.

[211] C. Li and F. Zeng, "Finite difference methods for fractional differential equations," *International Journal of Bifurcation and Chaos*, vol. 22, no. 04, p. 1230014, 2012.

[212] C. Deng, C. Wen, Y. Zou, W. Wang, and X. Li, "A hierarchical security control framework of nonlinear CPSs against DoS attacks with application to power sharing of AC microgrids," *IEEE Transactions on Cybernetics*, vol. 52, pp. 5255–5266, Jun. 2022.

[213] A. A. Cárdenas, S. Amin, and S. Sastry, "Research challenges for the security of control systems," in *Proceedings of the 3rd conference on Hot topics in security*, pp. 1–6, 2008.

[214] P. Ostalczyk, *Discrete Fractional Calculus: Applications in Control and Image Processing*, vol. 4. New Jersey: World Scientific, 2016.

[215] S. Chatterjee, O. Romero, and S. Pequito, "A separation principle for discrete-time fractional-order dynamical systems and its implications to closed-loop neurotechnology," *IEEE Control Systems Letters*, vol. 3, pp. 691–696, Jul. 2019.

[216] Z.-P. Jiang and Y. Wang, "Input-to-state stability for discrete-time nonlinear systems," *Automatica*, vol. 37, pp. 857–869, Jun. 2001.

[217] K. Zhao, F. L. Lewis, and L. Zhao, "Unifying performance specifications in tracking control of MIMO nonlinear systems with actuation faults," *Automatica*, vol. 155, p. 111102, Sep. 2023.

[218] C. P. Bechlioulis and G. A. Rovithakis, "Prescribed performance adaptive control of SISO feedback linearizable systems with disturbances," in *2008 16th Mediterranean Conference on Control and Automation*, pp. 1035–1040, 2008.

[219] W. Wang, D. Wang, Z. Peng, and T. Li, "Prescribed performance consensus of uncertain nonlinear strict-feedback systems with unknown control directions," *IEEE Transactions on Systems, Man, and Cybernetics: Systems*, vol. 46, pp. 1279–1286, Sep. 2016.

[220] S. Zhou and Y. Song, "Prescribed performance neuroadaptive fault-tolerant compensation for MIMO nonlinear systems under extreme actuator failures," *IEEE Transactions on Systems, Man, and Cybernetics: Systems*, vol. 51, pp. 5427–5436, Sep. 2021.

[221] W. Wang and C. Wen, "Adaptive actuator failure compensation control of uncertain nonlinear systems with guaranteed transient performance," *Automatica*, vol. 46, pp. 2082–2091, Dec. 2010.

[222] C. P. Bechlioulis and G. A. Rovithakis, "Adaptive control with guaranteed transient and steady state tracking error bounds for strict feedback systems," *Automatica*, vol. 45, pp. 532–538, Feb. 2009.

[223] W. Wang, J. Huang, and C. Wen, "Prescribed performance bound-based adaptive path-following control of uncertain nonholonomic mobile robots," *International Journal of Adaptive Control & Signal Processing*, vol. 31, pp. 805–822, May 2017.

[224] Y. Yang, J. Tan, and D. Yue, "Prescribed performance tracking control of a class of uncertain pure-feedback nonlinear systems with input saturation," *IEEE Transactions on Systems, Man, and Cybernetics: Systems*, vol. 50, pp. 1733–1745, May 2020.

[225] A. Oustaloup, F. Levron, B. Mathieu, and F. M. Nanot, "Frequency-band complex noninteger differentiator: Characterization and synthesis," *IEEE Transactions on Circuits and Systems I: Fundamental Theory and Applications*, vol. 47, pp. 25–39, Jan. 2000.

[226] X.-K. Liu, Y.-W. Wang, Z.-W. Liu, and Y. Huang, "On the stability of distributed secondary control for DC microgrids with grid-forming and grid-feeding converters," *Automatica*, vol. 155, p. 111164, Sep. 2023.

[227] H. Han, H. Zhang, J. Yang, and L. Yang, "Distributed containment control strategy for the dynamic stabilization of integrated energy system with multiple virtual leaders," *IEEE Transactions on Industrial Informatics*, vol. 20, pp. 6642–6651, Apr. 2024.

[228] C. Deng, F. Meng, X. Xie, D. Yue, W.-W. Che, and S. Fan, "Data-driven-based distributed fuzzy tracking control for nonlinear MASs under DoS attacks," *IEEE Transactions on Fuzzy Systems*, vol. 32, pp. 53–63, Jan. 2024.

[229] Y. Hong, J. Hu, and L. Gao, "Tracking control for multi-agent consensus with an active leader and variable topology," *Automatica*, vol. 42, pp. 1177–1182, Jul. 2006.

[230] M. Schatzman, *Numerical Analysis: A Mathematical Introduction.* Oxford University Press, USA, 2002.

Index